CHAPMAN & HALL/CRC
Monographs and Surveys in
Pure and Applied Mathematics 122

CANONICAL PROBLEMS IN SCATTERING AND POTENTIAL THEORY

PART I:

Canonical Structures in Potential Theory

CHAPMAN & HALL/CRC
Monographs and Surveys in
Pure and Applied Mathematics 122

CANONICAL PROBLEMS IN SCATTERING AND POTENTIAL THEORY

PART I:

Canonical Structures in Potential Theory

S. S. VINOGRADOV
P. D. SMITH
E. D. VINOGRADOVA

CHAPMAN & HALL/CRC
Boca Raton London New York Washington, D.C.

Library of Congress Cataloging-in-Publication Data

Vinogradov, Sergey S. (Sergey Sergeyevich)
Canonical problems in scattering and potential theory / Sergey S. Vinogradov, Paul D. Smith, Elena D. Vinogradova.
p. cm.— (Monographis and surveys in pure and applied mathematics ; 122)
Includes bibliographical references and index.
Contents: pt. 1. Canonical structures in potential theory
ISBN 1-58488-162-3 (v. 1 : alk. paper)
1. Potential theory (Mathematics) 2. Scattering (Mathematics) I. Smith, P.D. (Paul Denis), 1955- II. Vinogradova, Elena D. (Elena Dmitrievna) III. Title. IV. Chapman & Hall/CRC monographs and surveys in pure and applied mathematics ; 122.

QA404.7 . V56 2001
515′.9—dc21 2001028226

Direct all inquiries to CRC Press LLC, 2000 N.W. Corporate Blvd., Boca Raton, Florida 33431.

International Standard Book Number 1-58488-162-3
Library of Congress Card Number 2001028226
Printed in the United States of America 1 2 3 4 5 6 7 8 9 0
Printed on acid-free paper

To our children

Contents

Preface

Potential theory has its roots in the physical sciences and continues to find application in diverse areas including electrostatics and elasticity. From a mathematical point of view, the study of Laplace's equation has profoundly influenced the theory of partial differential equations and the development of functional analysis. Together with the wave operator and the diffusion operator, its study and application continue to dominate many areas of mathematics, physics, and engineering. Scattering of electromagnetic or acoustic waves is of widespread interest, because of the enormous number of technological applications developed in the last century, from imaging to telecommunications and radio astronomy.

The advent of powerful computing resources has facilitated numerical modelling and simulation of many concrete problems in potential theory and scattering. The many methods developed and refined in the last three decades have had a significant impact in providing numerical solutions and insight into the important mechanisms in scattering and associated static problems. However, the accuracy of present-day purely numerical methods can be difficult to ascertain, particularly for objects of some complexity incorporating edges, re-entrant structures, and dielectrics. An example is the open metallic cavity with a dielectric inclusion. The study of closed bodies with smooth surfaces is rather more completely developed, from an analytical and numerical point of view, and computational algorithms have attained a good degree of accuracy and generality. In contradistinction to highly developed analysis for closed bodies of simple geometric shape – which was the subject of Bowman, Senior, and Uslenghi's classic text on scattering [6] – structures with edges, cavities, or inclusions have seemed, until now, intractable to analytical methods.

Our motivation for this two-volume text on scattering and potential theory is to describe a class of analytic and semi-analytic techniques for accurately determining the diffraction from structures comprising edges and other complex cavity features. These techniques rely heavily on the solution of associated potential problems for these structures developed in Part I.

These techniques are applied to various classes of canonical scatterers, of particular relevance to edge-cavity structures. There are several reasons for focusing on such canonical objects. The exact solution to a potential theory problem or diffraction problem is interesting in its own right. As Bowman et al. [6] state, most of our understanding of how scattering takes place is obtained by detailed examination of such representative scatterers. Their study provides an exact quantification of the effects of edges, cavities, and inclusions. This is invaluable for assessing the relative importance of these effects in other, more general structures. Sometimes the solution developed in the text is in the form of a linear system of equations for which the solution accuracy can be determined; however, the same point about accurate quantification is valid. Such solutions thus highlight the generic difficulties that numerical methods must successfully tackle for more general structures. Reliable benchmarks, against which a solution obtained by such general-purpose numerical methods can be verified, are needed to establish confidence in the validity of these computational methods in wider contexts where analysis becomes impossible. Exact or semi-analytic solutions are valuable elsewhere: in inverse scattering, exact solutions may pinpoint special effects and distinguish between physically real effects and artefacts of the computational process. Moreover, many canonical structures are of direct technological interest, particularly where a scattering process is dominated by that observed in a related canonical structure.

Mathematically, we solve a class of mixed boundary value problems and develop numerical formulations for computationally stable, rapidly converging algorithms of guaranteed accuracy. The potential problems and diffraction problems are initially formulated as dual (or multiple) series equations, or dual (or multiple) integral equations. Central to the technique is the idea of regularisation. The general concept of regularisation is well established in many areas of mathematics. In this context, its main feature is the transformation of the *badly behaved* or singular part of the initial equations, describing a potential distribution or a diffraction process, to a *well behaved* set of equations (technically, second-kind Fredholm equations). Physically, this process of semi-inversion corresponds to solving analytically some associated potential problem, and utilising that solution to determine the full wave scattering.

The two volumes of this text are closely connected. Part I develops the theory of series equations and integral equations, and solves mixed boundary potential problems (mainly electrostatic ones) for structures with cavities and edges. The theory and structure of the dual equations that arise in this process reflect new developments and refinements since the major exposition of Sneddon [55]. In our unified approach, transformations connected with Abel's integral equation are employed to invert analytically the singular part of the operator defining the potential. Three-dimensional structures examined include shells and cavities obtained by opening apertures in canonically shaped closed surfaces; thus a variety of spherical and spheroidal cavities and toroidal and conical shells are considered. Although the main thrust of both volumes concerns three-dimensional effects, some canonical two-dimensional structures, such as slotted elliptical cylinders and various flat plates, are considered. Also, to illustrate how regularisation transforms the standard integral equations of potential theory and benefits subsequent numerical computations, the method is applied to a noncanonical structure, the singly-slotted cylinder of arbitrary cross-section.

Part II examines diffraction of acoustic and electromagnetic waves from similar classes of open structures with edges or cavities. The rigorous regularisation procedure relies on the techniques solutions developed in Part I to produce effective algorithms for the complete frequency range, quasi-static to quasi-optical. Physical interpretation of explicit mathematical solutions and relevant applications are provided.

The two volumes aim to provide an account of some mathematical developments over the last two decades that have greatly enlarged the set of soluble canonical problems of real physical and engineering significance. They gather, perhaps for the first time, a satisfactory mathematical description that accurately quantifies the physically relevant scattering mechanisms in complex structures. Our selection is not exhaustive, but is chosen to illustrate the types of structures that may be analysed by these methods, and to provide a platform for the further analysis of related structures.

In developing a unified treatment of potential theory and diffraction, we have chosen a concrete, rather than an abstract or formal style of analysis. Thus, constructive methods and explicit solutions from which practical numerical algorithms can be implemented, are obtained from an intensive and unified study of series equations and integral equations.

We hope this book will be useful to both new researchers and experienced specialists. Most of the necessary tools for the solution of series equations and integral equations are developed in the text; allied material on special functions and functional analysis is collated in an appendix so that the book is accessible to as wide a readership as possible. It is addressed to mathematicians, physicists, and electrical engineers. The text is suitable for postgraduate courses in diffraction and potential theory and related mathematical methods. It is also suitable for advanced-level undergraduates, particularly for project material.

We wish to thank our partners and families for their support and encouragement in writing this book. Their unfailing good humour and advice played a key role in bringing the text to fruition.

1

Laplace's Equation

Laplace's equation is one of the most important partial differential equations that arises in the application of mathematics to physical phenomena. It occurs in diverse contexts, including electrostatics, magnetostatics, elasticity, gravitation, steady-state heat conduction, incompressible fluid flow, and many related areas described in, for example, [44] and [13].

Common to these disciplines is the notion of a potential ψ, which is a scalar function of spatial position. We will be particularly interested in the electrostatic context, where the potential ψ is constant on equipotential surfaces, and the associated electric field vector $\overrightarrow{E}$ is expressed via the gradient

$$\overrightarrow{E} = -\nabla\psi. \tag{1.1}$$

This vector lies along the direction of most rapid decrease of ψ. Gauss' law states that the divergence of the electric field is proportional to charge density ρ at each point in space,

$$\nabla.\overrightarrow{E} = 4\pi\rho. \tag{1.2}$$

The proportionality factor in Equation (1.2) depends upon the choice of units. We employ Gaussian units [20] throughout; if Système International (SI) units are employed, the right-hand side of (1.2) is divided by $4\pi\varepsilon_0$ where ε_o denotes free space permittivity. (To convert capacitances from Gaussian to SI units, multiply by $4\pi\varepsilon_0$).

From (1.1) and (1.2), *Poisson's equation* follows,

$$\nabla.(\nabla\psi) = \nabla^2\psi = -4\pi\rho. \tag{1.3}$$

This equation describes how the potential is determined by the charge distribution in some region of space.

Now consider an electrostatic field with associated potential ψ. If a perfectly conducting surface S is immersed in this field, a charge distribution ρ^i is induced on the surface; it has an associated potential ψ^i satisfying (1.3). The total potential $\Psi = \psi + \psi^i$ is constant on S (an equipotential surface), the total electric field $-\nabla\Psi$ is normal to S (at each point), and because there are no charges except on S, the total potential satisfies *Laplace's equation,*

$$\nabla^2 \psi = 0, \tag{1.4}$$

at every point of space except on S.

In order to obtain a unique solution that is physically relevant, this partial differential equation must be complemented by appropriate boundary conditions; for example, the potential on one or more metallic conductors might be specified to be of unit value, and Laplace's equation is to be solved in the region excluding the conductors, but subject to this specification on the conductor surface. If one of the conductors encloses a (finite) region of interest, such boundary conditions may be sufficient to specify the required solution uniquely; however, in unbounded regions, some additional specification of the behaviour of the potential at infinity is required. Moreover, the presence of sharp edges on the bounding conducting surfaces may require that additional constraints, equivalent to the finiteness of energy, be imposed to ensure that a physically relevant solution is uniquely defined by Laplace's equation.

In this book we shall be interested in analytic and semi-analytic methods for solving Laplace's equation with appropriate boundary and other conditions. To make substantive progress, we shall consider orthogonal coordinate systems in which Laplace's equation is separable (i.e., it can be solved by the method of separation of variables), and the conductors occupy part or whole of a coordinate surface in these systems.

Laplace's equation can be solved by the method of separation of variables only when the boundary conditions are enforced on a *complete coordinate surface* (e.g., the surface of a sphere in the spherical coordinate system). As indicated in the preface, it is important to emphasize that the methods described in this book apply to a much wider class of surfaces, where the boundary conditions (describing, say, the electrostatic potential of a conductor) are prescribed on only part of a coordinate surface in the following way. Let u_1, u_2, and u_3 be a system of coordinates in which the three sets of coordinate surfaces, $u_1 =$ constant, $u_2 =$ constant, and $u_3 =$ constant, are mutually orthogonal. We shall consider portions of a coordinate surface typically specified by

$$u_1 = \text{constant}, \; a \leq u_2 \leq b \tag{1.5}$$

where a and b are fixed. For example, a spherical cap of radius a and subtending an angle θ_o (at the centre of the appropriate sphere) may be specified in the spherical coordinate system (r, θ, ϕ) by

$$r = a,\ 0 \leq \theta \leq \theta_0,\ 0 \leq \phi \leq 2\pi. \tag{1.6}$$

The determination of the electrostatic potential surrounding the cap can be posed as a *mixed boundary value problem*, and can be solved by the analytic methods of this book, despite its insolubility by the method of separation of variables.

Although the type of surface specified by (1.5) is somewhat restricted, it includes many cases not merely of mathematical interest, but of substantive physical and technological interest as well; the class of surfaces for which analytic solutions to the potential theory problem (of solving Laplace's equation) can be found is thus considerably enlarged, beyond the well-established class of solutions obtained by separation of variables (see, for example [54]). Since it will be central to later developments, Sections 1.1 and 1.2 briefly describe the form of Laplace's equation in some of these orthogonal coordinate systems, and the solutions generated by the classical method of separation of variables.

The formulation of potential theory for structures with edges is expounded in Section 1.3. For the class of surfaces described above, dual (or multiple) series equations arise naturally, as do dual (or multiple) integral equations. Various methods for solving such dual series equations are described in Section 1.4, including the Abel integral transform method that is the key tool employed throughout this text. It exploits features of Abel's integral equation (described in Section 1.5) and Abel-type integral representations of Legendre polynomials, Jacobi polynomials, and related hypergeometric functions (described in Section 1.6). In the final Section (1.7), the equivalence of the dual series approach and the more usual integral equation approach (employing single- or double-layer surface densities) to potential theory is demonstrated.

1.1 Laplace's equation in curvilinear coordinates

The study of Laplace's equation in various coordinate systems has a long history, generating, amongst other aspects, many of the special functions of applied mathematics and physics (Bessel functions, Legendre functions, etc.). In this section we gather material of a reference nature; for a greater depth of detail, we refer the interested reader to one of the numerous texts written on these topics, such as [44], [32] or [74].

Here we consider Laplace's equation in those coordinate systems that will be of concrete interest later in this book; in these systems the method of separation of variables is applicable. Let u_1, u_2, and u_3 be a system of

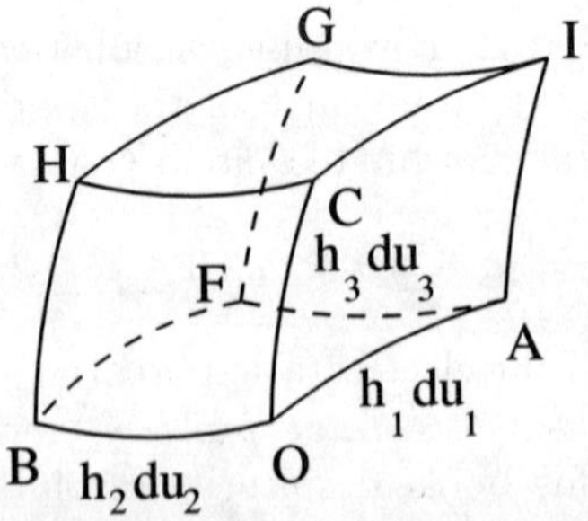

FIGURE 1.1. The elementary parallelepiped.

coordinates in which the coordinate surfaces $u_1 =$ constant, $u_2 =$ constant, and $u_3 =$ constant are mutually orthogonal (i.e., intersect orthogonally). Fix a point (u_1, u_2, u_3) and consider the elementary parallelepiped formed along the coordinate surfaces, as shown in Figure 1.1.

Thus O, A, B, and C have coordinates (u_1, u_2, u_3), $(u_1 + du_1, u_2, u_3)$, $(u_1, u_2 + du_2, u_3)$, and $(u_1, u_2, u_3 + du_3)$, respectively. The length ds of the diagonal line segment connecting (u_1, u_2, u_3) and $(u_1 + du_1, u_2 + du_2, u_3 + du_3)$ is given by

$$ds^2 = h_1^2 du_1^2 + h_2^2 du_2 + h_3^3 du_3^2 \tag{1.7}$$

where h_1, h_2, and h_3 are the metric coefficients (or Lamé coefficients, in recognition of the transformation of the Laplacian to general orthogonal coordinates first effected in [35]).

In terms of the Lamé coefficients, the lengths of the elementary parallelepiped edges equal $h_1 du_1, h_2 du_2$, and $h_3 du_3$, respectively, so that its volume is $h_1 h_2 h_3 du_1 du_2 du_3$. These coefficients depend, in general, upon the coordinates u_1, u_2, u_3 and can be calculated explicitly from the functional relationship between rectangular and curvilinear coordinates,

$$x = x(u_1, u_2, u_3), \;\; y = y(u_1, u_2, u_3), \;\; z = z(u_1, u_2, u_3). \tag{1.8}$$

It is useful to state the relationship between rectangular and curvilinear components of any vector $\overrightarrow{F}$. Designate by $\overrightarrow{i_x}, \overrightarrow{i_y}, \overrightarrow{i_z}$ the unit rectangular (Cartesian) coordinate vectors, and by $\overrightarrow{i_1}, \overrightarrow{i_2}, \overrightarrow{i_3}$ the unit coordinate vectors in the orthogonal curvilinear coordinate system; the unit vectors are defined by the relation (with $\overrightarrow{r} = x\,\overrightarrow{i_x} + y\,\overrightarrow{i_y} + z\,\overrightarrow{i_z}$):

$$\overrightarrow{i_i} = \frac{1}{h_i}\left(\frac{\partial x}{\partial u_i}\overrightarrow{i_x} + \frac{\partial y}{\partial u_i}\overrightarrow{i_y} + \frac{\partial z}{\partial u_i}\overrightarrow{i_z}\right) = \frac{\partial \overrightarrow{r}}{\partial u_i} \Big/ \left|\frac{\partial \overrightarrow{r}}{\partial u_i}\right| \tag{1.9}$$

where $i = 1, 2, 3$, and are mutually orthogonal. Then

$$\overrightarrow{F} = F_x \overrightarrow{i_x} + F_y \overrightarrow{i_y} + F_z \overrightarrow{i_z} = F_1 \overrightarrow{i_1} + F_2 \overrightarrow{i_2} + F_3 \overrightarrow{i_3}. \tag{1.10}$$

Taking inner products yields the following relations:

$$\begin{aligned} F_1 &= F_x(\vec{i_x}, \vec{i_1}) + F_y(\vec{i_y}, \vec{i_1}) + F_z(\vec{i_z}, \vec{i_1}) \\ F_2 &= F_x(\vec{i_x}, \vec{i_2}) + F_y(\vec{i_y}, \vec{i_2}) + F_z(\vec{i_z}, \vec{i_2}) \\ F_3 &= F_x(\vec{i_x}, \vec{i_3}) + F_y(\vec{i_y}, \vec{i_3}) + F_z(\vec{i_z}, \vec{i_3}) \end{aligned} \tag{1.11}$$

The differentials of the rectangular coordinates are linear functions of the curvilinear coordinates:

$$\begin{aligned} dx &= \frac{\partial x}{\partial u_1}du_1 + \frac{\partial x}{\partial u_2}du_2 + \frac{\partial x}{\partial u_3}du_3, \\ dy &= \frac{\partial y}{\partial u_1}du_1 + \frac{\partial y}{\partial u_2}du_2 + \frac{\partial y}{\partial u_3}du_3, \\ dz &= \frac{\partial z}{\partial u_1}du_1 + \frac{\partial z}{\partial u_2}du_2 + \frac{\partial z}{\partial u_3}du_3. \end{aligned} \tag{1.12}$$

Comparing the expression for elementary length $ds^2 = dx^2 + dy^2 + dz^2$ with (1.7), and using orthogonality of the coordinate basis vectors, we obtain

$$h_1^2 du_1^2 + h_2^2 du_2^2 + h_3^2 du_3^2 = dx^2 + dy^2 + dz^2; \tag{1.13}$$

substituting (1.12) into (1.13) and equating like coefficients shows that

$$h_i^2 = \left(\frac{\partial x}{\partial u_i}\right)^2 + \left(\frac{\partial y}{\partial u_i}\right)^2 + \left(\frac{\partial z}{\partial u_i}\right)^2 \quad (i = 1, 2, 3)\,. \tag{1.14}$$

Let $\psi = \psi(u_1, u_2, u_3)$ be a scalar function dependent upon spatial position, and let $\vec{A} = \vec{A}(u_1, u_2, u_3)$ be a vector function of position, the three components of which will be denoted $A_1 = A_1(u_1, u_2, u_3)$, $A_2 = A_2(u_1, u_2, u_3)$, and $A_3 = A_3(u_1, u_2, u_3)$. We wish to find the coordinate expression for the gradient of the scalar ψ (grad ψ) in this system, as well as the divergence (div $\vec{A}$) and circulation or curl (curl $\vec{A}$) of the vector $\vec{A}$.

It follows from Figure 1.1 that the first component of the gradient is

$$(\text{grad}\,\psi)_1 = \lim_{du_1 \to 0} \frac{\psi(u_1 + du_1, u_2, u_3) - \psi(u_1, u_2, u_3)}{h_1 du_1} = \frac{1}{h_1}\frac{\partial \psi}{\partial u_1}. \tag{1.15}$$

Analogously, the other two components are

$$(\text{grad}\,\psi)_2 = \frac{1}{h_2}\frac{\partial \psi}{\partial u_2}, \quad (\text{grad}\,\psi)_3 = \frac{1}{h_3}\frac{\partial \psi}{\partial u_3}. \tag{1.16}$$

To determine the divergence, let us calculate the total flux

$$\int_S \vec{A}.\vec{n}ds$$

of the vector $\vec{A}$ through the surface S of the elementary parallelepiped, the flux being calculated in the direction of the external unit normal $\vec{n}$. The flux through the surface OBHC is $A_1h_2h_3du_2du_3$, whereas the flux through surface AFGI is

$$A_1h_2h_3du_2du_3 + \frac{\partial}{\partial u_1}(A_1h_2h_3)du_1du_2du_3,$$

so the net flux through these two surfaces is

$$\frac{\partial}{\partial u_1}(A_1h_2h_3)du_1du_2du_3.$$

The net flux through the remaining two opposing pairs of surfaces is

$$\frac{\partial}{\partial u_2}(A_2h_3h_1)du_1du_2du_3 \text{ and } \frac{\partial}{\partial u_3}(A_3h_1h_2)du_1du_2du_3.$$

Thus the total flux through the complete parallelepiped surface is

$$\int_S \overrightarrow{A}.\overrightarrow{n}ds = \left[\frac{\partial}{\partial u_1}(A_1h_2h_3) + \frac{\partial}{\partial u_2}(A_2h_3h_1) + \frac{\partial}{\partial u_3}(A_3h_1h_2)\right] du_1du_2du_3.$$

According to the Gauss-Ostrogradsky theorem [74], [32]

$$\int_S \overrightarrow{A}.\overrightarrow{n}ds = \int_V \text{div}\, \overrightarrow{A}dV$$

where V is the volume enclosed by S. A comparison of the last two formulae shows that in curvilinear coordinates the divergence of $\overrightarrow{A}$ is (also denoted $\nabla.\overrightarrow{A}$),

$$\text{div}\,\overrightarrow{A} = \frac{1}{h_1h_2h_3}\left[\frac{\partial}{\partial u_1}(A_1h_2h_3) + \frac{\partial}{\partial u_2}(A_2h_3h_1) + \frac{\partial}{\partial u_3}(A_3h_1h_2)\right]. \tag{1.17}$$

To derive the circulation (curl $\overrightarrow{A}$) of the vector $\overrightarrow{A}$, consider the contour OBHC, which is denoted L. Observing that

$$\begin{aligned}
\int_0^B \overrightarrow{A}.\overrightarrow{dl} &= A_2h_2du_2, \\
\int_H^C \overrightarrow{A}.\overrightarrow{dl} &= -A_2h_2du_2 - \frac{\partial}{\partial u_3}(A_2h_2du_2)du_3, \\
\int_B^H \overrightarrow{A}.\overrightarrow{dl} &= A_3h_3du_3 + \frac{\partial}{\partial u_2}(A_3h_3du_3)du_2, \\
\int_C^O \overrightarrow{A}.\overrightarrow{dl} &= -A_3h_3du_3,
\end{aligned}$$

the circulation along this contour L is

$$\oint_L \overrightarrow{A}.\overrightarrow{dl} = \frac{\partial}{\partial u_2}(A_3h_3du_3)du_2 - \frac{\partial}{\partial u_3}(A_2h_2du_2)du_3.$$

According to Stokes' theorem [74], [32]

$$\oint_L \overrightarrow{A}.\overrightarrow{dl} = \int_S \text{curl}\,\overrightarrow{A}.\overrightarrow{n}ds$$

where S is the surface bounded by L, with the normal $\overrightarrow{n}$ defined above. A comparison of the last two formulae shows that the circulation curl $\overrightarrow{A} \equiv \nabla \times \overrightarrow{A}$ has first component

$$(\text{curl}\,\overrightarrow{A})_1 = \frac{1}{h_2h_3}\left[\frac{\partial}{\partial u_2}(h_3A_3) - \frac{\partial}{\partial u_3}(h_2A_2)\right]. \tag{1.18}$$

Considering the contours OCIA and OAFB, the other two components are

$$(\text{curl}\,\overrightarrow{A})_2 = \frac{1}{h_3h_1}\left[\frac{\partial}{\partial u_3}(h_1A_1) - \frac{\partial}{\partial u_1}(h_3A_3)\right], \tag{1.19}$$

$$(\text{curl}\,\overrightarrow{A})_3 = \frac{1}{h_1h_2}\left[\frac{\partial}{\partial u_1}(h_2A_2) - \frac{\partial}{\partial u_2}(h_1A_1)\right]. \tag{1.20}$$

The Laplacian can now be stated in curvilinear coordinate form, combining (1.15), (1.16), and (1.17) with the definition

$$\Delta\psi = \nabla^2\psi = \text{div}(\text{grad}\,\psi) \tag{1.21}$$

to obtain

$$\nabla^2\psi = \frac{1}{h_1h_2h_3}\left[\frac{\partial}{\partial u_1}(\frac{h_2h_3}{h_1}\frac{\partial\psi}{\partial u_1}) + \frac{\partial}{\partial u_2}\left(\frac{h_3h_1}{h_2}\frac{\partial\psi}{\partial u_2}\right) + \frac{\partial}{\partial u_3}\left(\frac{h_1h_2}{h_3}\frac{\partial\psi}{\partial u_3}\right)\right]. \tag{1.22}$$

Let us gather the explicit form of the metric coefficients, the volume element, and the Laplacian in the various coordinates systems of interest in this book.

1.1.1 Cartesian coordinates

The range of the coordinates is

$$-\infty < x < \infty, -\infty < y < \infty, -\infty < z < \infty.$$

The metric coefficients are $h_x = h_y = h_z = 1$, and the volume element is $dV = dx\,dy\,dz$. The forms of the Laplacian and gradient are, respectively,

$$\Delta\psi = \frac{\partial^2\psi}{\partial x^2} + \frac{\partial^2\psi}{\partial y^2} + \frac{\partial^2\psi}{\partial z^2} = 0, \tag{1.23}$$

$$\nabla\psi = \overrightarrow{i_x}\frac{\partial\psi}{\partial x} + \overrightarrow{i_y}\frac{\partial\psi}{\partial y} + \overrightarrow{i_z}\frac{\partial\psi}{\partial z}. \tag{1.24}$$

The coordinates surfaces ($x, y,$ or $z =$ constant) are planes.

1.1.2 Cylindrical polar coordinates

In terms of Cartesian coordinates, the cylindrical coordinates are

$$x = \rho \cos\phi, \; y = \rho \sin\phi, \; z = z,$$

and the range of the coordinates is $0 \leq \rho < \infty, 0 \leq \phi \leq 2\pi, -\infty < z < \infty$. The metric coefficients are

$$h_\rho = 1, \; h_\phi = \rho, \; h_z = 1,$$

and the volume element is $dV = \rho d\rho \, d\phi \, dz$. The forms of the Laplacian and gradient are, respectively,

$$\Delta\psi = \frac{1}{\rho}\frac{\partial}{\partial\rho}\left(\rho\frac{\partial\psi}{\partial\rho}\right) + \frac{1}{\rho^2}\frac{\partial^2\psi}{\partial\phi^2} + \frac{\partial^2\psi}{\partial z^2}, \tag{1.25}$$

$$\nabla\psi = \vec{i_\rho}\frac{\partial\psi}{\partial\rho} + \vec{i_\phi}\frac{1}{\rho}\frac{\partial\psi}{\partial\phi} + \vec{i_z}\frac{\partial\psi}{\partial z}. \tag{1.26}$$

The coordinates surfaces are cylinders ($\rho =$ constant), planes through the z-axis ($\phi =$ constant), or planes perpendicular to the z-axis ($z =$ constant).

1.1.3 Spherical polar coordinates

In terms of Cartesian coordinates, the spherical coordinates are

$$x = r\sin\theta\cos\phi, \; y = r\sin\theta\sin\phi, \; z = r\cos\theta,$$

and the range of the coordinates is $0 \leq r < \infty, 0 \leq \theta \leq \pi, 0 \leq \phi \leq 2\pi$. The metric coefficients are

$$h_r = 1, \; h_\theta = r, \; h_\phi = r\sin\theta.$$

The volume element is $dV = r^2 \sin\theta dr \, d\theta \, d\phi$ and the forms of the Laplacian and gradient are, respectively,

$$\Delta\psi = \frac{1}{r^2}\frac{\partial}{\partial r}\left(r^2\frac{\partial\psi}{\partial r}\right) + \frac{1}{r^2\sin\theta}\frac{\partial}{\partial\theta}\left(\sin\theta\frac{\partial\psi}{\partial\theta}\right) + \frac{1}{r^2\sin^2\theta}\frac{\partial^2\psi}{\partial\phi^2}, \tag{1.27}$$

$$\nabla\psi = \vec{i_r}\frac{\partial\psi}{\partial r} + \vec{i_\theta}\frac{1}{r}\frac{\partial\psi}{\partial\theta} + \vec{i_\phi}\frac{1}{r\sin\theta}\frac{\partial\psi}{\partial\phi}. \tag{1.28}$$

The coordinates surfaces are spheres ($r =$ constant), right circular cones ($\theta =$ constant), or azimuthal planes containing the z-axis ($\phi =$ constant).

1.1.4 Prolate spheroidal coordinates

There are two commonly used systems of spheroidal coordinates employing coordinates denoted (ξ, η, φ) and (α, β, φ), respectively. In terms of Cartesian coordinates, the first representation is

$$x = \frac{d}{2}\sqrt{(1-\eta^2)(\xi^2-1)}\cos\varphi,\; y = \frac{d}{2}\sqrt{(1-\eta^2)(\xi^2-1)}\sin\varphi,\; z = \frac{d}{2}\eta\xi,$$

where the parameter d will be identified as the interfocal distance; the range of coordinates is $1 \le \xi < \infty, -1 \le \eta \le 1, 0 \le \phi < 2\pi$.

The coordinate surface $\xi =$ constant > 1 is a prolate spheroid with foci at the points $(x, y, z) = (0, 0, \pm\frac{d}{2})$, with major semi-axis $b = \frac{d}{2}\xi$, and minor semi-axis $a = \frac{d}{2}\left(\xi^2 - 1\right)^{\frac{1}{2}}$,

$$\frac{x^2+y^2}{(\xi^2-1)} + \frac{z^2}{\xi^2} = \left(\frac{d}{2}\right)^2;$$

the degenerate surface $\xi = 1$ is the straight line segment $|z| \le \frac{d}{2}$. The coordinate surface $|\eta| =$ constant < 1 is a hyperboloid of revolution of two sheets with an asymptotic cone whose generating line passes through the origin and is inclined at an angle $\beta = \cos^{-1}(\eta)$ to the z–axis,

$$\frac{z^2}{\eta^2} - \frac{x^2+y^2}{(1-\eta^2)} = \left(\frac{d}{2}\right)^2;$$

the degenerate surface $|\eta| = 1$ is that part of the z–axis for which $|z| > \frac{1}{2}d$. The surface $\varphi =$ constant is a half-plane containing the z–axis and forming angle φ with the x, z–plane.

In the limit when the interfocal distance approaches zero and ξ tends to infinity, the prolate spheroidal system (ξ, η, φ) reduces to the spherical system $(r, \theta, \phi_{sphere})$ by making the identification

$$\frac{d}{2}\xi = r,\; \eta = \cos\theta,\; \varphi \equiv \phi_{sphere}$$

in such a way that the product $\frac{d}{2}\xi$ remains finite as $d \to 0, \xi \to \infty$.

The second representation (α, β, φ) of prolate spheroidal coordinates is obtained by setting $\xi = \cosh\alpha$ and $\eta = \cos\beta$ so that in terms of Cartesian coordinates

$$x = \frac{d}{2}\sinh\alpha\sin\beta\cos\varphi,\; y = \frac{d}{2}\sinh\alpha\sin\beta\sin\varphi,\; z = \frac{d}{2}\cosh\alpha\cos\beta.$$

The range of coordinates is $0 \le \alpha < \infty, 0 \le \beta \le \pi, 0 \le \phi < 2\pi$. Both representations are used equally in this book.

The metric coefficients are, respectively,

$$h_\xi = \frac{d}{2}\sqrt{\frac{\xi^2 - \eta^2}{\xi^2 - 1}},\ h_\eta = \frac{d}{2}\sqrt{\frac{\xi^2 - \eta^2}{1 - \eta^2}},\ h_\phi = \frac{d}{2}\sqrt{(\xi^2 - 1)(1 - \eta^2)}$$

and

$$h_\alpha = h_\beta = \frac{d}{2}\sqrt{\sinh^2\alpha + \sin^2\beta},\ h_\phi = \frac{d}{2}\sinh\alpha\sin\beta;$$

the volume element is

$$dV = \left(\frac{d}{2}\right)^3 (\xi^2 - \eta^2)\, d\xi\, d\eta\, d\phi$$
$$= \left(\frac{d}{2}\right)^3 (\sinh^2\alpha + \sin^2\beta)\sinh\alpha\sin\beta d\alpha d\beta d\phi.$$

The forms of the Laplacian and gradient are, respectively,

$$\left(\frac{d}{2}\right)^2 \Delta\psi = \frac{1}{(\xi^2 - \eta^2)}\left\{\frac{\partial}{\partial\xi}\left((\xi^2 - 1)\frac{\partial\psi}{\partial\xi}\right) + \frac{\partial}{\partial\eta}\left((1 - \eta^2)\frac{\partial\psi}{\partial\eta}\right)\right\} + \frac{1}{(\xi^2 - 1)(1 - \eta^2)}\frac{\partial^2\psi}{\partial\phi^2}, \quad (1.29)$$

$$\left(\frac{d}{2}\right)\nabla\psi = \vec{i_\xi}\sqrt{\frac{\xi^2 - 1}{\xi^2 - \eta^2}}\frac{\partial\psi}{\partial\xi} + \vec{i_\eta}\sqrt{\frac{1 - \eta^2}{\xi^2 - \eta^2}}\frac{\partial\psi}{\partial\eta} + \vec{i_\phi}\left[(\xi^2 - 1)(1 - \eta^2)\right]^{-\frac{1}{2}}\frac{\partial\psi}{\partial\phi}, \quad (1.30)$$

and

$$\left(\frac{d}{2}\right)^2 (\sinh^2\alpha + \sin^2\beta)\Delta\psi$$
$$= \frac{1}{\sinh\alpha}\frac{\partial}{\partial\alpha}\left(\sinh\alpha\frac{\partial\psi}{\partial\alpha}\right) + \frac{1}{\sin\beta}\frac{\partial}{\partial\beta}\left(\sin\beta\frac{\partial\psi}{\partial\beta}\right) + \left(\frac{1}{\sinh^2\alpha} + \frac{1}{\sin^2\beta}\right)\frac{\partial^2\psi}{\partial\phi^2}, \quad (1.31)$$

$$\left(\frac{d}{2}\right)\nabla\psi = \frac{1}{\sqrt{\sinh^2\alpha + \sin^2\beta}}\left\{\vec{i_\alpha}\frac{\partial\psi}{\partial\alpha} + \vec{i_\beta}\frac{\partial\psi}{\partial\beta}\right\} + \vec{i_\phi}\frac{1}{\sinh\alpha\sin\beta}\frac{\partial\psi}{\partial\phi}. \quad (1.32)$$

1.1.5 Oblate spheroidal coordinates

As with the prolate system, there are two commonly used systems of oblate spheroidal coordinates employing coordinates denoted (ξ, η, φ) and (α, β, φ), respectively. In terms of Cartesian coordinates, the first representation is

$$x = \frac{d}{2}\sqrt{(1-\eta^2)(\xi^2+1)}\cos\phi,\ y = \frac{d}{2}\sqrt{(1-\eta^2)(\xi^2+1)}\sin\phi,\ z = \frac{d}{2}\eta\xi$$

where the parameter d will be identified as interfocal distance; the range of the coordinates is $0 \le \xi < \infty, -1 \le \eta \le 1, 0 \le \phi < 2\pi$. The coordinate surface $\xi =$ constant is an oblate spheroid with foci at the points $(x, y, z) = \pm\left(\frac{d}{2}, \frac{d}{2}, 0\right)$,

$$\frac{x^2+y^2}{(\xi^2+1)} + \frac{z^2}{\xi^2} = \left(\frac{d}{2}\right)^2;$$

the degenerate surface $\xi = 0$ is the disk $x^2 + y^2 \le \left(\frac{d}{2}\right)^2$ in the plane $z = 0$. The coordinate surface $\eta =$ constant is a one-sheeted hyperboloid of revolution, with an asymptotic cone whose generating line passes through the origin and is inclined at the angle $\beta = \cos^{-1}(\eta)$ to the $z-$axis,

$$\frac{x^2+y^2}{(1-\eta^2)} - \frac{z^2}{\eta^2} = \left(\frac{d}{2}\right)^2.$$

The coordinate surface $\phi =$ constant is a half-plane containing the z-axis.

The second representation (α, β, φ) of oblate spheroidal coordinates is obtained by setting $\xi = \sinh\alpha$ and $\eta = \cos\beta$ so that in terms of Cartesian coordinates

$$x = \frac{d}{2}\cosh\alpha\sin\beta\cos\varphi,\ y = \frac{d}{2}\cosh\alpha\sin\beta\sin\varphi,\ z = \frac{d}{2}\sinh\alpha\cos\beta,$$

where the range of coordinates is $0 \le \alpha < \infty, 0 \le \beta \le \pi, 0 \le \phi < 2\pi$.

The metric coefficients are, respectively,

$$h_\xi = \frac{d}{2}\sqrt{\frac{\xi^2+\eta^2}{\xi^2+1}},\ h_\eta = \frac{d}{2}\sqrt{\frac{\xi^2+\eta^2}{1-\eta^2}},\ h_\phi = \frac{d}{2}\sqrt{(\xi^2+1)(1-\eta^2)},$$

and

$$h_\alpha = h_\beta = \frac{d}{2}\sqrt{\cosh^2\alpha - \sin^2\beta},\ h_\phi = \frac{d}{2}\cosh\alpha\sin\beta.$$

The forms of the Laplacian and gradient are, respectively,

$$\left(\frac{d}{2}\right)^2\Delta\psi = \frac{1}{(\xi^2+\eta^2)}\left\{\frac{\partial}{\partial\xi}\left((\xi^2+1)\frac{\partial\psi}{\partial\xi}\right) + \frac{\partial}{\partial\eta}\left((1-\eta^2)\frac{\partial\psi}{\partial\eta}\right)\right\} + \frac{1}{(\xi^2+1)(1-\eta^2)}\frac{\partial^2\psi}{\partial\phi^2}, \quad (1.33)$$

$$\left(\frac{d}{2}\right)\nabla\psi = \vec{i_\xi}\sqrt{\frac{\xi^2+1}{\xi^2+\eta^2}}\frac{\partial\psi}{\partial\xi} + \vec{i_\eta}\sqrt{\frac{1-\eta^2}{\xi^2+\eta^2}}\frac{\partial\psi}{\partial\eta} + \vec{i_\phi}\frac{1}{\sqrt{(\xi^2+1)(1-\eta^2)}}\frac{\partial\psi}{\partial\phi}, \quad (1.34)$$

and

$$\left(\frac{d}{2}\right)^2(\cosh^2\alpha - \sin^2\beta)\,\Delta\psi = \frac{1}{\cosh\alpha}\frac{\partial}{\partial\alpha}\left(\cosh\alpha\frac{\partial\psi}{\partial\alpha}\right) + \frac{1}{\sin\beta}\frac{\partial}{\partial\beta}\left(\sin\beta\frac{\partial\psi}{\partial\beta}\right) + \left(\frac{1}{\sin^2\beta} - \frac{1}{\cosh^2\alpha}\right)\frac{\partial^2\psi}{\partial\phi^2}, \quad (1.35)$$

$$\left(\frac{d}{2}\right)\nabla\psi = \frac{1}{\sqrt{\cosh^2\alpha - \sin^2\beta}}\left\{\vec{i_\alpha}\frac{\partial\psi}{\partial\alpha} + \vec{i_\beta}\frac{\partial\psi}{\partial\beta}\right\} + \vec{i_\phi}\frac{1}{\cosh\alpha\sin\beta}\frac{\partial\psi}{\partial\phi}. \quad (1.36)$$

1.1.6 *Elliptic cylinder coordinates*

In terms of Cartesian coordinates, the elliptic cylinder coordinates are

$$x = \frac{d}{2}\cosh\alpha\cos\beta,\ y = \frac{d}{2}\sinh\alpha\sin\beta,\ z = z,$$

where the range of the coordinates is $-\infty < \alpha < \infty$, $0 \le \beta \le \pi$, $-\infty < z < \infty$. The metric coefficients are

$$h_\alpha = h_\beta = \frac{d}{2}\sqrt{\cosh^2\alpha - \cos^2\beta}, h_z = 1,$$

and the volume element is $dV = \left(\frac{d}{2}\right)^3(\cosh^2\alpha - \cos^2\beta)$. The forms of the Laplacian and gradient are, respectively,

$$\Delta\psi = \frac{1}{\left(\frac{d}{2}\right)^2(\cosh^2\alpha - \cos^2\beta)}\left\{\frac{\partial^2\psi}{\partial\alpha^2} + \frac{\partial^2\psi}{\partial\beta^2}\right\} + \frac{\partial^2\psi}{\partial z^2}, \quad (1.37)$$

$$\nabla\psi = \frac{1}{\left(\frac{d}{2}\right)(\cosh^2\alpha - \cos^2\beta)^{\frac{1}{2}}}\left\{\vec{i_\alpha}\frac{\partial\psi}{\partial\alpha} + \vec{i_\beta}\frac{\partial\psi}{\partial\beta}\right\} + \vec{i_z}\frac{\partial\psi}{\partial z}.$$

An alternative representation employs $\xi = \cosh\alpha, \eta = \cos\beta$, so that

$$x = \frac{d}{2}\xi\eta,\ y = \frac{d}{2}\sqrt{(\xi^2-1)(1-\eta^2)},\ z = z,$$

where the range of the coordinates is $1 \leq \xi < \infty, -1 \leq \eta \leq 1, -\infty < z < \infty$. The metric coefficients are

$$h_\xi = \frac{d}{2}\sqrt{\frac{\xi^2 - \eta^2}{\xi^2 - 1}}, \quad h_\eta = \frac{d}{2}\sqrt{\frac{\xi^2 - \eta^2}{1 - \eta^2}}, \quad h_z = 1.$$

The volume element is $dV = \left(\frac{d}{2}\right)^3 \left(\xi^2 - \eta^2\right) \left\{\left(\xi^2 - 1\right)\left(1 - \eta^2\right)\right\}^{-\frac{1}{2}} d\xi \, d\eta \, dz$. The forms of the Laplacian and gradient are, respectively,

$$\Delta\psi = \frac{\sqrt{\xi^2 - 1}}{\left(\frac{d}{2}\right)^2 \left(\xi^2 - \eta^2\right)} \frac{\partial}{\partial \xi}\left(\sqrt{\xi^2 - 1}\frac{\partial \psi}{\partial \xi}\right) + \frac{\sqrt{1 - \eta^2}}{\left(\frac{d}{2}\right)^2 \left(\xi^2 - \eta^2\right)} \frac{\partial}{\partial \eta}\left(\sqrt{1 - \eta^2}\frac{\partial \psi}{\partial \eta}\right) + \frac{\partial^2 \psi}{\partial z^2} \quad (1.38)$$

$$\nabla\psi = \vec{i_\xi}\left(\frac{d}{2}\right)^{-1}\sqrt{\frac{\xi^2 - 1}{\xi^2 - \eta^2}}\frac{\partial \psi}{\partial \xi} + \vec{i_\eta}\left(\frac{d}{2}\right)^{-1}\sqrt{\frac{1 - \eta^2}{\xi^2 - \eta^2}}\frac{\partial \psi}{\partial \eta} + \vec{i_z}\frac{\partial \psi}{\partial z} \quad (1.39)$$

The coordinate surfaces are confocal elliptic cylinders with semi-focal distance $\frac{d}{2}$(when ξ or α is constant) or confocal, one-sheeted hyperbolic cylinders (when η or β is constant), or planes perpendicular to the z-axis ($z =$ constant).

1.1.7 *Toroidal coordinates*

In terms of Cartesian coordinates, the toroidal coordinates employ a scale factor $c > 0$ and

$$x = \frac{c \sinh \alpha \cos \phi}{\cosh \alpha - \cos \beta}, \quad y = \frac{c \sinh \alpha \sin \phi}{\cosh \alpha - \cos \beta}, \quad z = \frac{c \sin \beta}{\cosh \alpha - \cos \beta},$$

where the range of the coordinates is $0 \leq \alpha < \infty, -\pi \leq \beta \leq \pi, -\pi \leq \phi \leq \pi$. The metric coefficients are

$$h_\alpha = h_\beta = \frac{c}{\cosh \alpha - \cos \beta}, \quad h_\phi = \frac{c \sinh \alpha}{\cosh \alpha - \cos \beta},$$

and the volume element is $dV = c^3 \sinh \alpha \left(\cosh \alpha - \cos \beta\right)^{-3} d\alpha \, d\beta \, d\phi$. The form of the Laplacian and gradient can be expressed as

$$h_\alpha h_\beta h_\phi \Delta\psi = \frac{\partial}{\partial \alpha}\left(h_\phi \frac{\partial \psi}{\partial \alpha}\right) + \frac{\partial}{\partial \beta}\left(h_\phi \frac{\partial \psi}{\partial \beta}\right) + \frac{1}{\left(\cosh \alpha - \cos \beta\right)\sinh \alpha}\frac{\partial^2 \psi}{\partial \phi^2}, \quad (1.40)$$

$$\nabla\psi = \overrightarrow{i_\alpha} c^{-1} (\cosh\alpha - \cos\beta) \frac{\partial\psi}{\partial\alpha} + \overrightarrow{i_\beta} c^{-1} (\cosh\alpha - \cos\beta) \frac{\partial\psi}{\partial\beta} + \overrightarrow{i_z} c^{-1} \frac{(\cosh\alpha - \cos\beta)}{\sinh\alpha} \frac{\partial\psi}{\partial\phi}. \quad (1.41)$$

The coordinate surfaces corresponding to constant α are tori (with minor radius $r = c/\sinh\alpha$ and major radius $R = c\coth\alpha$, the tori are $\left(\sqrt{x^2+y^2} - R\right)^2 + z^2 = r^2$); for constant β, the coordinate surfaces are spheres of radius $a = c/\sin\beta$ and centre on the z-axis at $(x,y,z) = (0,0,b)$, where $b = c\cot\beta$; the coordinate surfaces of constant ϕ are azimuthal planes containing the z-axis. (See Figure 5.1.)

1.2 Solutions of Laplace's equation: separation of variables

In this section we describe the solutions to Laplace's equation generated by the classical method of separation of variables. A knowledge of these solutions is essential for the approach to the solution of mixed boundary value problems described in the next section, because it depends upon the formulation of an appropriate set of dual series equations with special function kernels.

1.2.1 Cartesian coordinates

We seek a solution to Laplace's equation in the form

$$\psi(x,y,z) = X(x)Y(y)Z(z). \quad (1.42)$$

Substitution in Equation (1.23) transforms it to

$$\frac{1}{X}\frac{d^2X}{dx^2} + \frac{1}{Y}\frac{d^2Y}{dy^2} + \frac{1}{Z}\frac{d^2Z}{dz^2} = 0. \quad (1.43)$$

Each term in this equation is a function of only one independent variable, so there are constants ("separation constants") ν and μ such that

$$\frac{1}{X}\frac{d^2X}{dx^2} = -\nu^2 \Rightarrow X'' + \nu^2 X = 0, \quad (1.44)$$

$$\frac{1}{Y}\frac{d^2Y}{dy^2} = -\mu^2 \Rightarrow Y'' + \nu^2 Y = 0, \quad (1.45)$$

and hence

$$\frac{1}{Z}\frac{d^2Z}{dz^2} - \left(\nu^2 + \mu^2\right) = 0 \Rightarrow Z'' - \left(\nu^2 + \mu^2\right) Z = 0. \quad (1.46)$$

Thus the original equation involving partial derivatives has been reduced to three ordinary differential equations.

The process just described is the classical process of separation of variables and leads to infinitely many solutions of the form (1.42), depending on the parameters ν and μ, which can take real or complex values. The solution of Equations (1.44)–(1.46) can be expressed in terms of elementary functions of form

$$X_\nu(x) = A_\nu \cos \nu x + B_\nu \sin \nu x, \tag{1.47}$$

$$Y_\mu(y) = C_\mu \cos \mu y + D_\mu \sin \mu y, \tag{1.48}$$

and

$$Z_{\nu,\mu}(z) = E_{\nu,\mu} e^{-\sqrt{\nu^2+\mu^2} z} + F_{\nu,\mu} e^{+\sqrt{\nu^2+\mu^2} z}, \tag{1.49}$$

where $A_\nu, B_\nu, C_\mu, D_\mu, E_{\nu,\mu}$, and $F_{\nu,\mu}$ are constants.

The required solution of the given physical problem is obtained by linear superposition of the particular solutions (1.42) formed from (1.47)–(1.49), of the form

$$\sum_{\nu,\mu} X_\nu(x) Y_\mu(y) Z_{\nu,\mu}(z) \quad \text{or} \quad \iint X_\nu(x) Y_\mu(y) Z_{\nu,\mu}(z) d\nu d\mu,$$

where the specific conditions of the problem dictate the range of parameters ν, μ used in the summation or integration as appropriate.

1.2.2 *Cylindrical polar coordinates*

Applying the method of separation of variables, the Laplace Equation (1.25) has particular solutions of the form

$$\psi(\rho, \phi, z) = R(\rho)\Phi(\phi)Z(z), \tag{1.50}$$

where

$$\frac{1}{\rho}\frac{d}{d\rho}(\rho\frac{dR}{d\rho}) + \left(\lambda^2 - \frac{\mu^2}{\rho^2}\right) R = 0, \tag{1.51}$$

$$\frac{d^2\Phi}{d\phi^2} + \mu^2\Phi = 0, \tag{1.52}$$

$$\frac{d^2 Z}{dz^2} - \lambda^2 Z = 0, \tag{1.53}$$

and λ and μ are the "separation constants." The solutions of the latter two equations are the same as those considered above in (1.44) and (1.46):

$$\Phi_\mu(\phi) = A_\mu \cos(\mu\phi) + B_\mu \sin(\mu\phi), \tag{1.54}$$

$$Z_\lambda(z) = C_\lambda e^{-\lambda z} + D_\lambda e^{+\lambda z}. \tag{1.55}$$

Equation (1.51) cannot be expressed in terms of elementary functions; rescaling $u = \lambda\rho$, we obtain Bessel's differential equation (see Appendix B.5),

$$u\frac{d}{du}(u\frac{dR}{du}) + (u^2 - \mu^2)R = 0. \tag{1.56}$$

Its solutions are linear combinations of Bessel functions,

$$R_{\lambda,\mu}(\rho) = E_{\lambda,\mu} J_\mu(\lambda\rho) + F_{\lambda,\mu} Y_\mu(\lambda\rho), \tag{1.57}$$

where $J_\mu(\lambda\rho)$ and $Y_\mu(\lambda\rho)$ are the Bessel functions of order μ, of first and second kind, respectively.

1.2.3 Spherical polar coordinates

In spherical polars, the Laplace Equation (1.27) has separated solutions

$$\psi(r, \theta, \phi) = R(r)\Theta(\theta)\Phi(\phi) \tag{1.58}$$

where

$$\frac{1}{r^2}\frac{d}{dr}(r^2\frac{dR}{dr}) - \frac{\nu(\nu+1)}{r^2}R = 0, \tag{1.59}$$

$$\frac{1}{\sin\theta}\frac{d}{d\theta}\left(\sin\theta\frac{d\Theta}{d\theta}\right) + \left[\nu(\nu+1) - \frac{\mu^2}{\sin^2\theta}\right]\Theta = 0, \tag{1.60}$$

$$\frac{d^2\Phi}{d\phi^2} + \mu^2\Phi = 0, \tag{1.61}$$

and μ, ν are the most conveniently chosen forms of the separation constants. The solutions of these equations are

$$R(r) = A_\nu r^\nu + B_\nu r^{-\nu-1}, \tag{1.62}$$

$$\Theta(\theta) = C_{\nu,\mu} P_\nu^\mu(\cos\theta) + D_{\nu,\mu} Q_\nu^\mu(\cos\theta), \tag{1.63}$$

$$\Phi(\phi) = E_\mu \cos \mu\phi + F_\mu \sin \mu\phi, \tag{1.64}$$

where $P_\nu^\mu(\cos\theta)$ and $Q_\nu^\mu(\cos\theta)$ are the associated Legendre functions (see Appendix B.4) of the first and second kind, respectively. When boundary conditions are applied on spherical coordinate surfaces, no boundaries of which lie along the planes ϕ = constant, enforcement of continuity and of periodicity upon Φ requires that μ be zero or a positive integer, i.e., $\mu = m\,(m = 0, 1, 2...)$. The Legendre functions $P_\nu^m(\cos\theta)$ are finite over the range $0 \leq \theta \leq \pi$ only when ν is an integer n, equal to m, or larger. These requirements, of periodicity of the solution over the range $0 \leq \theta \leq \pi$, and of its finiteness, restrict the separation constants so that the particular solutions of Laplace's equation in spherical coordinates are linear combinations of

$$r^n Y_{mn}^{(e)},\ r^n Y_{mn}^{(o)},\ r^{-n-1} Y_{mn}^{(e)},\ \text{and}\ r^{-n-1} Y_{mn}^{(o)},$$

where

$$Y_{mn}^{(e)} = \cos(m\phi) P_n^m(\cos\theta) \text{ and } Y_{mn}^{(o)} = \sin(m\phi) P_n^m(\cos\theta) \tag{1.65}$$

are the "spherical harmonics." Those harmonics with $m = 0$ are zonal harmonics (since these functions depend only on θ, the nodal lines divide the sphere into zones), those with $m = n$ are sectoral harmonics (since these functions depend only on ϕ, the nodal lines divide the sphere into sectors), and the rest, for $0 < m < n$, are known as tesseral harmonics. Their properties are described in the references in Appendix B.

1.2.4 Prolate spheroidal coordinates

The separated solutions of Laplace's equation in prolate spheroidal coordinates (1.29) are

$$\psi(\xi, \eta, \phi) = X(\xi) H(\eta) \Phi(\phi),$$

where

$$\frac{d}{d\xi}\left[(\xi^2 - 1)\frac{dX}{d\xi}\right] - \left[n(n+1) + \frac{m^2}{\xi^2 - 1}\right] X = 0, \tag{1.66}$$

$$\frac{d}{d\eta}\left[(1 - \eta^2)\frac{dH}{d\eta}\right] + \left[n(n+1) - \frac{m^2}{1 - \eta^2}\right] H = 0, \tag{1.67}$$

$$\frac{d^2\Phi}{d\phi^2} + m^2\Phi = 0. \tag{1.68}$$

The separation constants are n and m. Admissible solutions of the third equation, with periodic boundary conditions on Φ, are

$$\Phi_m(\phi) = E_m \cos(m\phi) + D_m \sin(m\phi), \tag{1.69}$$

where m is zero or a positive integer. The first and second equations have as solutions the associated Legendre functions P_n^m and Q_n^m of the first and second kind. For the second equation, if $\eta \in [-1, 1]$, the only finite solutions (at $\eta = \pm 1$) for H must be proportional to the Legendre function of the first kind, $P_n^m(\eta)$, where n is zero or a positive integer; if this restriction is removed

$$H(\eta) = C_n^m P_n^m(\eta) + D_n^m Q_n^m(\eta). \tag{1.70}$$

The maximum range of the variable ξ is $[1, \infty)$. For most values of n and m there is no solution to (1.66) which is finite over the whole of this interval, so we use whatever linear combination of $P_n^m(\xi)$ and $Q_n^m(\xi)$ that is finite inside the boundaries of the problem,

$$X(\xi) = A_n^m P_n^m(\xi) + B_n^m Q_n^m(\xi). \tag{1.71}$$

In this way, the partial solution of Laplace's equation $\psi_{nm}(\xi, \eta, \phi)$ is the product of (1.69)–(1.71).

In the alternative representation of Laplace's Equation (1.31), the separated solutions take the form

$$\psi(\alpha, \beta, \phi) = A(\alpha)B(\beta)\Phi(\phi),$$

where Φ satisfies (1.68); A satisfies

$$\frac{1}{\sinh\alpha}\frac{d}{d\alpha}\left(\sinh\alpha\frac{dA}{d\alpha}\right) - \left[n(n+1) + \frac{m^2}{\sinh^2\alpha}\right]A = 0, \tag{1.72}$$

so that it is a linear combination of $P_n^m(\cosh\alpha)$ and $Q_n^m(\cosh\alpha)$; and B satisfies

$$\frac{1}{\sin\beta}\frac{d}{d\beta}\left(\sin\beta\frac{dB}{d\beta}\right) + \left[n(n+1) - \frac{m^2}{\sin^2\beta}\right]B = 0, \tag{1.73}$$

so that it is a linear combination of $P_n^m(\cos\beta)$ and $Q_n^m(\cos\beta)$.

1.2.5 Oblate spheroidal coordinates

The separated equations for the θ- and η- coordinates are the same as for prolate spheroids, generating solutions $\sin m\theta, \cos m\theta$ and $P_n^m(\eta)$, where m and n are positive integers (or zero). The equation for the ξ- coordinate has solutions $P_n^m(i\xi)$ and $Q_n^m(i\xi)$. Thus, the partial solutions of Laplace's equation in this system have the form

$$\phi_{nm}(\xi,\eta,\theta) = [A_n^m P_n^m(i\xi) + B_n^m Q_n^m(i\xi)]\, P_n^m(\eta)\, [E_m \cos m\theta + F_m \sin m\theta]\,. \tag{1.74}$$

In the alternative form of Laplace's equation the separated equations have solutions $\sin m\theta, \cos m\theta, P_n^m(\cos\beta)$, and $P_n^m(i\sinh\alpha)$, $Q_n^m(i\sinh\alpha)$. The partial solutions are similar to the form of (1.74).

1.2.6 Elliptic cylinder coordinates

The separated solutions of Laplace's Equation (1.38) in elliptic cylinder coordinates are

$$\psi(\xi,\eta,z) = A(\alpha)B(\beta)Z(z)$$

where, in general, A and B satisfy Mathieu's equation and the modified Mathieu equation, respectively. For a full description of these functions and their properties, the reader is referred to [40] and [75]. If ψ is independent of z, Laplace's equation becomes

$$\frac{\partial^2\psi}{\partial\alpha^2} + \frac{\partial^2\psi}{\partial\beta^2} = 0,$$

which has separated solutions

$$B(\beta) = B_m^1 \cos m\beta + B_m^2 \sin m\beta,$$

$$A(\alpha) = A_m^1 e^{-m\alpha} + A_m^2 e^{m\alpha}.$$

1.2.7 Toroidal coordinates

Our treatment of the method of separation of variables in this system is based on that given by N.N. Lebedev [36]. Unlike the cases considered previously, we cannot directly separate variables in Equation (1.40). However, define a new function V by

$$\psi = V\sqrt{2\cosh\alpha - 2\cos\beta},$$

where $\sqrt{2\cosh\alpha - 2\cos\beta}$ may be called the "asymmetry factor;" Laplace's Equation (1.40) becomes

$$\frac{d^2V}{d\alpha^2} + \frac{d^2V}{d\beta^2} + \coth\alpha\frac{dV}{d\alpha} + \frac{1}{4}V + \frac{1}{\sinh^2\alpha}\frac{d^2V}{d\phi^2} = 0.$$

This admits separation of variables: setting $V = A(\alpha)B(\beta)\Phi(\phi)$, we find that

$$\sinh^2\alpha\left[\frac{1}{A}\frac{d^2A}{d\alpha^2}+\frac{1}{B}\frac{d^2\beta}{d\beta^2}+\frac{\coth\alpha}{A}\frac{dA}{d\alpha}+\frac{1}{4}\right]=-\frac{1}{\Phi}\frac{d^2\Phi}{d\phi^2}=\mu^2,$$

where μ^2 is a constant. This implies

$$\frac{d^2\Phi}{d\phi^2}+\mu^2\Phi=0,$$

$$\frac{1}{A}\frac{d^2A}{d\alpha^2}+\frac{\coth\alpha}{A}\frac{dA}{d\alpha}+\frac{1}{4}-\frac{\mu^2}{\sinh^2\alpha}=-\frac{1}{B}\frac{d^2B}{d\beta^2}=\nu^2,$$

where ν^2 is another constant, so that

$$\frac{d^2B}{d\beta^2}+\nu^2B=0,$$

$$\frac{1}{\sinh\alpha}\frac{d}{d\alpha}\left(\sinh\alpha\frac{dA}{d\alpha}\right)-\left(\nu^2-\frac{1}{4}+\frac{\mu^2}{\sinh^2\alpha}\right)A=0. \tag{1.75}$$

Thus Laplace's equation in toroidal coordinates has infinitely many particular solutions of the form

$$\phi=\sqrt{2\cosh\alpha-2\cos\beta}\,A_{\mu,\nu}(\alpha)B_\nu(\beta)\Phi_\mu(\phi),$$

where

$$B_\nu=C_\nu\cos(\nu\beta)+D_\nu\sin(\nu\beta),$$

$$\Phi_\mu(\phi)=E_\mu\cos(\mu\phi)+F_\mu\sin(\mu\phi),$$

and $A=A_{\mu,\nu}$ satisfies (1.75). The introduction of a new variable $z=\cosh\alpha$ into this equation transforms it to

$$\frac{d}{dz}\left[(1-z^2)\frac{dA}{dz}\right]+\left[(\nu-\frac{1}{2})(\nu+\frac{1}{2})-\frac{\mu^2}{1-z^2}\right]A=0,$$

which may be recognised as the differential equation for the associated Legendre functions $P^\mu_{\nu-\frac{1}{2}}$ or $Q^\mu_{\nu-\frac{1}{2}}$; thus

$$A_{\nu,\mu}(\alpha)=G_{\nu,\mu}P^\mu_{\nu-\frac{1}{2}}(\cosh\alpha)+H_{\nu,\mu}Q^\mu_{\nu-\frac{1}{2}}(\cosh\alpha).$$

1.3 Formulation of potential theory for structures with edges

The focus of this book is potential theory – the study of solutions of Laplace's equation – especially for structures in which edge effects are important. As already indicated, the boundary conditions must be supplemented by a decay condition at infinity as well as finite energy constraints near edges, so that a unique and physically relevant solution can be found.

Since edges introduce distinctive features into the theory, let us distinguish between closed surfaces, those possessing no boundary or edge, and open shells, which have one or more boundaries. A spherical surface is closed, whilst the hemispherical shell is open with a circular boundary. A more sophisticated distinction can be formulated in topological terms, but this is unnecessary for our purposes. The smoothness of the surface, including the presence of singularities such as corners or conical tips, is important in considering the existence and uniqueness of solutions. This topic has been extensively investigated by Kellogg [32]. However, the surfaces under investigation in this book are portions of coordinate surfaces as described in the Introduction, and both the surfaces and bounding curves are analytic or piecewise analytic. The smoothness conditions, which must be imposed on the closed or open surfaces in a more general formulation of potential theory, are automatically satisfied and will be omitted from further discussion except for two cases, the conical shells considered in Chapter 6, and the two-dimensional axially-slotted cylinders of arbitrary cross-sectional profile considered in Section 7.5; appropriate smoothness conditions are considered in the respective sections.

This section outlines generic aspects of potential theory applicable to both open and closed surfaces, together with those features that are distinctive for open shells. Let us begin with the conditions under which a uniqueness theorem, assuring existence of potentials for closed surfaces, can be asserted.

A closed surface separates space into two regions, namely *internal* and *external*; the internal region may be composed of two or more disconnected parts depending upon the topology of the closed surface. Thus, we can consider either the internal boundary value problem for Laplace's equation or the external boundary value problem. The term *boundary value problem* requires an explicit definition of the type of boundary condition imposed on solutions $U(\overrightarrow{r})$ of Laplace's equation on the closed surface S. Either U is specified everywhere on S (the Dirichlet problem) or its normal derivative

$$\frac{\partial U}{\partial n}$$

(in the direction of the outward normal $\overrightarrow{n}$ on S) is specified on S (the Neumann problem), or a linear combination of U and its normal derivative is specified. These three types, known as first-, second-, and third-kind boundary value problems, respectively, may be expressed as

$$U = f_1 \text{ on } S,$$

$$\frac{\partial U}{\partial n} = f_2 \text{ on } S,$$

or

$$\frac{\partial U}{\partial n} + h(U - f_3) = 0 \text{ on } S,$$

where f_1, f_2, f_3, and h are given functions on S. Thus the internal Dirichlet boundary value problem for Laplace's equation can be formulated as follows.

Problem 1 *Let V be a given region of space which is open, and is bounded by the closed surface S. Find the function U that (a) satisfies Laplace's equation $\Delta U = 0$ within the region V, (b) is continuous in the closed region $V \cup S$ including the boundary surface S, and (c) takes an assigned value on S.*

The external Dirichlet boundary value problem for an infinite open region V exterior to the closed surface S requires an additional constraint on the behaviour of the solution as the observation point tends to infinity.

Problem 2 *Let V be an infinite open region exterior to the closed surface S. Find the function U that (a) satisfies Laplace's equation $\Delta U = 0$ in the infinite region V, (b) is continuous in the closed region $V \cup S$ including the bounding surface S, (c) takes on assigned value on S, and (d) converges uniformly to zero at infinity: $U(\vec{r}) \to 0$ as $|\vec{r}| \to \infty$.*

It is proved in [32] and [60] that when these conditions are satisfied, a unique solution providing a potential can be guaranteed. The Kelvin transform

$$V(\vec{r}) = r^{-1} U(r^{-2} \vec{r})$$

of U is harmonic, except at $\vec{r} = \vec{0}$, if U is harmonic (see [17])). If we require that the function U be harmonic at infinity, i.e., the function V is harmonic at the origin, then condition (d) may be omitted; in either case, the radial derivative $\partial U/\partial r = O(r^{-2})$ as $r \to \infty$. Sometimes the conditions (a)–(c), or (a)–(d) above are referred to as "the conditions of the uniqueness theorem." If U is harmonic, and its value is prescribed on the surface S, then V solves the Dirichlet problem where its value is prescribed in the obvious way on the surface S', which is the image of S under the Kelvin transform $\vec{r} \longmapsto r^{-2} \vec{r}$ of inversion in a unit sphere centred at the origin.

The strict demarcation of internal and exterior regions is lost once a closed surface is punctured and the potentials in previously disconnected regions are coupled to one another across the aperture introduced in the closed surface. Whilst the conditions described above are satisfactory for closed bodies, open surfaces require a supplementary condition to deal appropriately with the singular behaviour of potentials near the edges or rims of the aperture boundary curve.

Physical motivation for the final form and choice of this condition can be found in the electrostatic example of an ideally conducting body with a point or edge. When charged, a high-level electrostatic field is created near the point or edge due to charge concentration in its vicinity; the field tends to infinity as the point of observation approaches the point or edge. By contrast, away from the edge, the surface charge density varies smoothly as does the potential. However, in the vicinity of the edge, the electrostatic field

$$\vec{E} = -\nabla U \tag{1.76}$$

exhibits extremely high values.

At first sight, this localized high-level electrostatic field might be considered an "equivalent source." Nevertheless, some care is needed in this interpretation because the energy integral attached to a real source occupying a volume V diverges:

$$\frac{1}{2}\iiint_V \varepsilon_0 \left|\vec{E}\right|^2 dV = \infty. \tag{1.77}$$

(As an illustration, consider a unit charge placed at the origin of a spherical coordinate frame. The potential is $V = r^{-1}$ and the electrostatic field is radically directed: $\vec{E} = \vec{r}/r^3$; the energy integral is clearly divergent.)

On the other hand, the energy associated with the charged conductor might reasonably be expected to be finite, so that the apparent or equivalent source in the vicinity of the edge possesses a weaker (integrable) singularity than that of a real source. The discussion of appropriate models for real physical sources has a long history; suffice it to say that in the absence of such localized sources, the energy associated with the structure must remain bounded.

This discussion provides a physical motivation for our additional "edge constraint," namely that the gradient of the potential (electrostatic or otherwise) must be square integrable over the whole volume V of space:

$$\iiint_V |\text{grad}\, U|^2 \, dV = \iiint_V |\nabla U|^2 \, dV < \infty. \tag{1.78}$$

Abstracting from the particular physical problem that the potential function $U(\vec{r})$ describes, we assume that the value $|\nabla U|^2$ is proportional to the volume density of the energy, and whereas this gradient may exhibit singular behaviour at various points of the region under consideration, the total energy within any bounded volume including the edges must be finite, as in (1.78). We will see later that this condition ensures that the potential is uniquely determined.

From a mathematical point of view, the condition (1.78) is important in establishing existence and uniqueness of solutions to Laplace's equation. One way of demonstrating existence of solutions is via the "Dirichlet principle," which asserts that any function U that minimises

$$\iiint_V |\text{grad}\, U|^2 \, dV, \tag{1.79}$$

subject to the constraint $U = f$ on S, where the continuous function f is prescribed, satisfies Laplace's equation $\Delta U = 0$ subject to the boundary condition $U = f$ on S. This principle has had a chequered career, which is traced in [43], but eventually it was placed on a rigorous basis for a large class of bounding surfaces S. The principle stimulated much careful analysis of surfaces (there are surfaces for which Laplace's equation cannot be solved uniquely) and lead to the development of functional analysis through the examination of the class of functions for which the minimum of (1.79) is actually attained.

Accepting that Laplace's equation, with the boundary condition $U = f$ on S, has at least one solution, uniqueness is established by considering the difference U_1 of any two such distinct solutions. U_1 is harmonic and vanishes on S, and the divergence theorem shows that

$$\iint_\Sigma U_1 \frac{\partial U}{\partial r} dS - \iint_S U_1 \frac{\partial U}{\partial n} dS = \iiint_V |\text{grad}\, U_1|^2 \, dV, \tag{1.80}$$

where Σ denotes a large spherical surface of radius R enclosing S, and $\overrightarrow{n}$ is the outward normal on S; the bounds, $U_1 = O(R^{-1})$ and $\frac{\partial U_1}{\partial r} = O(R^{-2})$ as $R \to \infty$, show that both sides of (1.80) vanish as $R \to \infty$, so that U_1 is identically zero, and the solution U is unique. This argument is not directly valid when S is an open surface with edges (the divergence theorem is not applicable); it may be modified by surrounding the open surface by a small open region with a smooth bounding surface S_ε whose volume ε contracts to zero; uniqueness holds for the surface S_ε, and by letting $\varepsilon \to 0$, the same result can be recovered for the surface S, provided the energy integral (1.79) is finite. The same identity can be employed to show that if S is a smooth surface bounding an open volume, the energy integral (1.79) is finite.

Examples of nontrivial solutions to Laplace's equation that decay at infinity (according to $U(\overrightarrow{r}) \to 0$ as $|\overrightarrow{r}| \to \infty$) yet vanish on an open surface S_0 may be constructed should the requirement of finiteness of the energy integral be disregarded. Consider, in cylindrical polars (ρ, ϕ, z), the half-plane $\phi = 0$. For any positive integer n, the functions $\psi_n = A_n \rho^{-\frac{n}{2}} \sin(n\phi/2)$ satisfy Laplace's equation (with arbitrary constants A_n)

and vanish on S. The image of S under inversion in a unit sphere located at $(\rho, \phi, z) = (1, \pi, 0)$ is a circular disc D. The Kelvin transform of ψ_n is harmonic on D, vanishes on D, and is $O(|\vec{r}|^{-1})$ as $|\vec{r}| \to \infty$.

Thus, in formulating the statement of boundary value problems for Laplace's equation, two differences between closed and open surfaces are apparent. First, the well-defined concept of internal and external boundary value problems for closed surfaces disappears, the determination of the potential for open surfaces becomes a *mixed* boundary value problem for Laplace's equation; secondly, as well as the conditions standardly imposed in the determination of the potential field associated with a closed body, an extra boundedness condition (1.78) must be imposed on the energy to determine uniquely the potential distribution associated with an open surface.

Later chapters examine potential theory for open shells that are portions of the orthogonal coordinate surfaces described in Section 1.1. By way of illustration, consider the particular example of a spherical shell S_0 of radius a subtending an angle θ_0 at the origin; it is defined in spherical coordinates by

$$r = a,\ 0 \le \theta \le \theta_0,\ 0 \le \phi \le 2\pi.$$

The spherical surface S of radius a may be regarded as the union of the shell S_0 and the "aperture" S_1 given by

$$r = a,\ \theta_0 < \theta \le \pi,\ 0 \le \phi \le 2\pi.$$

Problem 3 *Suppose the shell S is charged to unit potential. Find the potential $U(r, \theta, \phi)$ that satisfies the following conditions: (1) $\Delta U = 0$ at all points, except on the shell; (2) U is everywhere continuous, including all points on the surface $S = S_0 \cup S_1$; on S_0, U takes a prescribed value: $U(a, \theta, \phi) = \Phi(\theta, \phi)$, at all points of S_0; (3) the normal or radial derivative is continuous at all points of S_1:*

$$\lim_{r\to a+} \frac{\partial U}{\partial r}(r, \theta, \phi) = \lim_{r\to a-} \frac{\partial U}{\partial r}(r, \theta, \phi)\ \textit{for}\ \theta_0 \le \theta \le \pi, 0 \le \phi \le 2\pi;$$

(4) U converges uniformly to 0 at infinity: $U(r, \theta, \phi) \to 0$ as $r \to \infty$, and (5) the energy integral must be bounded in any volume V including the edges:

$$\iiint_V |\nabla U|^2 \, dV =$$
$$\iiint_V \left\{ \left|\frac{\partial U}{\partial r}\right|^2 + \frac{1}{r^2}\left|\frac{\partial U}{\partial \theta}\right|^2 + \frac{1}{r^2 \sin^2\theta}\left|\frac{\partial U}{\partial \phi}\right|^2 \right\} dV < \infty.$$

More generally, let us formally state the first-kind mixed boundary value problem (BVP) for Laplace's equation pertaining to an open surface S_0 that is a portion of a coordinate surface S in one of those coordinate systems in which Laplace's equation can be solved by the method of separation of variables (Section 1.2). The term *mixed* refers to the enforcement of different boundary conditions on the two portions comprising the surface S (namely the shell S_0 and the aperture S_1).

Let (q_1, q_2, q_3) be the curvilinear coordinates in this system, and suppose that S is the coordinate surface on which q_1 takes a fixed value, q_1^0. Let I_2 and I_3 be the intervals over which q_2 and q_3 range (in the spherical cap example, $I_2 = [0, \pi]$ and $I_3 = [0, 2\pi]$ where q_2 and q_3 are identified with θ and ϕ). Thus S is parametrised by $I = I_2 \times I_3$.

We consider shells S_0 which are parametrised by $I_0 = I_2^{(0)} \times I_3^{(0)}$ where $I_2^{(0)}$ is composed of one or more subintervals of I_2, and $I_3^{(0)}$ is a similar subset of I_3 ; however, as a rule, either $I_2^{(0)} = I_2$ or $I_3^{(0)} = I_3$. The "aperture" area S_1 may then be parametrised by I_1, the complement of I_0 in I $(I = I_0 \cup I_1)$.

Problem 4 *The first-kind mixed BVP for Laplace's equation. Find the potential $U = U(q_1, q_2, q_3)$ satisfying the following conditions: (1) $\Delta U = 0$ at all points, of space except on S; (2) U is everywhere continuous, including all points on the surface $S = S_0 \cup S_1$, that is*

$$\lim_{q_1 \to q_1^0 + 0} U(q_1, q_2, q_3) = \lim_{q_1 \to q_1^0 - 0} U(q_1, q_2, q_3) \tag{1.81}$$

where $(q_2, q_3) \in I$; (3) the value of U is prescribed on S_0, by a given continuous function F:

$$\lim_{q_1 \to q_1^0 + 0} U(q_1, q_2, q_3) = \lim_{q_1 \to q_1^0 - 0} U(q_1, q_2, q_3) = F(q_2, q_3) \tag{1.82}$$

where $(q_2, q_3) \in I_0$; (4) the normal derivative $\frac{\partial U}{\partial q_1}$ must be continuous on the aperture S_1:

$$\lim_{q_1 \to q_1^0 + 0} \frac{\partial U}{\partial q_1}(q_1, q_2, q_3) = \lim_{q_1 \to q_1^0 - 0} \frac{\partial U}{\partial q_1}(q_1, q_2, q_3) \tag{1.83}$$

where $(q_2, q_3) \in I_1$; (5) $U(q_1, q_2, q_3)$ converges uniformly to 0 at infinity:

$$U(q_1, q_2, q_3) \to 0 \text{ as } |(q_1, q_2, q_3)| \to \infty; \tag{1.84}$$

and (6) the energy integral must be bounded in any arbitrary volume V including the edges:

$$\iiint_V |\nabla U|^2 \, dV = \iiint_V \left\{ \left| \frac{1}{h_1} \frac{\partial U}{\partial q_1} \right|^2 + \left| \frac{1}{h_2} \frac{\partial U}{\partial q_2} \right|^2 + \left| \frac{1}{h_3} \frac{\partial U}{\partial q_3} \right|^2 \right\} dV < \infty \tag{1.85}$$

(where h_1, h_2, h_3 are the metric coefficients).

The condition (1.85) gives rise to most of the so called *edge conditions* appearing in the literature; these prescribe the singular behaviour of the potential close to an edge, determining, for example, the order of the singularity. It is worth noting that the normal derivative is continuous onto the surface S_0, but may take different values as its approaches a point on S_0 from one side or the other. Also, in the vicinity of the edge the normal derivative is generally unbounded. The jump in normal derivative across the surface S_0 is the single-layer density used in the standard integral representation of the field (see Section 1.7). Physically, it is proportional to surface charge density.

In contrast to first-kind mixed problems are those of second kind, in which the role of U and $\frac{\partial U}{\partial q_1}$ are interchanged in the boundary conditions (1.82) and (1.83).

Problem 5 *The second-kind mixed BVP for Laplace's equation. Find the potential* $U = U(q_1, q_2, q_3)$ *satisfying the following conditions: (1)* $\Delta U = 0$ *at all points of space, except on* S*; (2) the normal derivative*

$$\frac{\partial U}{\partial n} = \frac{\partial U}{\partial q_1}$$

is everywhere continuous, including all points on the surface $S = S_0 \cup S_1$*, that is*

$$\left[\frac{\partial U}{\partial q_1}\right]_{q_1=q_1^0-0} = \left[\frac{\partial U}{\partial q_1}\right]_{q_1=q_1^0+0} \tag{1.86}$$

where $(q_2, q_3) \in I$*; (3) the value of the normal derivative is prescribed on* S_0*, by a continuous function* G*:*

$$\lim_{q_1 \to q_1^0-0} \frac{\partial U}{\partial q_1}(q_1, q_2, q_3) = \lim_{q_1 \to q_1^0+0} \frac{\partial U}{\partial q_1}(q_1, q_2, q_3) = G(q_2, q_3), \tag{1.87}$$

where $(q_2, q_3) \in I_0$*; (4)* U *is continuous on the aperture* S_1*:*

$$\lim_{q_1 \to q_1^0-0} U(q_1, q_2, q_3) = \lim_{q_1 \to q_1^0+0} U(q_1, q_2, q_3), \tag{1.88}$$

where $(q_2, q_3) \in I_1$*; (5)* $U(q_1, q_2, q_3)$ *converges uniformly to* 0 *as* $|(q_1, q_2, q_3)| \to \infty$ *(cf. (1.84)); and (6) the energy integral (1.85) must be finite.*

Succeeding chapters provide constructive methods for uniquely solving both types of mixed boundary value problems for Laplace's equation. Our methods utilise the special functions associated with the orthogonal coordinate system of relevance to the particular problem at hand to obtain a pair of functional equations, which are enforced on S_0 and on the aperture S_1, respectively. A constructive and rigorously correct mathematical method – to be explained in the next chapter – may be applied to solve this pair, to determine completely the unique potential satisfying the appropriate six conditions listed above.

Let us describe generally how these functional equations arise, for the first-kind mixed boundary value problems for Laplace's equation, under the somewhat restrictive assumption that the solution is independent of one coordinate, say q_3, so that

$$\frac{\partial U}{\partial q_3}(q_1, q_2, q_3) = 0. \tag{1.89}$$

In this case the function F (see (1.82)) is independent of $q_3 : F(q_2, q_3) \equiv F(q_2)$.

Dual (or multiple) series equations arise when the eigenvalue spectrum of the Sturm-Liouville problem, originating from the ordinary differential equations obtained in application of the separation of variables technique applied to Laplace's equation, is discrete. Separated solutions are generated for the two regions separated by S (namely, the regions $q_1 < q_1^o$ and $q_1 > q_1^o$) in the form

$$U_n(q_1, q_2) = \begin{Bmatrix} x_n^{(1)} R_n^{(1)}(q_1) A_n(q_2), & q_1 < q_1^o \\ x_n^{(2)} R_n^{(2)}(q_1) A_n(q_2), & q_1 > q_1^o \end{Bmatrix} \tag{1.90}$$

(where the index $n = 0, 1, 2, \ldots$ labels the spectrum), and the corresponding total solution is the superposition

$$U(q_1, q_2) = \sum_{n=0}^{\infty} U_n(q_1, q_2) = \sum_{n=0}^{\infty} \left\{ \begin{array}{c} x_n^{(1)} R_n^{(1)}(q_1), q_1 < q_1^o \\ x_n^{(2)} R_n^{(2)}(q_1), q_1 > q_1^o \end{array} \right\} A_n(q_2). \tag{1.91}$$

The unknown Fourier coefficients $\left\{x_n^{(1)}\right\}_{n=0}^{\infty}$ and $\left\{x_n^{(2)}\right\}_{n=0}^{\infty}$ are to be determined; $R_n^{(1)}$, $R_n^{(2)}$ are *radial* functions, and A_n is an *angle* function by convention.

Both $R_n^{(1)}$ and $R_n^{(2)}$ satisfy the same ordinary differential equation and provide a basis for the set of all solutions of this differential equation; $R_n^{(1)}$ is chosen to be regular in the domain $q_1 \leq q_1^o$ (so determining it uniquely up to a constant factor), whereas $R_n^{(2)}$ is chosen to satisfy the condition (1.84); thus $R_n^{(2)}$ is regular in the domain $q_1 \geq q_1^o$ and determined uniquely up to a constant factor. The infinite set of angle functions $\{A_n\}_{n=0}^{\infty}$ is complete and orthogonal on I_2 with respect to a weight function, denoted h :

$$\int_{I_2} h(q_2) A_n(q_2) A_m(q_2) dq_2 = \alpha_n \delta_{nm}. \tag{1.92}$$

The constants α_n are necessarily positive, so that the normalised functions $\hat{A}_n = A_n / \alpha_n^{\frac{1}{2}}$ form a complete orthonormal set.

The continuity condition (1.81), together with (1.92), gives a relationship between $x_n^{(1)}$ and $x_n^{(2)}$,

$$x_n^{(2)} = \left(R_n^{(1)}(q_1^o)/R_n^{(2)}(q_1^o)\right) x_n^{(1)}, \tag{1.93}$$

so that (1.91) becomes

$$U(q_1,q_2) = \sum_{n=0}^{\infty} x_n^{(1)} \left\{ \begin{array}{ll} R_n^{(1)}(q_1), & q_1 < q_1^o \\ R_n^{(1)}(q_1^o) R_n^{(2)}(q_1)/R_n^{(2)}(q_1^o), & q_1 > q_1^o \end{array} \right\} A_n(q_2), \tag{1.94}$$

or, in symmetric form,

$$U(q_1,q_2) = \sum_{n=0}^{\infty} X_n \left\{ \begin{array}{l} R_n^{(2)}(q_1^o) R_n^{(1)}(q_1),\ q_1 < q_1^o \\ R_n^{(1)}(q_1^o) R_n^{(2)}(q_1),\ q_1 > q_1^o \end{array} \right\} A_n(q_2), \tag{1.95}$$

where we have rescaled $x_n^{(1)} = R_n^{(2)}(q_1^o) X_n$. Enforcing the boundary conditions (1.82) and (1.83) leads to the pair of functional equations

$$\sum_{n=0}^{\infty} X_n R_n^{(1)}(q_1^o) R_n^{(2)}(q_1^o) A_n(q_2) = F(q_2),\ q_2 \in I_2^{(0)}, \tag{1.96}$$

$$\sum_{n=0}^{\infty} X_n W\left(R_n^{(1)}(q_1^o), R_n^{(2)}(q_1^o)\right) A_n(q_2) = 0,\ q_2 \in I_2 \backslash I_2^{(0)}, \tag{1.97}$$

where the Wronskian

$$W\left(R_n^{(1)}(q_1), R_n^{(2)}(q_1)\right) = R_n^{(1)}(q_1)\frac{d}{dq_1}R_n^{(2)}(q_1) - R_n^{(2)}(q_1)\frac{d}{dq_1}R_n^{(1)}(q_1)$$

is evaluated at $q_1 = q_1^o$. These equations are referred to as *dual series equations* if the interval $I_2 \backslash I_2^{(0)}$ is a simply connected subset of I_2; otherwise, they are referred to as *triple-* or *multiple-*series equations depending on the total number of connected subintervals of I_2 appearing in Equations (1.96) and (1.97)

Enforcement of the finite energy condition (1.85) provides a unique solution to (1.96) and (1.97); essentially, it provides the correct functional space setting for the coefficients X_n. The simplest but most effective choice of the volume V of integration in (1.85) is the interior region ($q_1 \leq q_1^o, q_2 \in I_2, q_3 \in I_3$); it is bounded, finite, and involves the edges. Substitution of the relevant derivatives, obtained from term-by-term differentiation of (1.91) and (1.94), into the energy integral (1.85) gives a condition which the Fourier coefficients ($x_n^{(1)}$ or X_n) must satisfy. This condition will always take the form

$$\sum_{n=0}^{\infty} c_n \left|x_n^{(1)}\right|^2 < \infty, \tag{1.98}$$

where c_n is some explicitly known coefficient.

Conversely, as we will see in succeeding sections, the condition (1.98) ensures that the operations of term-by-term integration and differentiation, to be applied on the series (1.96) and (1.97), are justified and valid. If the angle functions are normalised, the condition (1.98) becomes

$$\sum_{n=0}^{\infty} |y_n|^2 < \infty, \tag{1.99}$$

where $\{y_n\}_{n=0}^{\infty}$ is a suitably rescaled sequence related to $\left\{x_n^{(1)}\right\}_{n=0}^{\infty}$ or $\{X_n\}_{n=0}^{\infty}$. Thus the sequence $\{y_n\}_{n=0}^{\infty}$ belongs to the set of square summable Fourier coefficients l_2.

When the spectrum of the relevant Sturm-Liouville problem is continuous, a similar argument produces dual (or multiple) integral equations. This schematic outline of the formulation and basic features of boundary value problems for structures with edges will be refined and analysed more carefully when concrete configurations are encountered.

1.4 Dual equations: a classification of solution methods

It is perhaps well known that a comprehensive theory to solve dual equations does not exist, and that many treatments have been developed to obtain solutions to such equations. Essentially, these treatments can be grouped into three basic methods: the definition-extension method described by W.E. Williams, [76] the substitution method described by B. Noble, [46] and the multiplying factor method also described by B. Noble [47].

The common and distinctive feature of all these methods is the utilization, in one form or another, of Abel's integral equation (or transform) technique. Let us illustrate these methods with the simple problem of determining the potential of a charged spherical cap when its surface is held at a constant unit value of potential. This problem produces the following dual series equations involving Legendre polynomials $P_n(\cos\theta)$,

$$\sum_{n=0}^{\infty} a_n P_n(\cos\theta) = 1, \ \theta \in (0, \theta_0), \tag{1.100}$$

$$\sum_{n=0}^{\infty} (2n+1) a_n P_n(\cos\theta) = 0, \ \theta \in (\theta_0, \pi), \tag{1.101}$$

where the unknown desired set of coefficients $\{a_n\}_{n=0}^{\infty}$ must belong to the Hilbert functional space l_2. The concrete form of condition (1.85) that imposes this constraint on the coefficients is

$$\sum_{n=0}^{\infty}\frac{n+1}{2n+1}\left|a_n\right|^2<\infty. \tag{1.102}$$

1.4.1 The definition method

To solve Equations (1.100) and (1.101), let us define a function g on $[0,\theta_0]$, which provides the extension of (1.101) to the complete interval $[0,\pi]$. That is, let

$$\sum_{n=0}^{\infty}(2n+1)a_nP_n(\cos\theta)=\begin{Bmatrix} g(\theta), & \theta\in[0,\theta_0) \\ 0, & \theta\in(\theta_0,\pi]\end{Bmatrix}. \tag{1.103}$$

In (1.103) the left-hand side is the Fourier-Legendre expansion for a certain function F; the right-hand side is the piecewise continuous expression of that function F on $[0,\pi]$.

The orthogonality property of the set of Legendre polynomials $\{P_n\}_{n=0}^{\infty}$ on $[0,\pi]$ allows us to express $\{a_n\}_{n=0}^{\infty}$ in terms of the function g :

$$a_n=\frac{1}{2}\int_0^{\theta_0}g(\theta)P_n(\cos\theta)\sin\theta d\theta. \tag{1.104}$$

Substitute this expression in (1.100), and invert the order of integration and summation; the original Equations (1.100) and (1.101) are then reduced to the first-kind Fredholm integral equation

$$\int_0^{\theta_0}g^*(\vartheta)K(\vartheta,\theta)d\vartheta=1,\ \theta\in(0,\theta_0) \tag{1.105}$$

where $g^*(\vartheta)=\sin\vartheta g(\vartheta)$, and the kernel is

$$K(\vartheta,\theta)=\frac{1}{2}\sum_{n=0}^{\infty}P_n(\cos\vartheta)P_n(\cos\theta). \tag{1.106}$$

Following the idea developed by W. E. Williams [76], we solve (1.105) by the successive solution of two Abel integral equations. To this end, represent the kernel (1.106) in the form

$$K(\vartheta,\theta)=\frac{1}{2\pi}\int_0^{\min(\vartheta,\theta)}\frac{d\phi}{\sqrt{(\cos\phi-\cos\vartheta)(\cos\phi-\cos\theta)}} \tag{1.107}$$

This representation is easily obtained from the Dirichlet-Mehler formula (see Appendix B.94),

$$P_n(\cos\theta)=\frac{\sqrt{2}}{\pi}\int_0^{\theta}\frac{\cos(n+\frac{1}{2})\phi}{\sqrt{\cos\phi-\cos\theta}}d\phi, \tag{1.108}$$

from which it follows that

$$\sum_{n=0}^{\infty} P_n(\cos\vartheta)\cos(n+\frac{1}{2})\phi = \begin{Bmatrix} [2(\cos\phi-\cos\vartheta)]^{-\frac{1}{2}}, & 0 \le \phi < \vartheta \\ 0, & \vartheta < \phi \le \pi \end{Bmatrix}. \quad (1.109)$$

Decomposing the integration domain in (1.105) into two parts, $(0,\theta) \cup (\theta,\theta_0)$, and using the expression (1.107), one obtains the repeated integral

$$\frac{1}{2\pi}\int_0^{\theta}\frac{d\phi}{\sqrt{\cos\phi-\cos\theta}}\int_{\phi}^{\theta_0}\frac{g^*(\vartheta)}{\sqrt{\cos\phi-\cos\vartheta}}d\vartheta = 1, \qquad \theta \in (0,\theta_0). \quad (1.110)$$

A double application of the inversion formulae to this iterated Abel integral equation (see the next section) yields the solution for the function g in closed form:

$$g(\vartheta) = \frac{\sqrt{2}}{\pi}\left\{\frac{2\cos\frac{1}{2}\theta_0}{\sqrt{\cos\vartheta-\cos\theta_0}} + \frac{\pi}{2} - \arcsin\left(\frac{\cos\frac{1}{2}\theta_0}{\cos\frac{1}{2}\theta}\right)\right\}. \quad (1.111)$$

The substitution of this expression for the function g in (1.104) gives, after elementary integration, the final solution for the Fourier coefficients:

$$a_n = \frac{1}{\pi}\left[\frac{\sin n\theta_0}{n} + \frac{\sin(n+1)\theta_0}{n+1}\right]. \quad (1.112)$$

The function g coincides with the surface charge density on the cap, and possesses the expected singularity of order $-\frac{1}{2}$ as $\vartheta \to \theta_0$ (Formula (1.111)). Moreover, by construction, the solution $\{a_n\}_{n=0}^{\infty}$ given in (1.112) lies in l_2.

1.4.2 The substitution method

This method is based upon expansion of a certain discontinuous function in a Fourier-Legendre series. Its analytic form permits us to find a representation of the solution that automatically satisfies one of the dual equations. Considering (1.108), let us represent the coefficients a_n in terms of an unknown function U, so that

$$a_n = \int_0^{\theta_0} U(t)\cos(n+\frac{1}{2})t\,dt. \quad (1.113)$$

The functional Equation (1.101) is automatically satisfied, but the companion Equation (1.100) is transformed, after an interchange of integration and summation, to the Abel integral equation

$$\int_0^{\theta}\frac{U(t)dt}{\sqrt{\cos t-\cos\theta}} = \sqrt{2}, \qquad \theta \in (0,\theta_0). \quad (1.114)$$

This possesses the obvious solution

$$U(t) = \frac{2}{\pi}\cos\frac{t}{2}, \qquad t \in (0, \theta_0)$$

and substitution in (1.113) immediately leads to the previously obtained solution (1.112).

1.4.3 Noble's multiplying factor method

The essence of the multiplying factor method is the following. Each of the Equations (1.100) and (1.101) is multiplied by a suitable functional factor and then an appropriate integral operator, or a combination of integral and differential operators is applied to transform the left-hand side of (1.100) or (1.101) to the same functional expression – a Fourier series (or similar) involving the coefficients a_n. The coefficients a_n are then obtained from the calculation of the Fourier coefficients of the piecewise continuous function obtained by the transform of the right-hand side of (1.100) and (1.101) under this process.

In our example problem, the operators are derived from the well-known identities arising from the inversion of the Dirichlet-Mehler formulae (B.94):

$$\cos(n+\frac{1}{2})\theta = \frac{1}{\sqrt{2}}\frac{d}{d\theta}\int_0^\theta \frac{P_n(\cos\phi)}{\sqrt{\cos\phi - \cos\theta}}\sin\phi d\phi, \tag{1.115}$$

$$\cos(n+\frac{1}{2})\theta = \frac{1}{\sqrt{2}}(n+\frac{1}{2})\int_\theta^\pi \frac{P_n(\cos\phi)}{\sqrt{\cos\theta - \cos\phi}}\sin\phi d\phi. \tag{1.116}$$

Let K_1 and K_2 denote operators defined by

$$(K_1 f)(\theta) = \frac{1}{\sqrt{2}}\frac{d}{d\theta}\int_0^\theta \frac{f(\phi)\sin\phi d\phi}{\sqrt{\cos\phi - \cos\theta}} \tag{1.117}$$

and

$$(K_2 f)(\theta) = \frac{1}{\sqrt{2}}\int_\theta^\pi \frac{f(\phi)\sin\phi d\phi}{\sqrt{\cos\theta - \cos\phi}}. \tag{1.118}$$

Applying K_1 to (1.100) and K_2 to (1.101) yields

$$\sum_{n=0}^\infty a_n \cos(n+\frac{1}{2})\theta = \begin{Bmatrix} \cos\frac{1}{2}\theta, & \theta \in (0, \theta_0) \\ 0, & \theta \in (\theta_0, \pi) \end{Bmatrix}. \tag{1.119}$$

A calculation of the coefficients a_n, using the orthogonality property

$$\int_0^\pi \cos(n+\frac{1}{2})\theta\cos(m+\frac{1}{2})\theta d\theta = \frac{\pi}{2}\delta_{nm},$$

leads to the previously obtained form (1.112) of the desired solution.

Thus all the three methods described above employ Abel's integral equation in some form or another. Noble's multiplying factor method can be seen as a direct application of fractional integration. The relationship between fractional integration and integral transforms of Abel type is discussed fully in [55].

1.4.4 The Abel integral transform method

A more direct and readily justified method of solving dual series is the *Abel integral transform method* which was developed in [67], [68], [69], [70] and [71]. It can be directly identified with the *integral representation method* described in [21], the only difference being that the mathematical validity of the operations in the first approach is properly established, whereas the analysis of the latter approach is purely formal in manner.

This is not to assert that the Abel integral transform method is a completely new method to solve dual, triple, and multiple series or integral equations of this class. It is clearly rooted in the integral representation method described in [21]. It is worth emphasizing that each of the sequences of mathematical operations associated with the Abel integral transform method is straightforwardly justified, so there is no doubt about the validity of solutions obtained by this approach. The name of the method highlights the transform at its core.

Various classes of dual and triple series equations are solved in a mathematically rigorous manner by the application of this method in Chapter 2. By way of illustration, let us apply the Abel integral transform method to the charged spherical cap problem described earlier in this section. The method transforms each functional equation (of the dual series equations) to an integral equation that is recognizable as Abel's integral equation with zero forcing term. It has a unique solution, namely zero, which provides the basis for the final solution step.

Considering (1.100), we replace the right-hand side with the expression, derived from (1.108), $n = 0$,

$$1 = \frac{\sqrt{2}}{\pi}\int_0^\theta \frac{\cos\frac{1}{2}\phi d\phi}{\sqrt{\cos\phi - \cos\theta}}.$$

The terms on the left-hand side are replaced by the representation (1.108). An interchange of integration and summation is permissible under the condition (1.102), and leads to the homogeneous Abel integral equation

$$\int_0^\theta \frac{f(\phi)d\phi}{\sqrt{\cos\phi - \cos\theta}} = 0,\ \theta \in [0,\theta_0), \tag{1.120}$$

where

$$f(\phi) = \sum_{n=0}^{\infty} a_n \cos(n + \frac{1}{2})\phi - \cos\frac{1}{2}\phi.$$

Because Equation (1.120) has the unique zero solution, we obtain

$$\sum_{n=0}^{\infty} a_n \cos(n + \frac{1}{2})\phi - \cos\frac{1}{2}\phi = 0, \ \phi \in [0, \theta_0). \tag{1.121}$$

Turning to (1.101), term-by-term integration of this series is also permitted, because the series is uniformly Abel-summable (this point is discussed in greater detail in Section 2.2). Multiplying by $\sin\theta$ and integrating over (θ, π) (when $\theta > \theta_0$), produces

$$\sum_{n=0}^{\infty} a_n \left[P_{n-1}(\cos\theta) - P_{n+1}(\cos\theta)\right] = 0, \ \theta \in (\theta_0, \pi]. \tag{1.122}$$

Here we have used the well-known formula (see Appendix, (B.58))

$$(2n+1)P_n(x) = \frac{d}{dx}\left[P_{n-1}(x) - P_{n+1}(x)\right], \tag{1.123}$$

and the Dirichlet-Mehler formula (see [55]) for the Legendre polynomials

$$P_n(\cos\theta) = \frac{\sqrt{2}}{\pi}\int_\theta^\pi \frac{\sin(n+\frac{1}{2})\phi d\phi}{\sqrt{\cos\theta - \cos\phi}}, \tag{1.124}$$

to derive the Abel transform representation for the difference

$$P_{n-1}(\cos\theta) - P_{n+1}(\cos\theta) = -\frac{2\sqrt{2}}{\pi}\int_\theta^\pi \frac{\cos(n+\frac{1}{2})\phi \sin\phi d\phi}{\sqrt{\cos\theta - \cos\phi}}. \tag{1.125}$$

Note that both Equations (1.122) and (1.100) possess a common feature. The asymptotics for the Legendre polynomials

$$P_n(x) = O(n^{-\frac{1}{2}}) \text{ as } n \to \infty$$

ensure that the asymptotic behaviour of the terms in each series is the same.

The substitution of the transform (1.125) into (1.122) produces the integral equation

$$\int_\theta^\pi \frac{g(\phi)\sin\phi d\phi}{\sqrt{\cos\theta - \cos\phi}} = 0, \ \theta \in (\theta_0, \pi], \tag{1.126}$$

where

$$g(\phi) = \sum_{n=0}^{\infty} a_n \cos(n + \frac{1}{2})\phi, \ \phi \in (\theta_0, \pi]; \tag{1.127}$$

it has the solution

$$\sum_{n=0}^{\infty} a_n \cos(n+\frac{1}{2})\phi = 0,\ \phi \in (\theta_0, \pi]. \tag{1.128}$$

The interchange of summation and integration is justified under condition (1.102).

Combining the results (1.121) and (1.128) produces the same result as given by Noble's multiplying factor method, and consequently the same closed form expression (1.112) for the coefficients a_n.

In spite of its simplicity, this example illustrates all the features that are characteristic of the Abel integral transform method. The main features that occur in a typical application to potential theory, which requires the solution of dual, triple, or multiple series equations, or integral equations of this type, the kernels of which involve hypergeometric functions, are as follows.

The very first step is to determine the solution class from the edge condition. This key point allows us to establish the validity of various mathematical operations on series or integrals. Next, the convergence rate of each member of the dual equations must be assessed. For example, the convergence rate of (1.100) is $O(n^{-\frac{3}{2}})$ as $n \to \infty$, whereas the rate of (1.101) is $O(n^{-\frac{1}{2}})$, as $n \to \infty$. The equation with the slower convergence is subjected to an integration operation that equilibrates the convergence rate of both equations (for example, see the transition from (1.101) to (1.122)).

Although both members of this pair of transformed functional equations now possess the same convergence rate, each involves different kernels ($P_n(\cos\theta)$ or $P_{n-1}(\cos\theta) - P_{n+1}(\cos\theta)$ in the example above). The third step represents the kernel (and right-hand side) of each equation as an Abel integral transform. One can then interchange the order of summation and integration (for dual series equations), or the order of double integrals (for dual integral equations), as appropriate. As a result, one obtains two independent integral equations of Abel type, each of which possesses a unique solution, namely zero (see (1.121) and (1.128) of the example above).

The final phase is to recognize that a Fourier series (or Fourier integral) in the unknown coefficients has been obtained by this process; the series or integral is equal to a known function, with a piecewise continuous representation on its complete interval of definition. Our example above produced the representation (1.119). Using properties of orthogonality and completeness of terms in the series – or an inverse Fourier integral transform as appropriate – we obtain the final solution for the unknown coefficients.

This general description provides a common formal structure to construct solutions of multiple series or integral equations. The mathematical tools to realise each step of this process are discussed in Chapter 2.

1.5 Abel's integral equation and Abel integral transforms

Most texts on linear integral equations invariably discuss Abel's integral equation in the first few pages because it is a precursor of the modern theory of linear integral equations. Originally, Abel's integral equation

$$f(x) = \int_0^x \frac{u(\xi)d\xi}{\sqrt{x-\xi}} \tag{1.129}$$

arose from the following problem in mechanics. A particle moving under the influence of gravity, along a smooth curve in a vertical plane, takes the time $f(x)$ to move from the vertical height x to a fixed point on the curve. The problem is to find the function u defining that curve, known as the tautochrone.

Instead of Equation (1.129), Abel set himself the problem of solving the more general equation

$$f(x) = \int_a^x \frac{u(\xi)d\xi}{(x-\xi)^\lambda}, \qquad (0 < \lambda < 1), \tag{1.130}$$

where f is a known function and u is the function to be determined. Details of the solution of this generalised Abel's equation can be found in several texts, including [63], [50], and [24]; we simply state the inversion formula for (1.130):

$$u(\xi) = \frac{\sin \lambda\pi}{\pi} \frac{d}{d\xi} \int_a^\xi \frac{f(x)dx}{(\xi - x)^{1-\lambda}}. \tag{1.131}$$

The companion form of the generalised Abel integral equation is

$$f(x) = \int_x^b \frac{u(\xi)d\xi}{(\xi - x)^\lambda}, \qquad (0 < \lambda < 1), \tag{1.132}$$

and has the solution

$$u(\xi) = -\frac{\sin \lambda\pi}{\pi} \frac{d}{d\xi} \int_\xi^b \frac{f(x)dx}{(x-\xi)^{1-\lambda}}. \tag{1.133}$$

In the deduction of (1.131) and (1.133), the following well known formula involving the beta function B (see Appendix, (B.8)) is used,

$$\int_\xi^z \frac{dx}{(z-x)^{1-\mu}(x-\xi)^\mu} = B(\mu, 1-\mu) = \frac{\pi}{\sin \mu\pi}, \qquad (0 < \mu < 1). \tag{1.134}$$

From the solution given in [24] we may formulate a theorem concerning solution existence.

Theorem 6 *Necessary and sufficient conditions that the integral Equation* (1.130) *should have a continuous solution on* (a, b) *are that* $f(x)$ *be continuous in* (a, b), *that* $f(a) = 0$, *and that*

$$\int_a^x \frac{f(\xi)d\xi}{(x-\xi)^{1-\lambda}}$$

have a continuous derivative on (a, b). *If these conditions are fulfilled,* (1.130) *has only one continuous solution, given by Formula* (1.131).

An analogous theorem may be stated for the integral Equation (1.132). Omitting their deduction (see [55]), let us state three results connected with Abel's integral equation, which will be used subsequently.

Theorem 7 *If* ϕ *is finite, and has only a finite number of discontinuities in* (a, b), *the function*

$$\Phi(x) = \int_a^x \frac{\phi(\xi)d\xi}{(x-\xi)^\lambda}, \qquad (\lambda < 1),$$

is continuous on (a, b), *including at the point* a, *where it vanishes.*

Theorem 8 *If* ϕ *is continuous on* (a, b), *and has a derivative that is finite except for a finite number of discontinuities in* (a, b), *and if* $\phi(a) = 0$, *the function*

$$\Phi(x) = \int_a^x \frac{\phi(\xi)}{(x-\xi)^\lambda}d\xi, \qquad (\lambda < 1),$$

has a derivative that is continuous on (a, b) *and is given by the formula*

$$\Phi'(x) = \int_a^x \frac{\phi'(\xi)}{(x-\xi)^\lambda}d\xi.$$

Theorem 9 *(Dirichlet's extended formula). Let* ϕ *be a function of two variables. If* ϕ *is finite in the region* $a \leq y \leq x \leq b$, *and its discontinuities (if any) are regularly distributed, and if* λ, μ, ν *are constants satisfying*

$$0 \leq \lambda < 1, 0 \leq \mu < 1, 0 \leq \nu < 1,$$

then

$$\int_a^b \int_a^x \frac{\phi(x,y)dydx}{(x-y)^\lambda (b-x)^\mu (y-a)^\nu} = \int_a^b \int_y^b \frac{\phi(x,y)dxdy}{(x-y)^\lambda (b-x)^\mu (y-a)^\nu}. \tag{1.135}$$

Let us consider a further generalisation of Abel's integral equation in the form

$$f(x) = \int_a^x \frac{U(\xi)d\xi}{\{h(x)-h(\xi)\}^\lambda}, \qquad x \in (a, b), \qquad 0 < \lambda < 1, \tag{1.136}$$

where h is a strictly monotonically increasing and continuously differentiable function on (a, b) (so $h' > 0$ in this interval). Differing terminology has been used for this generalisation in the literature. R. P. Kanwal [31] treated the generalised Abel integral Equation (1.130) or (1.132) as a *special case of the singular integral equation* (1.136). I. N. Sneddon [55] refers to (1.136) as an Abel-type integral equation, but usually in the context of some specific choices of the function h. We propose to use this terminology whatever choice for h is made.

The pairs (1.130) and (1.131), or (1.132) and (1.133), can be considered as companion integral transforms. For example, if the transform (1.130) is designated as the direct Abel integral transform, then integral transform (1.131) is its inverse. Similar terminology can be applied to the pair (1.132) and (1.133).

Let us solve Equation (1.136), following the treatments [50] and [55] closely. Consider the integral

$$\int_a^x \frac{h'(u)f(u)du}{\{h(x) - h(u)\}^{1-\lambda}},$$

and substitute for f from (1.136) to obtain

$$\int_a^x \int_a^u \frac{U(\xi)h'(u)d\xi du}{\{h(u) - h(\xi)\}^{\lambda}\{h(x) - h(u)\}^{1-\lambda}}.$$

By changing the order of integration, this becomes

$$\int_a^x U(\xi)d\xi \int_\xi^x \frac{h'(u)du}{\{h(u) - h(\xi)\}^{\lambda}\{h(x) - h(u)\}^{1-\lambda}}.$$

The inner integral reduces to (1.134) under the obvious change of variable $z = h(u)$, so that

$$\int_a^x \frac{h'(u)f(u)du}{\{h(x) - h(u)\}^{1-\lambda}} = \frac{\pi}{\sin \lambda\pi} \int_a^x U(\xi)d\xi. \tag{1.137}$$

Differentiation of both sides of (1.137) produces the solution

$$U(\xi) = \frac{\sin \lambda\pi}{\pi} \frac{d}{d\xi} \int_a^\xi \frac{h'(u)f(u)du}{\{h(\xi) - h(u)\}^{1-\lambda}}. \tag{1.138}$$

Similarly, the integral equation

$$f(x) = \int_x^b \frac{U(\xi)d\xi}{\{h(x) - h(u)\}^{\lambda}}, \qquad x \in (a, b), \qquad 0 < \lambda < 1 \tag{1.139}$$

has the solution

$$U(\xi) = -\frac{\sin \lambda\pi}{\pi} \frac{d}{d\xi} \int_\xi^b \frac{h'(u)f(u)du}{\{h(u) - h(\xi)\}^{1-\lambda}}. \tag{1.140}$$

Two special cases of (1.136) and (1.139) will be of further interest. First, let $h(\xi) = \xi^2$: the integral equation

$$f(x) = \int_a^x \frac{U(\xi)d\xi}{(x^2 - \xi^2)^\lambda}, \qquad (0 < \lambda < 1) \tag{1.141}$$

has the solution

$$U(\xi) = \frac{2\sin\lambda\pi}{\pi}\frac{d}{d\xi}\int_a^\xi \frac{uf(u)du}{(x^2 - \xi^2)^{1-\lambda}}, \tag{1.142}$$

while its companion

$$f(x) = \int_x^b \frac{U(\xi)d\xi}{(\xi^2 - x^2)^\lambda}, \qquad (0 < \lambda < 1) \tag{1.143}$$

has the solution

$$U(\xi) = -\frac{2\sin\lambda\pi}{\pi}\frac{d}{d\xi}\int_\xi^b \frac{uf(u)du}{(\mu^2 - \xi^2)^{1-\lambda}}. \tag{1.144}$$

Next, consider $h(\xi) = \cosh\xi$: the integral equation

$$f(x) = \int_a^x \frac{U(\xi)d\xi}{(\cosh x - \cosh\xi)^\lambda}, \qquad (0 < \lambda < 1) \tag{1.145}$$

has the solution

$$U(\xi) = \frac{\sin\lambda\pi}{\pi}\frac{d}{d\xi}\int_a^\xi \frac{\sinh u f(u)du}{(\cosh\xi - \cosh u)^{1-\lambda}}, \tag{1.146}$$

while the companion integral equation

$$f(x) = \int_x^b \frac{U(\xi)d\xi}{(\cosh\xi - \cosh u)^\lambda}, \qquad (0 < \lambda < 1) \tag{1.147}$$

has the solution

$$U(\xi) = -\frac{\sin\lambda\pi}{\pi}\frac{d}{d\xi}\int_\xi^b \frac{\sinh u f(u)du}{(\cosh u - \cosh\xi)^{1-\lambda}}. \tag{1.148}$$

1.6 Abel-type integral representations of hypergeometric functions

In Section 1.3 we encountered the representation of one type of hypergeometric function as an integral transform of Abel type. The Dirichlet-Mehler formulae provide an integral representation for the Legendre polynomials (see (1.124) and Appendix, (B.94)), expressed in trigonometric form as

$$P_n(\cos\theta) = \frac{\sqrt{2}}{\pi}\int_0^\theta \frac{\cos(n+\frac{1}{2})\phi}{\sqrt{\cos\phi - \cos\theta}}d\phi, \tag{1.149}$$

$$P_n(\cos\theta) = \frac{\sqrt{2}}{\pi}\int_\theta^\pi \frac{\sin(n+\frac{1}{2})\phi}{\sqrt{\cos\theta - \cos\phi}}d\phi. \tag{1.150}$$

At first glance it seems that representations (1.149) and (1.150) transform one class of functions (the Legendre polynomials $P_n(\cos\theta)$) to another, the trigonometric functions of form $\cos(n+\frac{1}{2})\theta$, and $\sin(n+\frac{1}{2})\theta$.

From a wider perspective, these functions may be regarded as members of one and the same class, namely the Jacobi polynomials $P_n^{(\alpha,\beta)}$. For each fixed (α,β), with $\alpha > 1, \beta > -1$, the Jacobi polynomials $P_n^{(\alpha,\beta)}$ are polynomials of degree $n(=0,1,2\ldots)$ and are orthogonal on $[-1,1]$ with respect to the weight function $w_{\alpha,\beta}(x) = (1-x)^\alpha(1+x)^\beta$. Their properties are discussed in Appendix B.3. In particular, the relations between the trigonometric functions and the Legendre polynomials are

$$\cos n\theta = \frac{\Gamma\left(\frac{1}{2}\right)\Gamma(n+1)}{\Gamma\left(n+\frac{1}{2}\right)}P_n^{(-\frac{1}{2},-\frac{1}{2})}(\cos\theta), \tag{1.151}$$

$$\cos(n+\frac{1}{2})\theta = \frac{\Gamma\left(\frac{1}{2}\right)\Gamma(n+1)}{\Gamma\left(n+\frac{1}{2}\right)}\cos\frac{1}{2}\theta P_n^{(-\frac{1}{2},\frac{1}{2})}(\cos\theta), \tag{1.152}$$

$$\sin n\theta = \frac{\Gamma\left(\frac{3}{2}\right)\Gamma(n+1)}{\Gamma\left(n+\frac{1}{2}\right)}\sin\theta P_{n-1}^{(\frac{1}{2},\frac{1}{2})}(\cos\theta), \tag{1.153}$$

$$\sin(n+\frac{1}{2})\theta = \frac{\Gamma\left(\frac{1}{2}\right)\Gamma(n+1)}{\Gamma\left(n+\frac{1}{2}\right)}\sin\frac{1}{2}\theta P_n^{(\frac{1}{2},-\frac{1}{2})}(\cos\theta), \tag{1.154}$$

and

$$P_n(\cos\theta) = P_n^{(0,0)}(\cos\theta). \tag{1.155}$$

On the other hand, the trigonometric functions $\cos(\nu x), \sin(\nu x)$ with continuous parameter ν, occur in the well-known representations [19] of the Bessel functions

$$J_0(\nu\rho) = \frac{2}{\pi}\int_0^\rho \frac{\cos\nu x}{\sqrt{\rho^2 - x^2}}dx, \tag{1.156}$$

$$J_0(\nu\rho) = \frac{2}{\pi}\int_\rho^\infty \frac{\sin \nu x}{\sqrt{x^2-\rho^2}}dx. \tag{1.157}$$

Echoing previous remarks about the representations (1.149) and (1.150), an Abel-type transform of the trigonometric functions ($\cos \nu x$, $\sin \nu x$) produces another functional class (J_0). However, upon recalling the well-known relationships [19]

$$\cos \nu x = \left(\frac{\pi \nu x}{2}\right)^{\frac{1}{2}} J_{-\frac{1}{2}}(\nu x), \tag{1.158}$$

$$\sin \nu x = \left(\frac{\pi \nu x}{2}\right)^{\frac{1}{2}} J_{\frac{1}{2}}(\nu x), \tag{1.159}$$

it becomes clear that the Abel transforms (1.156) and (1.157) should be considered in the wider context of Bessel functions.

In other words, the trigonometric functions $\cos n\theta$, $\sin n\theta$, $\cos(n+\frac{1}{2})\theta$, and $\sin(n+\frac{1}{2})\theta$, with integer or half-integer parameter, should be considered as a special subclass of the Jacobi polynomials $P_n^{(\alpha,\beta)}(\cos\theta)$ (for appropriate (α,β)); whereas the trigonometric functions $\cos \nu x$, $\sin \nu x$, with real parameter ν, should be considered as a special subclass of the Bessel functions $J_\mu(\nu x)$ (for appropriate μ).

In turn, both the class of Bessel functions J_μ and the class of Jacobi polynomials $P_n^{(\alpha,\beta)}$, with arbitrary values of the parameters (α,β) or μ, belong to the wider class of hypergeometric functions in a very simple manner. Both are particular examples of the generalised hypergeometric function [59]

$$_pF_q\left(a_1,\dots,a_p;b_1,\dots,b_q;z\right) \equiv \sum_{k=0}^{\infty} \frac{(a_1)_k(a_2)_k\dots(a_p)_k}{(b_1)_k(b_2)_k\dots(b_q)_k}\cdot\frac{z^k}{k!} \tag{1.160}$$

where the notation for the *Pochhammer symbol*

$$(a)_k \overset{def}{=} a\,(a+1)\dots(a+k-1)\,;\,(a)_0 \overset{def}{=} 1 \tag{1.161}$$

has been used; the *upper parameters* $\overrightarrow{a} = (a_1,\dots,a_p)$ are unrestricted, whereas the *lower parameters* $\overrightarrow{b} = (b_1,\dots,b_q)$ are restricted so that no b_j is zero or a negative integer. Note that when a is neither zero nor a negative integer,

$$(a)_k = \frac{\Gamma(a+k)}{\Gamma(a)}. \tag{1.162}$$

When $p \le q$, the series converges for all complex z; when $p = q+1$, the series has radius of convergence 1 (its convergence on the unit disc $|z| = 1$ is discussed in Appendix B.2). If the one of upper parameters is equal to zero or a negative integer, then the series terminates and is a *hypergeometric* polynomial.

The Jacobi polynomial $P_n^{(\alpha,\beta)}$ may be recognised as a generalised hypergeometric function (see Appendix, (B.25)); it is *hypergeometric* polynomial

$$P_n^{(\alpha,\beta)}(x) = \binom{n+\alpha}{n} {}_2F_1\left(-n, n+\alpha+\beta+1; \alpha+1; \frac{1-x}{2}\right). \quad (1.163)$$

From the symmetry property (see Appendix, (B.26))

$$P_n^{(\alpha,\beta)}(-x) = (-1)^n P_n^{(\beta,\alpha)}(x),$$

we deduce the alternative representation

$$P_n^{(\alpha,\beta)}(x) = (-1)^n \binom{n+\beta}{n} {}_2F_1\left(-n, n+\alpha+\beta+1; \beta+1; \frac{1+x}{2}\right). \quad (1.164)$$

Bessel functions of arbitrary order also have a hypergeometric representation in terms of the special *confluent* hypergeometric functions,

$$J_\mu(z) = \frac{(z/2)^\mu}{\Gamma(\mu+1)} {}_0F_1(\mu+1; -\frac{1}{4}z^2), \quad (1.165)$$

$$J_\mu(z) = \frac{(z/2)^\mu}{\Gamma(\mu+1)} e^{iz} {}_1F_1(\mu+\frac{1}{2}; 2\mu+1; 2iv). \quad (1.166)$$

Let us derive the integral representation of Abel type for the Jacobi polynomials. From (1.163) and (1.160) immediately follows the finite series representation:

$$P_n^{(\alpha,\beta)}(x) = \frac{\Gamma(n+\alpha+1)}{n!\Gamma(\alpha+1)} \sum_{m=0}^{n} \frac{(-n)_m (n+\alpha+\beta+1)_m}{m!(\alpha+1)_m} \left(\frac{1-x}{2}\right)^m. \quad (1.167)$$

Fix the parameter $\eta \in [0,1)$; multiply both sides of (1.167) by the factor $(1-x)^\alpha (x-t)^{-\eta}$ and integrate over the interval $(t,1)$ to obtain

$$\int_t^1 \frac{(1-x)^\alpha P_n^{(\alpha,\beta)}(x)}{(x-t)^\eta} dx = \frac{\Gamma(n+\alpha+1)}{n!\Gamma(\alpha+1)} \sum_{m=0}^{n} \frac{(-n)_m (n+\alpha+\beta+1)_m}{2^m m!(\alpha+1)_m} A_\eta^{m+\alpha}(t), \quad (1.168)$$

where

$$A^q_\eta(t) \stackrel{def}{=} \int_t^1 (1-x)^q (x-t)^{-\eta}\, dx. \tag{1.169}$$

The change of variable by $1-x=(1-t)y$ expresses $A^q_\eta(t)$ in terms of the beta function B (see Appendix, (B.8)):

$$\begin{aligned} A^q_\eta(t) &= (1-t)^{q+1-\eta} \int_0^1 y^q (1-y)^{-\eta} dy = (1-t)^{q+1-\eta} B(q+1, 1-\eta) \\ &= (1-t)^{q+1-\eta} \frac{\Gamma(q+1)\Gamma(1-\eta)}{\Gamma(q+2-\eta)}. \end{aligned} \tag{1.170}$$

Substituting (1.170) into (1.168), replacing (α, β) by $(\alpha+\eta-1, \beta-\eta+1)$, and bearing in mind Definition (1.167), one obtains, after some manipulation, the following *integral representation of Abel type*:

$$P_n^{(\alpha,\beta)}(t) = \frac{(1-t)^{-\alpha}\Gamma(n+1+\alpha)}{\Gamma(1-\eta)\Gamma(n+\alpha+\eta)} \int_t^1 \frac{(1-x)^{\alpha+\eta-1} P_n^{(\alpha+\eta-1,\beta-\eta+1)}(x)}{(x-t)^\eta} dx. \tag{1.171}$$

Interchanging the role of α and β in (1.171), changing the sign of x and t, and taking into account Identity (1.164), we obtain another such *integral representation:*

$$P_n^{(\alpha,\beta)}(t) = \frac{(1+t)^{-\beta}\Gamma(n+1+\beta)}{\Gamma(1-\eta)\Gamma(n+\beta+\eta)} \int_{-1}^t \frac{(1+x)^{\beta+\eta-1} P_n^{(\alpha-\eta+1,\beta+\eta-1)}(x)}{(t-x)^\eta} dx. \tag{1.172}$$

Formulae (1.171) and (1.172) have an interpretation in terms of fractional integration operators [55]. When $\eta=0$, the following two notable identities corresponding to integration in conventional sense result:

$$(1-t)^{\alpha+1}\, P_n^{(\alpha+1,\beta-1)}(t) = (n+\alpha+1) \int_t^1 (1-x)^\alpha P_n^{(\alpha,\beta)}(x) dx, \tag{1.173}$$

$$(1+t)^{\beta+1}\, P_n^{(\alpha-1,\beta+1)}(t) = (n+\beta+1) \int_{-1}^t (1+x)^\beta P_n^{(\alpha,\beta)}(x) dx. \tag{1.174}$$

When expressed in algebraic form, the Dirichlet-Mehler Formulae (1.149) and (1.150) are special cases of the integral representations (1.171) and (1.172) with $\alpha=\beta=0, \eta=\frac{1}{2}$ (setting $t=\cos\theta$, and $x=\cos\phi$):

$$P_n(t) = \pi^{-\frac{1}{2}} \frac{\Gamma(n+1)}{\Gamma(n+\frac{1}{2})} \int_t^1 \frac{(1-x)^{-\frac{1}{2}} P_n^{(-\frac{1}{2},\frac{1}{2})}(x)}{(x-t)^{\frac{1}{2}}} dx, \tag{1.175}$$

$$P_n(t) = \pi^{-\frac{1}{2}} \frac{\Gamma(n+1)}{\Gamma(n+\frac{1}{2})} \int_{-1}^{t} \frac{(1+x)^{-\frac{1}{2}} P_n^{(\frac{1}{2},-\frac{1}{2})}(x)}{(t-x)^{\frac{1}{2}}} dx. \tag{1.176}$$

Let us now obtain the integral representations of Abel kind for the Bessel functions. The well-known Sonine's integrals provide a simple starting point. Sonine's first integral [14] is

$$J_{\nu+\xi+1}(z) = \frac{z^{\xi+1}}{2^{\xi}\Gamma(\xi+1)} \int_0^{\frac{\pi}{2}} J_\nu(z\sin\theta)\sin^{\nu+1}\theta\cos^{2\xi+1}\theta d\theta, \tag{1.177}$$

where $\nu > -1, \xi > -1$. The trivial transformation $z = xt, \rho = x\sin\theta$ produces the desired *integral representation of Abel kind*:

$$t^{-\xi-1} J_{\nu+\xi+1}(xt) = \frac{x^{-\xi-\nu-1}}{2^{\xi}\Gamma(\xi+1)} \int_0^x J_\nu(\rho t)\rho^{\nu+1}(x^2-\rho^2)^{\xi} d\rho. \tag{1.178}$$

A limiting form of Sonine's second integral [55] is

$$t^{-\eta-1} J_{\nu-\eta-1}(xt) = \frac{x^{\nu-\eta-1}}{2^{\xi}\Gamma(\eta+1)} \int_0^{\infty} J_\nu[t(s^2+x^2)^{\frac{1}{2}}](s^2+x^2)^{-\frac{\nu}{2}} s^{2\eta+1} ds, \tag{1.179}$$

where $\frac{\nu}{2} - \frac{1}{4} > \eta > -1$. The substitution $s^2 + x^2 = \rho^2$ transforms (1.179) to the second *integral representation of Abel kind* for Bessel functions:

$$t^{-\eta-1} J_{\nu-\eta-1}(xt) = \frac{x^{\nu-\eta-1}}{2^{\eta}\Gamma(\eta+1)} \int_x^{\infty} J_\nu(\rho t)\rho^{-\nu+1}(\rho^2-x^2)^{\eta} d\rho. \tag{1.180}$$

Special cases of (1.178) and (1.180) with $\xi = 0$ and $\eta = 0$, respectively are

$$\frac{1}{t} J_{\nu+1}(xt) = x^{-\nu-1} \int_0^x J_\nu(\rho t)\rho^{\nu+1} d\rho, \tag{1.181}$$

$$\frac{1}{t} J_{\nu-1}(xt) = x^{\nu-1} \int_x^{\infty} J_\nu(\rho t)\rho^{-\nu+1} d\rho. \tag{1.182}$$

The comments about fractional integration directly following Formula (1.172) are of equal pertinence to the representations (1.178) and (1.180) and their confluent forms (1.181) and (1.182).

These basic integral representations of Abel kind will be extensively exploited in later chapters. Other useful relationships can be found in [55].

1.7 Dual equations and single- or double-layer surface potentials

Let S_0 be an open surface, which is a portion of a larger closed surface S; let S_1 be the complementary part of S_0 in S (thus $S = S_0 \cup S_1$) so that

S_1 may be regarded as an "aperture" in S. Given S_0, the choice of S (and hence S_1) may be made arbitrarily, but we shall require that it satisfies the hypotheses for the application of Green's theorem (see [32]).

Classical potential theory represents the solution of Laplace's equation by means of single- or double-layer surface potentials [32]. In Section 1.3, the formulation of mixed boundary value problems for S_0 and the Laplace equation was discussed. This apparently alternative approach (which produces dual series equations or dual integral equations) is in fact entirely equivalent, at least in the context of the class of coordinate surfaces S discussed in Section 1.3.

Let P be an arbitrary point on S, and M be an observation point. Introduce an origin O; let $\overrightarrow{r'}$ and $\overrightarrow{r}$ denote the position vectors $\overrightarrow{OP}$ and $\overrightarrow{OM}$, and denote the distance between P and M by $R_{PM} = R\left(\overrightarrow{r'}, \overrightarrow{r}\right) = \left|\overrightarrow{r} - \overrightarrow{r'}\right|$. At P, we shall also consider the inward- and outward-pointing unit normal vectors $\overrightarrow{n_i}$ and $\overrightarrow{n_e}$.

Let us commence by considering the first boundary value problem for Laplace's equation, assuming that the value of potential U is specified on the open surface $S_0 : U(\overrightarrow{r'}) = F(\overrightarrow{r'})$ for some continuous function F. As already mentioned, classical potential theory presents the solution of Laplace's equation in terms of surface potentials. As a consequence of Green's fundamental theorem [32], the value of the harmonic function U at any interior point $\overrightarrow{r}$ of the region bounded by S is given by

$$U^{(i)}(\overrightarrow{r}) = \frac{1}{4\pi}\iint_S \left[\frac{1}{R(\overrightarrow{r}, \overrightarrow{r'})}\frac{\partial U^{(i)}}{\partial n_i} - U^{(i)}(\overrightarrow{r'})\frac{\partial}{\partial n_i}\left(\frac{1}{R(\overrightarrow{r}, \overrightarrow{r'})}\right)\right] ds. \tag{1.183}$$

When $\overrightarrow{r}$ lies outside S, the integral in (1.183) vanishes. In the exterior region, the solution at any point $\overrightarrow{r}$ exterior to S satisfies

$$U^{(e)}(\overrightarrow{r}) = \frac{1}{4\pi}\iint_S \left[\frac{1}{R(\overrightarrow{r}, \overrightarrow{r'})}\frac{\partial U^{(e)}}{\partial n_e} - U^{(e)}(\overrightarrow{r'})\frac{\partial}{\partial n_e}\left(\frac{1}{R(\overrightarrow{r}, \overrightarrow{r'})}\right)\right] ds. \tag{1.184}$$

When $\overrightarrow{r}$ lies inside S, the integral in (1.184) vanishes.

When the surface is open, the distinction between internal and external regions disappears (see Section 1.3) and the solution at any point $\overrightarrow{r}$ not on S must be considered as a sum of (1.183) and (1.184),

$$U(\overrightarrow{r}) = U^{(i)}(\overrightarrow{r}) + U^{(e)}(\overrightarrow{r}). \tag{1.185}$$

The solution and its normal derivative must be continuous at any point $\overrightarrow{r'}$ of the aperture surface S_1 so that

$$U^{(i)}(\overrightarrow{r'}) - U^{(e)}(\overrightarrow{r'}) = 0, \tag{1.186}$$

$$\frac{\partial}{\partial n}U^{(i)}(\vec{r'}) - \frac{\partial}{\partial n}U^{(e)}(\vec{r'}) = 0, \tag{1.187}$$

where $\vec{n} \equiv \vec{n_e} = -\vec{n_i}$.

Thus the solution U of the first-kind boundary value problem is given by

$$U(\vec{r}) = -\frac{1}{4\pi}\iint_{S_0}\left[\frac{\partial U^{(i)}}{\partial n} - \frac{\partial U^{(e)}}{\partial n}\right]\frac{1}{R(\vec{r}, \vec{r'})}ds, \tag{1.188}$$

whereas the solution of the second-kind boundary value problem (in which the normal derivative is specified on S_0) is represented by

$$U(\vec{r}) = -\frac{1}{4\pi}\iint_{S_0}\left[U^{(e)}(\vec{r'}) - U^{(i)}(\vec{r'})\right]\frac{\partial}{\partial n}\left(\frac{1}{R(\vec{r}, \vec{r'})}\right)ds. \tag{1.189}$$

Introducing the notations for the *jump functions* occurring in (1.188) and (1.189),

$$\sigma_D(\vec{r'}) \stackrel{def}{=} \frac{\partial U^{(i)}}{\partial n} - \frac{\partial U^{(e)}}{\partial n}, \tag{1.190}$$

$$\sigma_N(\vec{r'}) \stackrel{def}{=} U^{(e)}(\vec{r'}) - U^{(i)}(\vec{r'}), \tag{1.191}$$

the integral formulae become

$$U(\vec{r}) = -\frac{1}{4\pi}\iint_{S_0}\sigma_D(\vec{r'})\frac{1}{R(\vec{r}, \vec{r'})}ds, \tag{1.192}$$

and

$$U(\vec{r}) = -\frac{1}{4\pi}\iint_{S_0}\sigma_N(\vec{r'})\frac{\partial}{\partial n}\left(\frac{1}{R(\vec{r}, \vec{r'})}\right)ds. \tag{1.193}$$

The first integral (1.192) is the potential associated with a simple or single-layer distribution on S ; the second integral (1.193) is the potential of a double-layer distribution on S [32].

Thus the first-kind boundary value problem, in which the Dirichlet boundary condition (prescribing the value of U on S_0) is given by $U|_{S_0} = F$, gives rise to the following Fredholm integral equation of the first kind for the unknown single-layer distribution σ_D:

$$F(\vec{r_s}) = -\frac{1}{4\pi}\iint_{S_0}\sigma_D(\vec{r'})\frac{1}{R(\vec{r_s}, \vec{r'})}ds, \qquad \vec{r_s} \in S_0. \tag{1.194}$$

In a similar way, the second-kind boundary value problem in which the Neumann boundary condition (prescribing the value of $\frac{\partial U}{\partial n}$ on S_0) is given by $\frac{\partial U}{\partial n}|_{S_0} = G$ produces a Fredholm integral equation of the first kind for the unknown double-layer distribution σ_N:

$$G(\overrightarrow{r_s}) = -\frac{1}{4\pi}\iint_{S_0} \sigma_N(\overrightarrow{r'})\frac{\partial^2}{\partial n_s \partial n'}\left[\frac{1}{R(\overrightarrow{r_s},\overrightarrow{r'})}\right] ds \qquad \overrightarrow{r_s} \in S_0 \tag{1.195}$$

where $\overrightarrow{n_s}$ denotes the outward-pointing unit normal at $\overrightarrow{r_s}$.

The distance function, between any two arbitrary points in space $\overrightarrow{r}$ and $\overrightarrow{r'}$,

$$R(\overrightarrow{r},\overrightarrow{r'}) \equiv \left|\overrightarrow{r} - \overrightarrow{r'}\right|$$

plays an important part in classical potential theory since the Green's function for Laplace's equation in three-dimensional free space is

$$G(\overrightarrow{r},\overrightarrow{r'}) = \frac{1}{4\pi}\frac{1}{R(\overrightarrow{r},\overrightarrow{r'})}. \tag{1.196}$$

The reciprocal of the distance function, $R^{-1}(\overrightarrow{r},\overrightarrow{r'})$, is often called the source function of Laplace's equation; it is the potential function associated with the positive unit charge in electrostatics. It solves the nonhomogeneous Laplace's equation (Poisson's equation)

$$\nabla^2 U(\overrightarrow{r}) = -\delta(\overrightarrow{r} - \overrightarrow{r'}), \tag{1.197}$$

where $\delta(\overrightarrow{r} - \overrightarrow{r'})$ is the delta-function [28]; the differentiation in (1.197) is performed with respect to the primed variables.

We wish to investigate potential problems in coordinate systems that admit separation of variables for Laplace's equation. Accordingly, let us consider Poisson's equation in generalised curvilinear coordinates (q_1, q_2, q_3):

$$\nabla^2 U(q_1, q_2, q_3) = -h_{q_1}^{-1}h_{q_2}^{-1}h_{q_3}^{-1}\delta(q_1 - q_1')\delta(q_2 - q_2')\delta(q_3 - q_3') \tag{1.198}$$

where $h_{q_i}(i = 1, 2, 3)$ are the metric coefficients (see Section 1.1) and, as before, the differentiation in (1.198) is performed with respect to the primed variables. The metric coefficients perform a normalising function in (1.198) because

$$\int_V \delta(\overrightarrow{r} - \overrightarrow{r'})dV = \iiint_{\text{all } q_1,q_2,q_3} \delta(\overrightarrow{r} - \overrightarrow{r'})h_{q_1}h_{q_2}h_{q_3}dq_1dq_2dq_3 = 1, \tag{1.199}$$

which follows from the fundamental property of the δ-function,

$$\int_{x'=x-\varepsilon}^{x'=x+\varepsilon} \delta(x' - x)dx = 1, \text{ for } \varepsilon > 0.$$

We wish to obtain the Fourier series, or Fourier integral representation as appropriate, for the source function or for the Green's function. We consider in detail the spherical coordinate context, and simply state the final results for other coordinate systems. In spherical coordinates (r, θ, ϕ) the Green's function $G_0(\vec{r}, \vec{r'})$ of free space must satisfy

$$\Delta G_0(r, \theta, \phi, r', \theta', \phi') = -\frac{1}{r^2 \sin\theta}\delta(r - r')\delta(\theta - \theta')\delta(\phi - \phi'), \qquad (1.200)$$

where the Laplacian operator Δ is given by (1.27). Since $G_0(\vec{r}, \vec{r'})$ satisfies the homogeneous Laplace's equation when $\vec{r} \neq \vec{r'}$, and is a symmetric function of the primed and unprimed coordinates, we may expand $G_0(r, \theta, \phi; r', \theta', \phi')$ in terms of eigenfunctions of the Laplacian as

$$\frac{1}{r'}\sum_{m=0}^{\infty} \cos m(\phi - \phi') \sum_{n=m}^{\infty} A_{nm} P_n^m(\cos\theta) P_n^m(\cos\theta') \begin{Bmatrix} (r/r')^n, & r < r' \\ (r/r')^{-n-1}, & r > r' \end{Bmatrix}. \qquad (1.201)$$

This function is finite at $r = 0$ and satisfies the regularity condition at infinity.

The value A_{nm} is determined by the inhomogeneous term of (1.200). Multiply both sides of this equation by r^2, and integrate with respect to r over a small interval $(r' - \epsilon, r' + \epsilon)$ about r'. Remembering the continuity of the terms at $r = r'$ and passing to the limit $\epsilon \to 0$, we obtain

$$r^2 \frac{\partial}{\partial r} G_o(r, \theta, \phi; r', \theta', \phi')\Big|_{r=r'-0}^{r=r'+0} = -\frac{1}{\sin\theta}\delta(\theta - \theta')\delta(\phi - \phi'). \qquad (1.202)$$

Substituting (1.201) in this expression, and utilising the Wronskian relation for the independent solutions of (1.59), we find

$$\sum_{m=0}^{\infty} \cos m(\phi - \phi') \sum_{n=m}^{\infty} A_{nm}(2n+1) P_n^m(\cos\theta) P_n^m(\cos\theta') = \frac{\delta(\theta - \theta')\delta(\phi - \phi')}{\sin\theta}. \qquad (1.203)$$

Multiplying both sides of this equation by $P_l^k(\cos\theta)\cos k\phi$ and integrating over the full range of the variables θ and ϕ produces

$$A_{nm} = \frac{1}{4\pi}(2 - \delta_{m0})\frac{(n-m)!}{(n+m)!}. \qquad (1.204)$$

The representation (1.201) of the free space Green's function with coefficients (1.204) is not unique in spherical coordinates. It is a representation that is *discontinuous in the coordinate* r. A representation that is discontinuous in the coordinate θ will be derived in Chapter 6.

Similar representations of the free-space Green's function may be deduced by this method for those coordinate systems where the method of separation of variables is applicable. In particular, let us now state the Green's functions of this type for the Laplace equation in Cartesian, cylindrical polar, and spherical coordinates.

Cartesian coordinates.

The distance function is

$$R(\overrightarrow{r}, \overrightarrow{r'}) = \left\{(x-x')^2 + (y-y')^2 + (z-z')^2\right\}^{\frac{1}{2}}, \tag{1.205}$$

and the Green's function $G_o(x, y, z; x', y', z')$, which is discontinuous in z, is

$$\frac{2}{\pi^2}\int_0^\infty d\nu \cos[\nu(x-x')] \int_0^\infty \frac{\cos[\mu(y-y')]}{\sqrt{\nu^2+\mu^2}} \left\{\begin{matrix} e^{-\sqrt{\nu^2+\mu^2}(z-z')}, & z > z' \\ e^{\sqrt{\nu^2+\mu^2}(z-z')}, & z < z' \end{matrix}\right\}. \tag{1.206}$$

Cylindrical polar coordinates.

The distance function is

$$R(\overrightarrow{r}, \overrightarrow{r'}) = \left\{\rho^2 + (\rho')^2 - 2\rho\rho' \cos(\phi-\phi') + (z-z')^2\right\}^{\frac{1}{2}}, \tag{1.207}$$

and the Green's function $G_o(\rho, \phi, z; \rho', \phi', z')$, which is discontinuous in ρ, is

$$\frac{1}{\pi^2}\int_0^\infty d\nu \cos[\nu(z-z')] \times \sum_{m=0}^{\infty}(2-\delta_{0m}) \cos m(\phi-\phi') \left\{\begin{matrix} I_m(\nu\rho)K_m(\nu\rho'), & \rho < \rho' \\ I_m(\nu\rho')K_m(\nu\rho), & \rho > \rho' \end{matrix}\right\}. \tag{1.208}$$

Spherical polar coordinates.

The distance function is

$$R(\overrightarrow{r}, \overrightarrow{r'}) = \left\{r^2 + (r')^2 - 2rr'[\cos\theta\cos\theta' + \sin\theta\sin\theta'\cos(\phi-\phi')]\right\}^{\frac{1}{2}}, \tag{1.209}$$

and the Green's function $G_o(r, \theta, \phi; r', \theta', \phi')$, which is discontinuous in r, is given by Formulae (1.201) and (1.204).

Let us now establish the equivalence of the "dual series approach" and the method of single- or double-layer potentials in solving mixed boundary value problems for Laplace's equation. A constructive proof is not very complicated, requiring the three steps outlined below.

First, the free-space Green's function for Laplace's equation is expanded as a Fourier series, or represented as a Fourier integral, as in (1.201), (1.206), or (1.208). Secondly, the unknown distributions $\sigma_D(\overrightarrow{r'})$ or $\sigma_N(\overrightarrow{r'})$ are also expanded in a Fourier series or as a Fourier integral. On the surface, $S = S_0 \cup S_1$, the jump functions introduced in (1.190) and (1.191) satisfy

$$\frac{\partial U^{(i)}}{\partial n} - \frac{\partial U^{(e)}}{\partial n} = \begin{Bmatrix} \sigma_D(\overrightarrow{r'}), & \text{on } S_0 \\ 0, & \text{on } S_1 \end{Bmatrix} \tag{1.210}$$

and

$$U^{(e)}(\overrightarrow{r'}) - U^{(i)}(\overrightarrow{r'}) = \begin{Bmatrix} \sigma_N(\overrightarrow{r'}), & \text{on } S_0 \\ 0, & \text{on } S_1 \end{Bmatrix}. \tag{1.211}$$

These expansions are substituted in the integral Equations (1.194) and (1.195); because of the relationships of (1.210) and (1.211), the surface of integration is extended to the whole of S, which we may suppose is the coordinate surface corresponding to one coordinate (say q_1) being held constant, whilst the remaining two coordinates q_2, q_3 are varied over their full interval of definition. On the surface S, the harmonic functions (which are the separated solutions of Laplace's equation) are orthogonal and form a complete basis. Multiplying both sides of these equations by such a surface harmonic, and integrating over S (i.e., over the complete interval of variation of q_2, q_3), we obtain functional equations in matrix or integral form; these are valid for those values of q_2, q_3 such that $(q_1, q_2, q_3) \in S_o$. Additional equations are derived from Formulae (1.210) and (1.211) defining the jump functions on the aperture surface S_1.

Let us illustrate this abstractly described process with a concrete example. Consider the first-kind boundary value problem for Laplace's equation posed on an open spherical surface S_0 (or spherical cap) of radius a, subtending an angle θ_0 at the origin, with the boundary condition on S_0 being given as $U|_{S_0} = F$. Let S and S_1 denote, respectively, the complete spherical surface of radius a, and the aperture $r = a$, $\theta < \theta_0 \le \pi$, $0 \le \phi \le 2\pi$.

On S, the potential function $U(a, \theta, \phi)$ given by the function $F(\theta, \phi)$ is expressible as a Fourier series

$$F(\theta, \phi) = \sum_{m=0}^{\infty} (2 - \delta_{0m}) \cos m\phi \sum_{n=m}^{\infty} a_n^m P_n^m(\cos\theta), \tag{1.212}$$

where

$$a_n^m = \frac{1}{2\pi} \int_0^{2\pi} d\phi \int_0^{\pi} d\theta \sin\theta . F(\theta, \phi) P_n^m(\cos\theta) \cos m\phi \tag{1.213}$$

are known Fourier coefficients. We shall find the solution of the Laplace equation in this case as single-layer potential (1.192). Expand the jump function (1.210) in spherical surface harmonics

$$\left[\frac{\partial U^{(i)}}{\partial r} - \frac{\partial U^{(e)}}{\partial r}\right]_{r=a} = \frac{1}{a}\sum_{m=0}^{\infty}(2-\delta_{0m})\cos m\phi' \sum_{n=m}^{\infty} x_n^m P_n^m(\cos\theta'), \tag{1.214}$$

where $\theta' \in [0,\pi], \phi' \in [0,2\pi]$, and $\{x_n^m\}_{m=0,n=m}^{\infty,\infty}$ denotes its unknown Fourier coefficients. We substitute the Green's function, $G_0(a,\theta,\phi;a,\theta',\phi')$ given by (1.201) with $r=r'=a$, into (1.194) to find

$$F(\theta,\phi) = -a^2\int_0^{2\pi} d\phi' \int_0^{\infty} d\theta' \sin\theta' \sigma_D(\theta',\phi') G_o(a,\theta,\phi;a,\theta',\phi'). \tag{1.215}$$

This is valid for $\theta \in [0,\theta_0)$, $\phi \in [0,2\pi]$.

Using (1.210) and expansions (1.212) and (1.214), we obtain from (1.215) the Fredholm integral equation of first kind

$$\begin{aligned}\frac{1}{4\pi}\int_0^{2\pi} d\phi' \int_0^{\pi} d\theta' \sin\theta' &\left\{\sum_{m=0}^{\infty}(2-\delta_{m0})\cos m\phi' \sum_{n=m}^{\infty} x_n^m P_n^m(\cos\theta')\right\} \\ \times \sum_{s=0}^{\infty}(2-\delta_{s0})\cos s(\phi-\phi') &\sum_{l=s}^{\infty}\frac{(l-s)!}{(l+s)!}P_l^s(\cos\theta')P_l^s(\cos\theta) \\ &= -\sum_{m=0}^{\infty}(2-\delta_{m0})\cos m\phi \sum_{n=m}^{\infty} a_n^m P_n^m(\cos\theta).\end{aligned} \tag{1.216}$$

This is valid for $\theta \in [0,\theta_0)$, $\phi \in [0,2\pi]$.

Exploiting the orthogonality of spherical surface harmonics, the left-hand side of this equation simplifies to a double series of the same format as the right-hand side, leading finally to the series equations

$$\begin{aligned}\sum_{m=0}^{\infty}(2-\delta_{0m})\cos m\phi &\sum_{n=m}^{\infty}\frac{x_n^m}{2n+1}P_n^m(\cos\theta) \\ &= -\sum_{m=0}^{\infty}(2-\delta_{0m})\cos m\phi \sum_{n=m}^{\infty} a_n^m P_n^m(\cos\theta).\end{aligned} \tag{1.217}$$

This is also valid for $\theta \in [0,\theta_0)$, $\phi \in [0,2\pi]$.

A companion equation follows directly from the definition of jump function (1.210) and its expansion in spherical surface harmonics:

$$\sum_{m=0}^{\infty}(2-\delta_{0m})\cos m\phi \sum_{n=m}^{\infty} x_n^m P_n^m(\cos\theta) = 0. \tag{1.218}$$

This is valid for the range $\theta \in (\theta_0,\pi]$, $\phi \in [0,2\pi]$.

Multiplication of both sides of Equations (1.217) and (1.218) by the factor $\cos k\phi$, followed by integration with respect to ϕ on $[0, 2\pi]$, produces a pair of dual series equations for the unknown coefficients x_n^m :

$$\sum_{n=m}^{\infty} \frac{x_n^m}{2n+1} P_n^m(\cos\theta) = -\sum_{n=m}^{\infty} a_n^m P_n^m(\cos\theta), \theta \in [0, \theta_0), \quad (1.219)$$

$$\sum_{n=m}^{\infty} x_n^m P_n^m(\cos\theta) = 0, \qquad \theta \in (\theta_o, \pi]. \quad (1.220)$$

Conversely, it is evident that transformation of the dual series equations (1.217) and (1.218) to an integral equation of Fredholm type can be easily realised in the following way. Apply the formula (1.214) in reverse order, i.e., for $\phi' \in [0, 2\pi]$,

$$\sum_{m=0}^{\infty} (2 - \delta_{m0}) \cos m\phi' \sum_{n=m}^{\infty} x_n^m P_n^m(\cos\theta') = \begin{cases} a\sigma_D(\theta', \phi'), & \theta' \in [0, \theta_0) \\ 0, & \theta' \in (\theta_0, \pi] \end{cases}, \quad (1.221)$$

from which it immediately follows that

$$x_n^m = \frac{a(2n+1)}{4\pi} \frac{(n-m)!}{(n+m)!} \times \int_0^{2\pi} d\phi' \cos m\phi' \int_0^{\theta_0} \sigma_D(\theta', \phi') P_n^m(\cos\theta') \sin\theta' d\theta'. \quad (1.222)$$

Substitution of (1.222) in (1.217) and an interchange of the order of summation and integration produces the original integral Equation (1.215), as desired.

Thus we have demonstrated the equivalence of the integral equation formulation ((1.194) or (1.215)) and the dual series equations formulation ((1.219) and (1.220)) for determining the potential.

[illegible] equation and [illegible] or double-layer surface potentials.

Multiplication of both sides of Equations [illegible] (1.2.41) [illegible] factor [illegible] follows. By integration with respect to φ, we [illegible] terms [illegible] the integral coefficients.

$$[illegible]$$

$$[illegible]$$

Conversely, [illegible] that the dual formulation of the dual series equations (1.2.[illegible]) and ([illegible]) [illegible] to an integral that a [illegible] solution [illegible] can be easily [illegible] satisfied [illegible] case. Apply the formula (1.2.[illegible]) [illegible] series [illegible]

$$[illegible] \qquad (1.2.21)$$

from which it immediately follows that

$$[illegible]$$

$$[illegible] \qquad (1.2.[illegible])$$

Substitution of (1.2.[illegible]) and (1.2.[illegible]) [illegible] the [illegible] and integral [illegible] the original integral equation (1.2.[illegible]) [illegible].

Thus we have demonstrated the equivalence of the integral equation formulation (1.2.[illegible]) or (1.2.[illegible]) and the dual series equations formulations (1.2.[illegible]) and (1.2.[illegible]) for double-layer potentials.

2

Series and Integral Equations

The spatial distribution of the electrostatic potential surrounding a conducting surface (open or closed) is determined at the most fundamental level by Laplace's equation, together with the appropriate boundary conditions, decay conditions at infinity, and, if necessary, edge conditions. The precise formulation of these conditions was described in Section 1.3.

An alternative but equivalent formulation utilizes integral representations for the potential in terms of the surface charge density (corresponding to the jump in the normal derivative of the potential across the surface); in turn, this density is determined as the solution of an integral equation holding at each point of the conducting surface (see Section 1.7).

These two formulations are the basis of all analytical and numerical methods devised to solve the potential problem for bodies of arbitrary or general shape. Certain classes of surfaces, including those that are portions of the orthogonal coordinate surfaces described in Chapter 1, admit another formulation of the potential problem, in terms of dual- (or triple- or multiple-) series equations, or dual- (or triple- or multiple-) integral equations. Although it is formally equivalent, this alternative approach has the benefit that, in many cases of physical interest, these equations can be solved analytically (in closed form), so that a direct assessment of the effect of edges and cavities in these geometries is possible. In other cases, the analytical solution process transforms or regularises the series (or integral) equations to a matrix (or integral) Fredholm equation of the second kind. Once converted, these equations provide a basis for approximate analytical solution techniques (such as successive approximation), or for a numerical solution procedure which is simple to implement, well conditioned, rapidly converging, and of guaranteed accuracy. Thus, edge effects and cavity contributions to the potential distribution can be accurately quantified.

Beyond the electrostatic context, this approach finds general application to mixed boundary value problems (of first-, second-, or third-kind) for the Laplace equation. It also provides a basis for assessing the scattering and diffraction by the class of bodies described above, of acoustic and electromagnetic waves, where the interest is in accurate quantification of the scattering process by edges, or of entrapment of wave energy by cavities.

This chapter considers various classes of series and integral equations. The core idea is to convert the set of equations to a second-kind Fredholm matrix or integral equation. The Abel integral transform method provides a unified and constructive treatment of this process. In some cases these equations can be solved explicitly, in closed form; in the remaining cases, the transformed system is well suited to either approximate analytical solution methods or to numerical methods. When the second-kind matrix system is truncated to a finite number N_{tr} of linear equations, the solution of the truncated system converges to the exact solution as $N_{tr} \to \infty$. It is possible to estimate accuracy as a function of truncation number N_{tr} and so produce solutions of specified accuracy. Precise treatments of the behaviour of second-kind systems under truncation are given in [2] and [30].

Proofs of the validity of this method, and of the uniqueness of solutions, are sketched in Section 2.1; readers with a deeper interest in the details are recommended to consult the paper [64].

The problem typified by the determination of the electrostatic potential surrounding a charged spherical cap (Section 1.3) leads to dual series equations involving the Jacobi polynomials $P_n^{(\alpha,\beta)}$ as kernels. This general class of equations is the first to be considered in the next section. They have the form

$$\sum_{n=0}^{\infty} c_n x_n P_n^{(\alpha,\beta)}(t) = F(t), \; t \in (-1, t_0), \tag{2.1}$$

$$\sum_{n=0}^{\infty} x_n P_n^{(\alpha,\beta)}(t) = G(t), \; t \in (t_0, 1), \tag{2.2}$$

where the functions F, G and coefficients c_n are known, t_0 is fixed in $(-1, 1)$, and the unknown coefficients x_n are to be determined. Typically,

$$c_n = n^{2\eta}\left(1 + O(n^{-1})\right), \text{ as } n \to \infty.$$

The regularisation generally obtained by the Abel transform method is outlined, and where possible, explicit solutions are found.

Two special subclasses which merit some separate consideration are examined in the following two sections (2.2 and 2.3), dual series with trigonometric kernels or with associated Legendre function kernels (these are closely related to ultraspherical polynomials).

Triple series equations provide a natural generalisation of dual series equations; the kernel class examined in Section 2.4 is restricted to those kernels of interest in subsequent chapters.

Preparatory to considering dual integral equations in their own right, the relationship between series and integral equations is explored in Section 2.5. The following Section (2.6) demonstrates how to apply the Abel integral transform to solve some dual integral equations with Bessel function kernels; this allows us to regularise a wide class of such dual integral equations.

The subdivision of the interval of definition for triple series equations examined in Section 2.4 is assumed to be symmetric; this restriction is removed to cover asymmetric subdivisions in Section 2.7.

Coupled systems of series equations are treated in the Section 2.8, and some general remarks on so-called *integro-series* equations are provided in the concluding section of the chapter.

2.1 Dual series equations involving Jacobi polynomials

This section considers dual series equations of the form (2.1) and (2.2). Since the function $y = P_n^{(\alpha,\beta)}$ satisfies the differential equation

$$\frac{1}{w(x)}\frac{d}{dx}\left(\left(1-x^2\right)w(x)\frac{dy}{dx}\right)+n(n+\alpha+\beta+1)y=0,$$

with weight function $w(x) = (1-x)^{\alpha}(1+x)^{\beta}$, the parameter η may be assumed to lie in the interval $[0,1)$; for if $\eta \geq 1$, we may replace $P_n^{(\alpha,\beta)}(t)$ by

$$\frac{-1}{n(n+\alpha+\beta+1)}\frac{1}{w(t)}\frac{d}{dt}\left(\left(1-t^2\right)w(t)\frac{dP_n^{(\alpha,\beta)}}{dt}(t)\right)$$

and integrate twice to obtain an equation similar to (2.1), but with a new coefficient c_n satisfying

$$c_n = n^{2(\eta-1)}\left(1+O(n^{-1})\right), \text{ as } n \to \infty.$$

It is convenient to employ the quantity $\lambda_n(\alpha,\beta;\eta)$ given by

$$\lambda_n(\alpha,\beta;\eta) = \frac{\Gamma(n+\alpha+1)\,\Gamma(n+\beta+1+\eta)}{\Gamma(n+\alpha+1-\eta)\,\Gamma(n+\beta+1)} \tag{2.3}$$

where Γ denotes the Gamma-function; Field's formula (see Appendix, (B.7)) shows that

$$\lambda_n(\alpha,\beta;\eta) = n^{2\eta}\left[1+O\left(n^{-1}\right)\right].$$

We consider the slightly more general form of (2.1) and (2.2):

$$\sum_{n=0}^{\infty} \lambda_n(\alpha,\beta;\eta)\, x_n (1-r_n)\, P_n^{(\alpha,\beta)}(t) = F(t),\ t \in (-1,t_0) \quad (2.4)$$

$$\sum_{n=0}^{\infty} x_n (1-q_n)\, P_n^{(\alpha,\beta)}(t) = G(t),\ t \in (t_0,1). \quad (2.5)$$

The infinite set of unknown coefficients $\{x_n\}_{n=0}^{\infty}$ are to be determined. The parameters α, β, η are constrained to satisfy $\alpha - \eta > -1, \beta > -1$, and for our applications we may always suppose that $\eta \in (0,1)$. The reason for this constraint will become clear once the method of regularisation is described below. The quantities $\{r_n\}_{n=0}^{\infty}, \{q_n\}_{n=0}^{\infty}$ are assumed to be known sequences, in general, of complex quantities satisfying

$$\lim_{n\to\infty} q_n = \lim_{n\to\infty} r_n = 0. \quad (2.6)$$

The right-hand sides of Equations (2.4) and (2.5) are assumed to be expandable in Fourier-Jacobi series of the form

$$F(t) = \sum_{n=0}^{\infty} \lambda_n(\alpha,\beta;\eta)\, f_n P_n^{(\alpha,\beta)}(t), \quad (2.7)$$

$$G(t) = \sum_{n=0}^{\infty} g_n P_n^{(\alpha,\beta)}(t). \quad (2.8)$$

We seek solutions to (2.4) and (2.5) in an appropriate functional space. Denote by $l_2(\mu)$ the space of sequences $\{x_n\}_{n=0}^{\infty}$ satisfying

$$\sum_{n=0}^{\infty} n^{\mu} \mid x_n \mid^2 < \infty. \quad (2.9)$$

We suppose that the coefficients f_n, g_n belong to $l_2(2\eta - 1)$, and the solution will be sought in the same class:

$$\{x_n\}_{n=0}^{\infty} \in l_2(2\eta-1), \quad \{f_n, g_n\}_{n=0}^{\infty} \in l_2(2\eta-1). \quad (2.10)$$

The specification (2.9) arises very naturally in connection with the edge condition of the uniqueness theorem for an open surface (see Section 1.3, (1.85)). Thus (2.5) and (2.8) contain series that converge to their sums in the weighted mean square sense with weight w.

Some care is needed in the interpretation of convergence of the series occurring in (2.4) and (2.7). In our applications, Equation (2.4) invariably arises from enforcing the continuity of either the potential or of its normal derivative across the aperture surface of the structure under consideration. Thus (2.4) is summable in the sense of Abel (see Appendix D.2), and moreover because the coefficients in (2.4) are $O(n^r)$ for some r dependent only on α, β, and uniformly on $[-1, t_0]$, the series is *uniformly Abel-summable* on $[-1, t_0]$.

Let us now describe the general scheme to solve Equations (2.4) and (2.5) by the process of regularisation. This was briefly outlined at the end of Section 1.4. First we integrate, with weight function $(1+t)^\beta$, both sides of (2.4) over the interval $(-1, t)$, using the integration formula (1.174) to obtain a more rapidly converging series:

$$\sum_{n=0}^{\infty} \frac{\lambda_n(\alpha, \beta; \eta)}{(n+1+\beta)} \{x_n(1-r_n) - f_n\} P_n^{(\alpha-1,\beta+1)}(t) = 0, t \in (-1, t_0) \quad (2.11)$$

This process is justified because the series is uniformly Abel-summable (on closed subintervals of $(-1, t_0)$).

Next we use the integral representation (1.172) of Abel kind for Jacobi polynomials $P_n^{(\alpha-1,\beta+1)}$, replacing index α by $\alpha - 1$ and β by $\beta + 1$:

$$P_n^{(\alpha-1,\beta+1)}(t) = \frac{(1+t)^{-\beta-1}\Gamma(n+\beta+2)}{\Gamma(1-\eta)\Gamma(n+\beta+1+\eta)} \int_{-1}^{t} \frac{(1+x)^{\beta+\eta} P_n^{(\alpha-\eta,\beta+\eta)}(x)}{(t-x)^\eta} dx \quad (2.12)$$

Substituting this representation for $P_n^{(\alpha-1,\beta+1)}$ in (2.11) and interchanging the order of summation and integration, we obtain the functional equation

$$\int_{-1}^{t} (t-x)^{-\eta} U(x) dx = 0, t \in (-1, t_0) \quad (2.13)$$

where

$$U(x) = (1+x)^{\beta+\eta} \sum_{n=0}^{\infty} \frac{\Gamma(n+\alpha+1)}{\Gamma(n+\alpha+1-\eta)} [x_n(1-r_n) - f_n] P_n^{(\alpha-\eta,\beta+\eta)}(x). \quad (2.14)$$

In obtaining the last equation, definition (2.3) was used. The interchange is justified by the weighted mean square convergence of the series (2.14) (see Appendix D.2). The reason for the constraint $\alpha - \eta > -1, \beta > -1$ is now clear.

Equation (2.13) is the homogenous form of Abel's integral equation. The inverse formula (1.131) shows that (2.13) has the unique trivial solution, and we obtain the functional equation

$$\sum_{n=0}^{\infty} \frac{\Gamma(n+\alpha+1)}{\Gamma(n+\alpha+1-\eta)} [x_n(1-r_n) - f_n] P_n^{(\alpha-\eta,\beta+\eta)}(x) = 0, x \in (-1, t_0). \quad (2.15)$$

To obtain a second equation over the interval $(t_0, 1)$, involving the same Jacobi polynomials as in (2.15), it is necessary to utilise the integral representation (1.171), replacing η by $1-\eta$:

$$P_n^{(\alpha,\beta)}(t) = \frac{(1-t)^{-\alpha}\,\Gamma(n+1+\alpha)}{\Gamma(\eta)\,\Gamma(n+\alpha+1-\eta)} \int_t^1 \frac{(1-x)^{\alpha-\eta}\,P_n^{(\alpha-\eta,\beta+\eta)}(x)}{(x-t)^{1-\eta}}dx. \tag{2.16}$$

Repeating the mathematical operations used to obtain equation (2.15) we find

$$\sum_{n=0}^{\infty} \frac{\Gamma(n+\alpha+1)}{\Gamma(n+\alpha+1-\eta)} \{x_n(1-q_n) - g_n\}\, P_n^{(\alpha-\eta,\beta+\eta)}(x) = 0, x \in (t_0, 1). \tag{2.17}$$

Combining Equations (2.15) with (2.17) we obtain

$$\sum_{n=0}^{\infty} c_n x_n P_n^{(\alpha-\eta,\beta+\eta)}(x) = \begin{Bmatrix} F_1(x), & x \in (-1, t_0) \\ F_2(x), & x \in (t_0, 1) \end{Bmatrix} \tag{2.18}$$

where

$$\begin{aligned} F_1(x) &= \sum_{n=0}^{\infty} c_n (x_n r_n + f_n) P_n^{(\alpha-\eta,\beta+\eta)}(x), \\ F_2(x) &= \sum_{n=0}^{\infty} c_n (x_n q_n + g_n) P_n^{(\alpha-\eta,\beta+\eta)}(x), \end{aligned}$$

and

$$c_n = \frac{\Gamma(n+\alpha+1)}{\Gamma(n+\alpha+1-\eta)}.$$

We recall that the coefficients $\{x_n\}_{n=0}^{\infty}$ lie in a space dependent upon η (2.10). It simplifies the solution to modify the Fourier coefficients so that they are square-summable sequences in $l_2 = l_2(0)$. Introducing the orthonormal Jacobi polynomials $\hat{P}_n^{(\alpha,\beta)}$, defined by

$$\hat{P}_n^{(\alpha,\beta)}(z) = P_n^{(\alpha,\beta)}(z) / \parallel P_n^{(\alpha,\beta)} \parallel \tag{2.19}$$

where the square norm $\parallel P_n^{(\alpha,\beta)} \parallel^2 \equiv h_n^{(\alpha,\beta)}$ is given by Formula (B.20) (see Appendix), we may normalise the coefficients x_n, f_n, g_n so that

$$\{y_n, \hat{f}_n, \hat{g}_n\} = \frac{\Gamma(n+1+\alpha)}{\Gamma(n+1+\alpha-\eta)} \left[h_n^{(\alpha-\eta,\beta+\eta)}\right]^{\frac{1}{2}} \{x_n, f_n, g_n\}; \tag{2.20}$$

these sequences are square-summable:

$$\left\{y_n, \hat{f}_n, \hat{g}_n\right\}_{n=0}^{\infty} \in l_2(0) \equiv l_2. \tag{2.21}$$

Equation (2.18) becomes

$$\sum_{n=0}^{\infty} y_n \hat{P}_n^{(\alpha-\eta,\beta+\eta)}(t) = \begin{Bmatrix} G_1(t), & t \in (-1, t_0) \\ G_2(t), & t \in (t_0, 1) \end{Bmatrix}, \tag{2.22}$$

where

$$\begin{aligned} G_1(t) &= \sum_{n=0}^{\infty} (y_n r_n + \hat{f}_n) \hat{P}_n^{(\alpha-\eta,\beta+\eta)}(t), \\ G_2(t) &= \sum_{n=0}^{\infty} (y_n q_n + \hat{g}_n) \hat{P}_n^{(\alpha-\eta,\beta+\eta)}(t). \end{aligned}$$

Conditions (2.7), (2.8), and (2.10) dictate that all series in (2.22) are Fourier-Jacobi series, so that we can exploit completeness and orthogonality of the orthonormal set $\{\hat{P}_n^{(\alpha-\eta,\beta+\eta)}\}_{n=0}^{\infty}$ on $[-1,1]$. After multiplication of both sides of (2.22) by the factor $(1-t)^{\alpha-\eta}(1+t)^{\beta+\eta}\hat{P}_s^{(\alpha-\eta,\beta+\eta)}(t)$ and integration over $(-1,1)$, we obtain the following infinite system of linear algebraic equations (i.s.l.a.e.)

$$\begin{aligned} (1-r_s)\, y_s + \sum_{n=0}^{\infty} y_n (r_n - q_n)\, \hat{Q}_{sn}^{(\alpha-\eta,\beta+\eta)}(t_0) \\ = \hat{f}_s + \sum_{n=0}^{\infty} \left(\hat{g}_n - \hat{f}_n\right) \hat{Q}_{sn}^{(\alpha-\eta,\beta+\eta)}(t_0), \end{aligned} \tag{2.23}$$

where $s = 0, 1, 2, \ldots$, and

$$\hat{Q}_{sn}^{(\alpha,\beta)}(t) = \int_t^1 (1-x)^{\alpha}(1+x)^{\beta}\, \hat{P}_s^{(\alpha,\beta)}(x)\, \hat{P}_n^{(\alpha,\beta)}(x)\, dx. \tag{2.24}$$

The function $\hat{Q}_{sn}^{(\alpha,\beta)}(t)$ is termed an *incomplete scalar product* of normalised Jacobi polynomials with weight function $(1-x)^{\alpha}(1+x)^{\beta}$ for the following reason. The conventional (weighted) scalar product of $\hat{P}_s^{(\alpha,\beta)}$ and $\hat{P}_n^{(\alpha,\beta)}$ is

$$\hat{Q}_{sn}^{(\alpha,\beta)}(-1) = \int_{-1}^1 (1-x)^{\alpha}(1+x)^{\beta}\, \hat{P}_s^{(\alpha,\beta)}(x)\, \hat{P}_n^{(\alpha,\beta)}(x)\, dx \tag{2.25}$$

and the "incompleteness" of (2.24) refers to the fact that integration is performed over the subinterval $[t,1]$. We shall also employ the *unnormalised* incomplete scalar product

$$Q_{sn}^{(\alpha,\beta)}(t) = \int_t^1 (1-x)^{\alpha}(1+x)^{\beta}\, P_s^{(\alpha,\beta)}(x)\, P_n^{(\alpha,\beta)}(x)\, dx \tag{2.26}$$

of unnormalised Jacobi polynomials. Some useful properties incomplete scalar product are stated in Appendix B.6.

It can be shown that $\{\hat{Q}_{sn}^{(\alpha,\beta)}(t)\}_{s,n=0}^{\infty}$ is the matrix of a projection operator $K(t)$ in l_2, therefore satisfying $K(t)^2 = K(t)$. Using this property and that of the diagonal operators d_r and d_q which correspond to the diagonal matrices $\text{diag}\{r_n\}_{n=0}^{\infty}$ and $\text{diag}\{q_n\}_{n=0}^{\infty}$, one can prove that the matrix operator of (2.23) is a completely continuous (or compact) perturbation H of the identity operator I in l_2. Thus Equation (2.23) is a Fredholm equation of the second kind (see Appendix C.3), which we may represent in the form

$$(I - H)\,y = b \tag{2.27}$$

where the vector $b \in l_2$ may be readily identified; the solution vector $y = \{y_n\}_{n=0}^{\infty}$ lies in l_2. Since projection operators have norm at most unity, the norm of the operator H is bounded by

$$\| H \| \leq \max_n | r_n | + \max_n | q_n | . \tag{2.28}$$

The Fredholm alternative (see Appendix C.3) is valid for (2.23) or (2.27); the equations can be solved by the truncation method or, in certain cases, by an iterative method of successive approximations. The truncation method replaces the infinite system by a finite number (those indexed by $s = 0, 1, \ldots, N_{tr}$) of linear algebraic equations, in which all infinite sums are truncated to retain only the variables $y_0, y_1, \ldots, y_{N_{tr}}$. Note that the solution is explicitly obtained in closed analytic form when $r_n = q_n = 0$. The mixed boundary value problems considered later will either have analytic solutions of this type or have solutions which can, in principle, be obtained by the method of successive approximations. In any case the system (2.23) is solvable numerically in a satisfactory manner via the truncation method. A detailed discussion of the rate of convergence of the solution to the truncated system to the exact (infinite) system is given in [30]; this makes it possible to estimate and guarantee accuracy of numerical solutions generated in this fashion.

A companion pair to the Equations (2.4) and (2.5) is the related set of dual series

$$\sum_{n=0}^{\infty} x_n (1 - q_n) P_n^{(\alpha,\beta)}(t) = G(t), t \in (-1, t_0) \tag{2.29}$$

$$\sum_{n=0}^{\infty} \lambda_n (\beta, \alpha, \eta)\, x_n (1 - r_n) P_n^{(\alpha,\beta)}(t) = F(t), t \in (t_0, 1). \tag{2.30}$$

The indices α, β, η are now constrained to satisfy $\alpha > -1, \beta - \eta > -1$. Essentially, the subintervals on which (2.4) and (2.5) are enforced are interchanged, and the factor $\lambda_n(\beta, \alpha, \eta)$ replaces $\lambda_n(\alpha, \beta, \eta)$. In contrast to

(2.7), F is assumed to be expandable in a Fourier-Jacobi series of the form

$$F(t) = \sum_{n=0}^{\infty} \lambda_n (\beta, \alpha, \eta) f_n P_n^{(\alpha,\beta)}(t), \tag{2.31}$$

but G is assumed to possess the same expansion (2.8).

Applying the same method used above to solve (2.4) and (2.5), we find

$$\sum_{n=0}^{\infty} c_n x_n P_n^{(\alpha+\eta,\beta-\eta)}(t) = \begin{Bmatrix} F_1(t) & , x \in (-1, t_0) \\ F_2(t) & , x \in (t_0, 1) \end{Bmatrix}, \tag{2.32}$$

where

$$\begin{aligned} F_1(t) &= \sum_{n=0}^{\infty} c_n (x_n q_n + g_n) P_n^{(\alpha+\eta,\beta-\eta)}(x), \\ F_2(t) &= \sum_{n=0}^{\infty} c_n (x_n r_n + f_n) P_n^{(\alpha+\eta,\beta-\eta)}(x), \end{aligned}$$

and

$$c_n = \frac{\Gamma(n+\beta+1)}{\Gamma(n+\beta+1-\eta)}.$$

After rescaling both known and unknown coefficients via

$$\{y_n, \hat{f}_n, \hat{g}_n\} = \frac{\Gamma(n+1+\beta)}{\Gamma(n+1+\beta-\eta)} \left[h_n^{(\alpha+\eta,\beta-\eta)}\right]^{\frac{1}{2}} \{x_n, f_n, g_n\}, \tag{2.33}$$

we obtain

$$\sum_{n=0}^{\infty} y_n \hat{P}_n^{(\alpha+\eta,\beta-\eta)}(t) = \begin{Bmatrix} G_1(t), & t \in (-1, t_0) \\ G_2(t), & t \in (t_0, 1) \end{Bmatrix}, \tag{2.34}$$

where

$$\begin{aligned} G_1(t) &= \sum_{n=0}^{\infty} (y_n q_n + \hat{g}_n) \hat{P}_n^{(\alpha+\eta,\beta-\eta)}(t), \\ G_2(t) &= \sum_{n=0}^{\infty} (y_n r_n + \hat{f}_n) \hat{P}_n^{(\alpha+\eta,\beta-\eta)}(t). \end{aligned}$$

From this, we finally obtain the i.s.l.a.e. of the second kind

$$\begin{aligned} (1 - q_s) y_s &+ \sum_{n=0}^{\infty} y_n (q_n - r_n) \hat{Q}_{sn}^{(\alpha+\eta,\beta-\eta)}(t_0) \\ &= \hat{g}_s + \sum_{n=0}^{\infty} \left(\hat{f}_n - \hat{g}_s\right) \hat{Q}_{sn}^{(\alpha+\eta,\beta-\eta)}(t_0), \end{aligned} \tag{2.35}$$

where $s = 0, 1, 2, ...$ This i.s.l.a.e. possesses very similar properties to those of (2.23).

It is not possible, in general, to solve the regularised systems (2.23) or (2.35) explicitly in closed form, except for certain choices of q_n and r_n. Without loss of generality we may suppose that $q_n = 0$. As an example, consider

$$r_n = \frac{-A}{n(n+\alpha+\beta+1)}$$

for some constant A. Then (2.23) implies that

$$Y(x) = \sum_{n=0}^{\infty} y_n \hat{P}_{sn}^{(\alpha+\eta,\beta-\eta)}(x)$$

satisfies

$$\frac{1}{w(x)}\frac{d}{dx}\left[\left(1-x^2\right)w(x)\frac{dY}{dx}(x)\right] + AY(x) =$$
$$\frac{1}{w(x)}\frac{d}{dx}\left[\left(1-x^2\right)w(x)\frac{d\hat{F}}{dx}(x)\right], \quad x \in (-1, t_0),$$

$$Y(x) = \hat{G}(x), \qquad x \in (t_0, 1)$$

where

$$\begin{aligned}\hat{F}(x) &= \sum_{n=0}^{\infty} \hat{f}_n \hat{P}_{sn}^{(\alpha+\eta,\beta-\eta)}(x), \\ \hat{G}(x) &= \sum_{n=0}^{\infty} \hat{g}_n \hat{P}_{sn}^{(\alpha+\eta,\beta-\eta)}(x).\end{aligned}$$

The differential equation may be solved to yield, when $x \in (-1, t_0)$,

$$Y(x) = \sum_{n=0}^{\infty} \frac{n(n+\alpha+\beta+1)}{n(n+\alpha+\beta+1)-A}\hat{f}_n \hat{P}_{sn}^{(\alpha+\eta,\beta-\eta)}(x) + CH_1(x) + DH_2(x),$$

where H_1, H_2 are a pair of linearly independent solutions of

$$\frac{1}{w(x)}\frac{d}{dx}\left[\left(1-x^2\right)w(x)\frac{dY}{dx}(x)\right] + AY(x) = 0,$$

and C, D are constants. The constants are explicitly determined by enforcing continuity of Y and its derivative at the point t_0, and the expansion coefficients of Y are then explicitly calculated.

More generally, the same argument can be applied when

$$r_n = \frac{A_1}{n(n+\alpha+\beta+1)} + \frac{A_2}{n^2(n+\alpha+\beta+1)^2} + \ldots + \frac{A_r}{n^r(n+\alpha+\beta+1)^r}$$

to produce a differential equation of order $2r$ that may be solved provided the corresponding homogeneous differential equation is solved. Again Y is made fully determinate by enforcing continuity on Y and its first $2r-1$ derivatives at t_0.

This idea lies behind various methods to improve the convergence of (2.23) under truncation, by replacing it with a more rapidly convergent system. An example of this technique will be given in Chapter 4.

Sometimes mixed-boundary value problems in potential theory or wave-scattering theory lead to dual series equations for which the parameter constraints (namely $\alpha-\eta > -1, \beta > -1$ on the pair (2.4) and (2.5), or $\alpha > -1, \beta-\eta > -1$ on the pair (2.29) and (2.30)) do not hold. We may overcome this difficulty by transforming the initial equations to an equivalent set which involve Jacobi polynomials with increased values of the indices.

There are two ways to effect such a transformation. One may apply the formula deduced from Rodrigues' formula [59] for Jacobi polynomials:

$$-2n(1-x)^{\alpha}(1+x)^{\beta} P_n^{(\alpha,\beta)}(x) = \frac{d}{dx}\{(1-x)^{\alpha+1}(1+x)^{\beta+1} P_{n-1}^{(\alpha+1,\beta+1)}(x)\}. \quad (2.36)$$

A second way successively applies the integration formulae (1.173) and (1.174). With completely arbitrary values of the parameters α or β, this construction is rather cumbersome, so that a completely general solution of this problem will not be presented here. However, we will treat specific examples in the following sections solving Equations (2.4) and (2.5) or Equations (2.29) and (2.30), to illustrate the merits and applicability of the abovementioned methods.

This completes our examination of dual series with Jacobi polynomial kernels. The functions $\hat{Q}_{sn}^{(\alpha,\beta)}$ that appear in the final regularised system play an extremely important role both in the analysis of and establishing the validity of the solution, as well as a wider role in the general investigation of the single (or double) layer potential density.

2.2 Dual series equations involving trigonometrical functions

Dual equations with trigonometric kernels have been investigated by a great many authors (see, for example, the bibliography in [55]). Apparently, Tranter [62] was the first to solve equations of this type by the definition method described in Section 1.4. In this section we present the original solution, placing it in the context of the general theory developed in the previous section for dual series equations involving Jacobi polynomials $P_n^{(\alpha,\beta)}$.

The fundamental connection arises from the relationships (1.151)–(1.154) between trigonometric functions and the Jacobi polynomials with indices $\alpha = \pm\frac{1}{2}$ and $\beta = \pm\frac{1}{2}$. In applications the parameter η invariably takes the value $\frac{1}{2}$, so as noted at the end of the previous section, the case when $\alpha = \beta = -\frac{1}{2}$ must be considered separately, since the solution described for the pair (2.4) and (2.5) requires $\alpha - \eta > -1$, whilst that for the pair (2.29) and (2.30) requires $\beta - \eta > -1$; an initial transformation as described at the end of the previous section must be effected. On the other hand, when $\alpha = \beta = \frac{1}{2}$, the solution described in the previous section is valid.

Let us consider the following dual series equations with kernels $e^{in\vartheta}$:

$$bx_0 - g_0 + \sum_{n\neq 0}\{z_n(1-q_n) - \xi_n\}e^{in\vartheta} = 0, \quad |\,\vartheta\,| < \vartheta_0 \tag{2.37}$$

$$ax_0 - f_0 + \sum_{n\neq 0}|\,n\,|\{z_n(1-r_n) - \zeta_n\}e^{in\vartheta} = 0, \quad |\,\vartheta\,| > \vartheta_0 \tag{2.38}$$

where the unknown coefficient sequence $\{z_n\}_{n\neq 0}$ will be assumed to lie in $l_2(1)$. The coefficients a, b, g_0, f_0 and the sequence coefficients ξ_n, ζ_n, q_n, r_n are assumed to be known; in addition, we suppose that $q_{-n} = q_n, r_n = r_{-n}$ and

$$\lim_{|n|\to\infty} q_n = \lim_{|n|\to\infty} r_n = 0.$$

Introduce the following notation:

$$\begin{Bmatrix} x_n \\ y_n \end{Bmatrix} = \begin{Bmatrix} z_n + z_{-n} \\ z_n - z_{-n} \end{Bmatrix}; \begin{Bmatrix} g_n \\ e_n \end{Bmatrix} = \begin{Bmatrix} \xi_n + \xi_{-n} \\ \xi_n - \xi_{-n} \end{Bmatrix}; \begin{Bmatrix} f_n \\ h_n \end{Bmatrix} = \begin{Bmatrix} \zeta_n + \zeta_{-n} \\ \zeta_n - \zeta_{-n} \end{Bmatrix}.$$

Then the pair of equations (2.37) and (2.38) is equivalent to the two pairs of functional equations in which the unknowns x_n, y_n are decoupled:

$$bx_0 - g_0 + \sum_{n=1}^{\infty}\{x_n(1-q_n) - g_n\}\cos n\vartheta = 0, \vartheta \in (0, \vartheta_0) \tag{2.39}$$

$$ax_0 - f_0 + \sum_{n=1}^{\infty} n\{x_n(1-r_n) - f_n\}\cos n\vartheta = 0, \vartheta \in (\vartheta_0, \pi) \tag{2.40}$$

and

$$\sum_{n=1}^{\infty}\{y_n(1-q_n) - e_n\}\sin n\vartheta = 0, \ \vartheta \in (0, \vartheta_0) \tag{2.41}$$

$$\sum_{n=1}^{\infty} n\{y_n(1-r_n) - h_n\}\sin n\vartheta = 0, \ \vartheta \in (\vartheta_0, \pi). \tag{2.42}$$

Let us consider first the pair (2.41) and (2.42) with sine function kernels; the pair with cosine function kernels will be treated later. Set $z = \cos\varphi$, $z_0 = \cos\varphi$ and use (1.153) to obtain

$$\sum_{n=1}^{\infty} n\{A_n(1-r_n) - d_n\}P_{n-1}^{\left(\frac{1}{2},\frac{1}{2}\right)}(z) = 0, \; z \in (-1, z_0) \tag{2.43}$$

$$\sum_{n=1}^{\infty} \{A_n(1-q_n) - c_n\}P_{n-1}^{\left(\frac{1}{2},\frac{1}{2}\right)}(z) = 0, \; z \in (z_0, 1) \tag{2.44}$$

where

$$\{A_n, d_n, c_n\} = \frac{\sqrt{\pi}}{2}\frac{\Gamma(n+1)}{\Gamma\left(n+\frac{1}{2}\right)}\{y_n, h_n, e_n\}. \tag{2.45}$$

The rescaled unknowns $\{A_n\}_{u=1}^{\infty}$ lie in l_2. Equations (2.43), (2.44) are of the form (2.4), (2.5) because $\lambda_{n-1}\left(\frac{1}{2},\frac{1}{2},\frac{1}{2}\right) = n$, and we may conclude that

$$(1-r_s)\hat{A}_s + \sum_{n=1}^{\infty}\hat{A}_n(r_n - g_n)\hat{Q}_{n-1,s-1}^{(0,1)}(z_0)$$
$$= \hat{d}_s + \sum_{n=1}^{\infty}\left(\hat{c}_n - \hat{d}_n\right)\hat{Q}_{n-1,s-1}^{(0,1)}(z_0), \tag{2.46}$$

where $s = 1, 2, ...,$ and

$$\{\hat{A}_s, \hat{d}_s, \hat{c}_s\} = \sqrt{\frac{2}{n}}\frac{\Gamma\left(n+\frac{1}{2}\right)}{\Gamma(n)}\{A_n, d_s, c_s\} = \sqrt{\frac{n\pi}{2}}\{y_n, h_n, e_n\}.$$

Notice that $\{\hat{A}_s\}_{s=1}^{\infty} \in l_2$.

We now turn to Equations (2.39) and (2.40) with cosine function kernels; an initial transformation of the parameter values is needed. First, replace the cosine terms by their Jacobi polynomial representation (1.151). Then integrate both sides of these equations using Formula (2.36). (This term-by-term integration is justified in the same way as in the previous section, using results in Appendix D.2.) We then integrate using Formula (1.174) to obtain

$$(1+t)^{\frac{3}{2}}\sum_{n=1}^{\infty}\frac{\Gamma(n+1)}{\Gamma\left(n+\frac{1}{2}\right)}\{x_n(1-r_n) - f_n\}P_{n-1}^{\left(-\frac{1}{2},\frac{3}{2}\right)}(t) =$$
$$\frac{2}{\sqrt{\pi}}(ax_0 - f_0)\{4(1+t)^{\frac{1}{2}} - (1-t)^{\frac{1}{2}}[\pi + 2\arcsin t]\}, \; t \in (-1, t_0), \tag{2.47}$$

$$(1-t)^{\frac{1}{2}}\sum_{n=1}^{\infty}\frac{\Gamma(n)}{\Gamma\left(n+\frac{1}{2}\right)}\{x_n(1-q_n) - g_n\}P_{n-1}^{\left(\frac{1}{2},\frac{1}{2}\right)}(t) =$$
$$-\frac{2}{\sqrt{\pi}}(bx_0 - g_0)(1+t)^{-\frac{1}{2}}\left\{\frac{\pi}{2} - \arcsin t\right\}, \; t \in (t_0, 1), \tag{2.48}$$

where $t = \cos\theta$, $t_0 = \cos\theta_0$.

Following the standard scheme described in Section 2.1, we obtain the dual series equations

$$(1+t)\sum_{n=1}^{\infty}\{x_n(1-r_n)-f_n\}P_{n-1}^{(0,1)}(t)$$
$$= -2(ax_0-f_0)\ln[\frac{1}{2}(1-t)],\ t\in(-1,t_0), \quad (2.49)$$

$$(1+t)\sum_{n=1}^{\infty}\{x_n(1-q_n)-g_n\}P_{n-1}^{(0,1)}(t) = -2(bx_0-g_0),\ t\in(t_0,1), \quad (2.50)$$

where the unknowns $\{x_n\}_{n=1}^{\infty}\in l_2(1)$. The following definite integral

$$\int_{-1}^{z}\frac{\frac{\pi}{2}+\arcsin x}{\sqrt{1-x}\sqrt{z-x}}dx = -\pi\ln\frac{1-z}{2} \quad (2.51)$$

which occurs in this process may be evaluated from the transform

$$-\sqrt{2}\pi\ln\left(\cos\frac{\phi}{2}\right) = \int_0^{\phi}\frac{\theta\sin\frac{1}{2}\theta\ d\theta}{\sqrt{\cos\theta-\cos\phi}}. \quad (2.52)$$

Introducing new coefficients

$$\{\hat{x}_n,\hat{f}_n,\hat{g}_n\} = \sqrt{\frac{2}{n}}\{x_n,f_n,g_n\} \quad (2.53)$$

we transform (2.49) and (2.50) to

$$F(t) = \begin{cases} F_1(t), & t\in(-1,t_0) \\ F_2(t), & t\in(t_0,1) \end{cases} \quad (2.54)$$

where

$$F(t) = (1+t)\sum_{n=1}^{\infty}\hat{x}_n\hat{P}_{n-1}^{(0,1)}(t), \quad (2.55)$$

$$F_1(t) = -2(ax_0-f_0)\ln\left[\frac{1}{2}(1-t)\right]+(1+t)\sum_{n=1}^{\infty}\left(\hat{x}_n r_n+\hat{f}_n\right)\hat{P}_{n-1}^{(0,1)}(t), \quad (2.56)$$

and

$$F_2(t) = -2(bx_0-g_0)+(1+t)\sum_{n=1}^{\infty}(\hat{x}_n q_n+\hat{g}_n)\hat{P}_{n-1}^{(0,1)}(t). \quad (2.57)$$

The rescaled solution $\{\hat{x}_n\}_{n=1}^{\infty}$ belongs to $l_2(2)$. Multiplying both sides of (2.54) by $\hat{P}_{m-1}^{(0,1)}(t)$ and integrating over $[-1,1]$, and employing the properties of the incomplete scalar product (see Appendix (B.6)), we obtain

$$
\begin{aligned}
&(1-r_m)\hat{x}_m - \sum_{n=1}^{\infty}\{\hat{x}_n(q_n-r_n)+\hat{g}_n-\hat{f}_n\}\hat{Q}_{n-1,m-1}^{(0,1)}(t_0) = \\
&\hat{f}_m + 2x_0\left\{-\frac{1-t_0}{m}\hat{P}_{m-1}^{(1,0)}(t_0)\left[b-a\ln\left(\frac{1-t_0}{2}\right)\right]+a\frac{1+t_0}{m^2}\hat{P}_{m-1}^{(0,1)}(t_0)\right\} \\
&\quad +2\left\{\frac{1-t_0}{m}\hat{P}_{m-1}^{(1,0)}(t_0)\left[g_0-f_0\ln\left(\frac{1-t_0}{2}\right)\right]-f_0\frac{1+t_0}{m^2}\hat{P}_{m-1}^{(0,1)}(t_0)\right\},
\end{aligned}
\tag{2.58}
$$

where $m=1,2,\ldots$.

Whatever the value of the constant x_0, the solution $\{\hat{x}_m\}_{m=1}^{\infty}$ of the system (2.58) lies in l_2; however, the value x_0 must be chosen so that it also lies in $l_2(2)$. This depends upon the smoothness of the function F, which is related to the rate of decrease of its Fourier coefficients [49, 79]. F is continuous everywhere on the interval $[-1,1]$ because (2.55) is a uniformly convergent series. The functions F_1 and F_2 are continuous on the sub-intervals $[-1,t_0)$ and $(t_0,1]$ respectively, so the only point where the function F may lose continuity is at t_0; observing that F is continuous at this point gives an equation for the constant x_0, namely,

$$F_1(t_0)=F_2(t_0). \tag{2.59}$$

From this condition we find

$$
\begin{aligned}
x_0 = c\left[g_0-f_0\ln\left(\frac{1-t_0}{2}\right)\right] &+ \\
&\frac{1+t_0}{2}c\sum_{n=1}^{\infty}\{\hat{x}_n(q_n-r_n)+\hat{g}_n-\hat{f}_n\}\hat{P}_{n-1}^{(0,1)}(t_0).
\end{aligned}
\tag{2.60}
$$

where

$$c=\left[b-a\ln\left(\frac{1-t_0}{2}\right)\right]^{-1}$$

Combined with (2.58), the relationship (2.60) gives the solution of the dual series equations involving trigonometric functions $\cos n\vartheta$. Let us substitute the expression (2.60) for x_0 in Equation (2.58), keeping in mind the relationship (see Appendix, (B.171))

$$\hat{Q}_{n-1,m-1}^{(0,1)}(t_0)=\frac{(1-t_0)^2}{m}\hat{P}_{n-1}^{(0,1)}(t_0)\hat{P}_{m-1}^{(1,0)}(t_0)+\frac{n}{m}\hat{Q}_{n-1,m-1}^{(1,0)}(t_0). \tag{2.61}$$

As a result we obtain

$$(1-r_m)\,X_m - \sum_{n=1}^{\infty}\{X_n\,(q_n-r_n)+G_n-F_n\}\times$$
$$\left\{\hat{Q}^{(1,0)}_{n-1,m-1}(t_0)+\frac{a\,(1+t_0)^2}{b-a\ln\left(\frac{1}{2}(1-t_0)\right)}\frac{\hat{P}^{(0,1)}_{n-1}(t_0)}{n}\frac{\hat{P}^{(0,1)}_{m-1}(t_0)}{m}\right\}$$
$$=F_m+2\frac{ag_0-f_0b}{b-a\ln\left(\frac{1}{2}(1-t_0)\right)}\frac{1+t_0}{m}\hat{P}^{(0,1)}_{m-1}(t_0) \quad (2.62)$$

where $m=1,2,\ldots$, and

$$\{X_m,G_m,F_m\}=m\{\hat{x}_m,\hat{g}_m,\hat{f}_m\}. \quad (2.63)$$

Because $\{X_m\}_{m=1}^{\infty}$ lies in l_2, the solution $\{\hat{x}_m\}$ lies in $l_2(2)$ as required.

This completes the regularisation of the dual series (2.39) and (2.40) or (2.41) and (2.42) , and hence of the original system (2.37) and (2.38) . There is a companion set of dual series, in which the sub-intervals on which the individual equations are interchanged. It is easily shown that they reduce to the same equations as (2.39) and (2.40) or (2.41) and (2.42) via the replacements $t_0\to -t_1$ ($\vartheta_1=\pi-\vartheta_0$, $t_1=\cos\vartheta_1=-\cos\vartheta_0=-t_0$), $\hat{A}_s\to(-1)^s\hat{A}_s$, $\{X_m,G_m,F_m\}\to(-1)^m\{X_m,G_m,F_m\}$.

To complete our consideration of dual series equations involving trigonometric kernels, we now consider the pairs of functional equations

$$\begin{cases}\sum_{n=0}^{\infty}\{x_n\,(1-q_n)-g_n\}\cos\left(n+\frac{1}{2}\right)\vartheta=0, & \vartheta\in(0,\vartheta_0)\\ \sum_{n=0}^{\infty}\left(n+\frac{1}{2}\right)\{x_n\,(1-r_n)-f_n\}\cos\left(n+\frac{1}{2}\right)\vartheta=0, & \vartheta\in(\vartheta_0,\pi)\end{cases} \quad (2.64)$$

and

$$\begin{cases}\sum_{n=0}^{\infty}\{y_n\,(1-q_n)-e_n\}\sin\left(n+\frac{1}{2}\right)\vartheta=0, & \vartheta\in(0,\vartheta_0)\\ \sum_{n=0}^{\infty}\left(n+\frac{1}{2}\right)\{y_n\,(1-r_n)-h_n\}\sin\left(n+\frac{1}{2}\right)\vartheta=0, & \vartheta\in(\vartheta_0,\pi).\end{cases} \quad (2.65)$$

In addition, we consider the companion equations in which the sub-intervals of definition of these equations are interchanged:

$$\begin{cases}\sum_{n=0}^{\infty}\left(n+\frac{1}{2}\right)\{x_n\,(1-r_n)-f_n\}\cos\left(n+\frac{1}{2}\right)\vartheta=0, & \vartheta\in(0,\vartheta_0)\\ \sum_{n=0}^{\infty}\{x_n\,(1-q_n)-g_n\}\cos\left(n+\frac{1}{2}\right)\vartheta=0, & \vartheta\in(\vartheta_0,\pi)\end{cases} \quad (2.66)$$

and

$$\begin{cases}\sum_{n=0}^{\infty}\left(n+\frac{1}{2}\right)\{y_n\,(1-r_n)-h_n\}\sin\left(n+\frac{1}{2}\right)\vartheta=0, & \vartheta\in(0,\vartheta_0)\\ \sum_{n=0}^{\infty}\{y_n\,(1-q_n)-e_n\}\sin\left(n+\frac{1}{2}\right)\vartheta=0, & \vartheta\in(\vartheta_0,\pi).\end{cases} \quad (2.67)$$

However, from the elementary relationships

$$\cos\left(n+\tfrac{1}{2}\right)(\pi-\theta)=(-1)^n\sin\left(n+\tfrac{1}{2}\right)\theta,$$
$$\sin\left(n+\tfrac{1}{2}\right)(\pi-\theta)=(-1)^n\cos\left(n+\tfrac{1}{2}\right)\theta,$$

it is evident that the pair (2.67) is of the same type as the pair (2.64), and also that the pair (2.66) is of the same type as (2.65). Thus, we shall consider only the pairs (2.64) and (2.65) and find solutions with $\{x_n, y_n\}_{n=0}^{\infty}\in l_2(1)$.

Using the identities (1.152) and (1.154), and setting $t=\cos\vartheta, t_0=\cos\vartheta_0$, we reformulate these equations in terms of Jacobi polynomials as

$$\sum_{n=0}^{\infty}\frac{\left(n+\frac{1}{2}\right)\Gamma(n+1)}{\Gamma\left(n+\frac{1}{2}\right)}\{x_n(1-r_n)-f_n\}P_n^{\left(-\frac{1}{2},\frac{1}{2}\right)}(t)=0,\ t\in(-1,t_0), \tag{2.68}$$

$$\sum_{n=0}^{\infty}\frac{\Gamma(n+1)}{\Gamma\left(n+\frac{1}{2}\right)}\{x_n(1-q_n)-g_n\}P_n^{\left(-\frac{1}{2},\frac{1}{2}\right)}(t)=0,\ t\in(t_0,1) \tag{2.69}$$

and

$$\sum_{n=0}^{\infty}\frac{\left(n+\frac{1}{2}\right)\Gamma(n+1)}{\Gamma\left(n+\frac{1}{2}\right)}\{y_n(1-r_n)-h_n\}P_n^{\left(\frac{1}{2},-\frac{1}{2}\right)}(t)=0,\ t\in(-1,t_0), \tag{2.70}$$

$$\sum_{n=0}^{\infty}\frac{\Gamma(n+1)}{\Gamma\left(n+\frac{1}{2}\right)}\{y_n(1-q_n)-e_n\}P_n^{\left(\frac{1}{2},-\frac{1}{2}\right)}(t)=0,\ t\in(t_0,1). \tag{2.71}$$

The general theory, developed in Section 2.1, is applicable to the second pair of dual equations, (2.70) and (2.71). We set $\eta=\alpha=\frac{1}{2}, \beta=-\frac{1}{2}$, so that $\lambda_n(\alpha,\beta;\eta)=n+\frac{1}{2}$, and then represent these equations in the standard form

$$\sum_{n=0}^{\infty}\lambda_n\left(\frac{1}{2},-\frac{1}{2};\frac{1}{2}\right)\{y_n^*(1-r_n)-h_n^*\}P_n^{\left(\frac{1}{2},-\frac{1}{2}\right)}(t)=0,\ t\in(-1,t_0) \tag{2.72}$$

$$\sum_{n=0}^{\infty}\{y_n^*(1-q_n)-e_n^*\}P_n^{\left(\frac{1}{2},-\frac{1}{2}\right)}(t)=0,\ t\in(t_0,1) \tag{2.73}$$

where

$$\{y_n^*, h_n^*, e_n^*\}=\frac{\Gamma(n+1)}{\Gamma\left(n+\frac{1}{2}\right)}\{y_n, h_n, e_n\}. \tag{2.74}$$

The regularised system from (2.65) is thus directly obtained from (2.23): the rescaled coefficients and unknowns

$$\{Y_s, H_s, E_s\} = \left(s + \frac{1}{2}\right)^{\frac{1}{2}} \{y_s, h_s, e_s\} \tag{2.75}$$

satisfy

$$(1 - r_s) Y_s + \sum_{n=0}^{\infty} Y_n (r_n - q_n) \hat{Q}_{sn}^{(0,0)} (t_0) = H_s + \sum_{n=0}^{\infty} (E_s - H_s) \hat{Q}_{sn}^{(0,0)} (t_0) \tag{2.76}$$

where $s = 0, 1, 2, \ldots$. Notice that in this case the incomplete inner product $\hat{Q}_{sn}^{(0,0)}$ is simply an incomplete inner product of normalised Legendre polynomials $\hat{P}_n = \hat{P}_n^{(0,0)} = (n + \frac{1}{2})^{\frac{1}{2}} P_n$:

$$\hat{Q}_{sn}^{(0,0)} (t_0) = \int_{t_0}^{1} \hat{P}_s (t) \hat{P}_n (t) \, dt. \tag{2.77}$$

Let us now consider the remaining dual series equations, (2.64). Instead of applying the variant (2.36) of Rodrigues' formula as was done previously (cf. (2.47),(2.48)), we apply the integration Formulae (1.173) and (1.174). First we use the relationship (1.173) with $\alpha = -\frac{1}{2}, \beta = \frac{1}{2}$,

$$\int_t^1 (1 - x)^{-\frac{1}{2}} P_n^{\left(-\frac{1}{2}, \frac{1}{2}\right)} (x) \, dx = \frac{(1 - t)^{\frac{1}{2}}}{n + \frac{1}{2}} P_n^{\left(\frac{1}{2}, -\frac{1}{2}\right)} (t) \tag{2.78}$$

and integrate both parts of Equations (2.68) and (2.69). (The term-by-term integration of a square-summable Fourier series is justified.) As a result, we obtain

$$\sum_{n=0}^{\infty} \frac{\Gamma(n+1)}{\Gamma\left(n + \frac{1}{2}\right)} \{x_n (1 - r_n) - f_n\} P_n^{\left(\frac{1}{2}, -\frac{1}{2}\right)} (t) = C (1 - t)^{-\frac{1}{2}}, \; t \in (-1, t_0), \tag{2.79}$$

$$\sum_{n=0}^{\infty} \frac{\Gamma(n+1)}{\Gamma\left(n + \frac{3}{2}\right)} \{x_n (1 - q_n) - g_n\} P_n^{\left(\frac{1}{2}, -\frac{1}{2}\right)} (t) = 0, \; t \in (-1, t_0), \tag{2.80}$$

where C is a constant to be determined later. This is in standard form for the application of the Abel integral transform method outlined in the previous section (with $\alpha = \eta = \frac{1}{2}, \beta = -\frac{1}{2}$). The first step is to integrate (2.79) again, but using Formula (1.174) with $\alpha = \frac{1}{2}, \beta = -\frac{1}{2}$:

$$\int_{-1}^{t} (1 + x)^{-\frac{1}{2}} P_n^{\left(\frac{1}{2}, -\frac{1}{2}\right)} (x) \, dx = \frac{(1 + t)^{\frac{1}{2}}}{n + \frac{1}{2}} P_n^{\left(-\frac{1}{2}, \frac{1}{2}\right)} (t). \tag{2.81}$$

We find

$$(1+t)^{\frac{1}{2}}\sum_{n=0}^{\infty}\frac{\Gamma(n+1)}{\Gamma\left(n+\frac{3}{2}\right)}\{x_n(1-r_n)-f_n\}P_n^{\left(-\frac{1}{2},\frac{1}{2}\right)}(t)$$
$$=C\left(\frac{\pi}{2}+\arcsin t\right),\quad t\in(-1,t_0). \quad (2.82)$$

Repeating the steps of the method described in the previous section converts Equations (2.82) and(2.80) to the equivalent pair

$$\sum_{n=0}^{\infty}x_nP_n(t)=\begin{Bmatrix}F_1(t), & t\in(-1,t_0)\\ F_2(t), & t\in(t_0,1)\end{Bmatrix}, \quad (2.83)$$

where

$$\begin{aligned} F_1(t) &= \left(\frac{2}{\pi}\right)^{\frac{1}{2}}K\left(\sqrt{\frac{1+t}{2}}\right)C+\sum_{n=0}^{\infty}(x_nr_n+f_n)P_n(t), \\ F_2(t) &= \sum_{n=0}^{\infty}(x_nq_n+g_n)P_n(t), \end{aligned}$$

and K denotes the complete elliptic integral of first kind (see Appendix, (B.78)). The value of the constant is determined by

$$C=\left(\frac{\pi}{2}\right)^{\frac{1}{2}}\left\{K\left(\sqrt{\frac{1+t_0}{2}}\right)\right\}^{-1}\sum_{n=0}^{\infty}\{x_n(q_n-r_n)+g_n-f_n\}P_n^{(0,0)}(t_0) \quad (2.84)$$

and the coefficients $\{x_n\}_{n=0}^{\infty}$ satisfy

$$(1-r_s)x_s-\sum_{n=1}^{\infty}\{x_n(q_n-r_n)Q_{ns}^{(0,0)}(t_0)$$
$$=\left(\frac{2}{\pi}\right)^{\frac{1}{2}}C\int_{-1}^{t_0}K\left(\sqrt{(1+t)/2}\right)P_n(t)\,dt$$
$$+f_s+\sum_{n=1}^{\infty}(g_n-f_n)Q_{ns}^{(0,0)}(t_0), \quad (2.85)$$

where $s=0,1,2,\ldots$. Note that $Q_{ns}^{(0,0)}$ is the unnormalised incomplete scalar product. The integral appearing in (2.85) may be simply expressed in terms of complete elliptic integrals (see later, (5.50)).

This completes the solution of the dual series (2.64) and concludes our regularisation of dual series equations with various types of trigonometric kernels.

2.3 Dual series equations involving associated Legendre functions

The associated Legendre functions P_n^m provide another interesting and special set of kernels for dual series equations, worthy of examination in their own right. Because P_n^m is essentially the mth derivative of the Legendre polynomial P_n, m-fold integration of the dual series equations immediately produces dual series equations with Legendre polynomial kernels that are readily solvable. This section examines the solution obtained by this simple process. For large m, however, the resulting scheme is numerically unstable; two stable modifications are therefore described. The advantages and limitations of the original and modified systems are discussed. These results were obtained jointly with Yu. A. Tuchkin; some of them appear in [72].

We therefore consider dual series equations involving associated Legendre functions $P_n^m(\cos\theta)$, and exploit the solution already obtained in Section 2.1. The index m is a fixed nonnegative integer. The dual series equations

$$\sum_{n=m}^{\infty} x_n^m (1-\varepsilon_n) P_n^m(\cos\theta) = G(\theta),\ \theta \in (0,\theta_0) \tag{2.86}$$

$$\sum_{n=m}^{\infty} (2n+1) x_n^m (1-\mu_n) P_n^m(\cos\theta) = F(\theta),\ \theta \in (\theta_0,\pi) \tag{2.87}$$

are to be solved for the unknown coefficients $\{x_n^m\}_{n=m}^{\infty}$. The quantities $\{\varepsilon_n\}_{n=0}^{\infty}, \{\mu_n\}_{n=0}^{\infty}$ are assumed to be known sequences of, in general, complex quantities decreasing at least as fast as $O(n^{-2})$ as $n \to \infty$:

$$\varepsilon_n = O(n^{-2})\,;\, \mu_n = O(n^{-2})\,. \tag{2.88}$$

The functions G, F are assumed to be expandable in Fourier-Legendre series

$$\begin{aligned} G(\theta) &= \sum_{n=m}^{\infty} g_n^m P_n^m(\cos\theta)\,, \\ F(\theta) &= \sum_{n=m}^{\infty} (2n+1) f_n^m P_n^m(\cos\theta)\,. \end{aligned} \tag{2.89}$$

where the coefficients g_n^m and f_n^m are of the form $g_n^m = \alpha_n n^{-m}$ and $f_n^m = \beta_n n^{-m-1}$ and satisfy

$$\sum_{n=m}^{\infty} |\alpha_n|^2 < \infty;\ \sum_{n=m}^{\infty} |\beta_n|^2 < \infty. \tag{2.90}$$

Furthermore, all the series contained in (2.86), (2.87), and (2.89) are assumed to be the Fourier series of their respective sums, i.e., are convergent in the weighted mean square sense with weight $w_{m,m}(x) = (1-x^2)^m$.

When $n \geqslant m$, the relationship between the associated Legendre functions P_n^m and the Jacobi polynomials $P_{n-m}^{(m,m)}$ is (see Appendix, (B.48))

$$P_n^m(\cos\theta) = 2^{-m}\sin^m\theta \frac{\Gamma(n+m+1)}{\Gamma(n+1)} P_{n-m}^{(m,m)}(\cos\theta). \tag{2.91}$$

The connection with Legendre polynomials is

$$P_n^m(x) = (-1)^m \left(1-x^2\right)^{\frac{m}{2}} \frac{d^m}{dx^m} P_n(x). \tag{2.92}$$

In terms of the parameters of the dual equations considered in Section 2.1, $\alpha = \beta = m$, and $\eta = \frac{1}{2}$.

Because of this connection, it is natural to seek the solution of the pair (2.86) and (2.87) in the class $l_2(2m)$ defined by (2.9):

$$\{x_n^m\}_{n=m}^{\infty} \in l_2(2m). \tag{2.93}$$

This condition which appears naturally in both potential theory and wave-scattering theory for open spherical surfaces, is equivalent to the boundedness condition for the energy integral, which is taken over a finite volume including the edges.

Defining

$$R_n(x) = P_{n+1}(x) - P_{n-1}(x),$$

the Legendre polynomials obey (see (1.123)),

$$(2n+1) P_n(x) = \frac{d}{dx} R_n(x). \tag{2.94}$$

All series in (2.86), (2.87) are (generalised) Fourier series, so they can be integrated term-by-term (see Appendix D.2). Set $x = \cos\theta$. Divide (2.86) and (2.87) by $\left(1-x^2\right)^{\frac{m}{2}}$; then integrate each equation m-times; a further integration of (2.87) is made using (2.94). As a result of this process, polynomials (in x) of degree $m-1$ and m, respectively appear on the right-hand sides of these equations with coefficients deriving from integration constants. Express each polynomial as a finite sum in terms of Legendre polynomials, with some undetermined coefficients to obtain

$$\sum_{n=m}^{\infty} x_n^m (1-\varepsilon_n) P_n(\cos\theta) = \sum_{n=0}^{m-1} C_n^m P_n(\cos\theta) + \sum_{n=m}^{\infty} f_n^m P_n(\cos\theta), \quad \theta \in (0,\theta_0), \tag{2.95}$$

$$\sum_{n=m}^{\infty} x_n^m (1-\mu_u) R_n(\cos\theta) =$$

$$\sum_{n=0}^{m-1} (C_n^m + A_n^m) R_n(\cos\theta) + \sum_{n=m}^{\infty} g_n^m R_n(\cos\theta), \theta \in (\theta_0, \pi) \quad (2.96)$$

where coefficients A_n^m, C_n^m denote arbitrary constants of integration.

This system has a solution in l_2, i.e.,

$$\{x_n^m\}_{n=m}^{\infty} \in l_2(0) \equiv l_2. \quad (2.97)$$

Each solution of (2.86) and (2.87) is a solution of (2.95) and (2.96) whatever the values of the coefficients A_n^m, C_n^m may be. However, any solution of (2.95) and (2.96) depends on the $2m$ arbitrary constants A_n^m, C_n^m and so in general is not a solution of (2.86) and (2.87). We now show how to determine A_n^m, C_n^m so that (2.93) is satisfied; the solution of (2.95) and (2.96) will also then be the solution of (2.86) and (2.87). This depends upon differentiating (2.95) and (2.96) the required number of times.

We now use the Dirichlet-Mehler integral representations for Legendre polynomials (1.149) and (1.150) and observe that

$$\begin{aligned} R_n(\cos\theta) &= -\frac{2\sqrt{2}}{\pi}\int_0^{\theta} \frac{\sin\left(n+\frac{1}{2}\right)\varphi \sin\varphi}{(\cos\varphi - \cos\theta)^{\frac{1}{2}}} d\varphi \\ &= \frac{2\sqrt{2}}{\pi}\int_{\theta}^{\pi} \frac{\cos\left(n+\frac{1}{2}\right)\varphi \sin\varphi}{(\cos\varphi - \cos\theta)^{\frac{1}{2}}} d\varphi. \end{aligned} \quad (2.98)$$

Transfer all terms in (2.95) and (2.96) to the left-hand sides of these equations, use the Dirichlet-Mehler integrals, and substitute the expression (2.98). Invert the order of summation and integration (the validity of this operation is ensured by (2.88) and (2.97)) to obtain two integral relationships, each of which is a homogenous Abel integral equation with a unique zero solution. As a result we obtain the following dual series equations.

$$\sum_{n=m}^{\infty} x_n^m \cos\left(n+\frac{1}{2}\right)\theta =$$

$$\sum_{n=0}^{m-1} C_n^m \cos\left(n+\frac{1}{2}\right)\theta + \begin{Bmatrix} F_1(\theta), & \theta \in (0,\theta_0) \\ F_2(\theta), & \theta \in (\theta_0,\pi) \end{Bmatrix}, \quad (2.99)$$

where

$$\begin{aligned} F_1(\theta) &= \sum_{n=m}^{\infty} (x_n^m \varepsilon_n + f_n^m) \cos\left(n+\frac{1}{2}\right)\theta, \\ F_2(\theta) &= \sum_{n=m}^{\infty} (x_n^m \mu_n + g_n^m) \cos\left(n+\frac{1}{2}\right)\theta + \sum_{n=0}^{m-1} A_n^m \cos\left(n+\frac{1}{2}\right)\theta. \end{aligned}$$

The set $\{\cos\left(n+\frac{1}{2}\right)\theta\}_{n=0}^{\infty}$ is orthogonal, so multiplying both sides of Equation (2.99) by $\cos\left(s+\frac{1}{2}\right)\theta$ and integrating term-by-term over $[0,\pi]$, we find

$$C_s^m + \sum_{n=m}^{\infty}\{x_n^m(\varepsilon_n-\mu_n)+(f_n^m-g_n^m)\}Q_{sn}(\theta_0) = -\sum_{n=0}^{\infty}A_n^m\{\delta_{sn}-Q_{sn}(\theta_0)\}, \quad (2.100)$$

where $s=0,1,2,...,m-1$, and

$$x_s^m(1-\mu_s)+\sum_{n=m}^{\infty}\{x_n^m(\mu_n-\varepsilon_n)Q_{sn}(\theta_0) = g_s^m+\sum_{n=m}^{\infty}(f_n^m-g_n^m)Q_{sn}(\theta_0)-\sum_{n=0}^{m-1}A_n^mQ_{sn}(\theta_0), \quad (2.101)$$

where $s=m,m+1,...$, and $Q_{sn}(\theta_0)=\hat{Q}_{sn}^{\left(-\frac{1}{2},\frac{1}{2}\right)}(\cos\theta_0)$.

Equation (2.101) is an infinite system of the linear algebraic equations of the second kind for the unknowns $\{x_n^m\}_{n=m}^{\infty}$; its solution depends on the m constants $A_0^m,\ldots,A_{m-1}^m$.

Let us introduce the formal notation $D_m^k(\vartheta)$ for the k-th derivative (with respect to ϑ) of

$$\sum_{n=m}^{\infty}\{x_n^m(\varepsilon_n-\mu_n)+(f_n^m-g_n^m)\}\cos\left(n+\frac{1}{2}\right)\vartheta-\sum_{n=0}^{m-1}A_n^m\cos\left(n+\frac{1}{2}\right)\vartheta. \quad (2.102)$$

Recollect our assumption that the solution of (2.101) belongs to the class $l_2(2m)$. From standard results, which connect the smoothness of a function with the rate of decrease of its Fourier coefficients [49, 79], the enforcement of the aggregate of m conditions

$$D_m^k(\vartheta_0)=0, \qquad k=0,1,2,...,m-1 \quad (2.103)$$

on Equations (2.99) is necessary and sufficient for the solution (2.101) to belong to the class $l_2(2m)$. Assuming this, one can differentiate the Equations (2.102) term-by-term. Combining (2.101) with (2.103) (the result of term-by-term differentiation of (2.102) at the point $\theta=\theta_0$), we are led to an infinite system of the linear algebraic equations for the aggregate of unknowns $\{A_n^m\}_{n=0}^{m-1}$ and $\{x_n^m\}_{n=m}^{\infty}\in l_2(2m)$.

It can be shown that Equations (2.86) and (2.87) are equivalent to the set of Equations (2.101)and (2.103); thus, we have successfully converted the original dual series equations (2.86) and (2.87) to an infinite system of linear algebraic equations, which can be solved by various numerical methods. The solution has asymptotic behaviour

$$x_s^m = \frac{D_m^k(\theta_0)}{\left(s+\frac{1}{2}\right)^{m+1}} \cdot \frac{2}{\pi}\Psi_s^m(\theta_0) + O\left(s^{-m-2}\right), \tag{2.104}$$

as $s \to \infty$, where $\Psi_s^m(\theta_0) = \sin\left(s+\frac{1}{2}\right)\theta_0$ or $\cos\left(s+\frac{1}{2}\right)\theta_0$ according as m is even or odd.

The simplicity in calculating the matrix elements of the system (2.100), (2.101) and the condition (2.103) is attractive: only trigonometric functions are used. However, it can be shown that for large m this scheme is unstable, and leads to significant errors in the calculation of the coefficients A_n^m. But provided m is not large, this system is very suitable for numerical calculation. Let us therefore modify the system to improve its stability. Write (2.103) as m equations for the unknown values A_n^m:

$$\left\{\frac{d^k}{d\theta^k}\left[\sum_{n=0}^{m-1} A_n^m \cos\left(n+\frac{1}{2}\right)\theta - \sum_{n=m}^{\infty} W_n^m \cos\left(n+\frac{1}{2}\right)\theta\right]\right\}_{\theta=\theta_0} = 0, \tag{2.105}$$

where $k = 0, 1, 2, \ldots m-1$, and

$$W_n^m = x_n^m(\varepsilon_n - \mu_n) + f_n^m - g_n^m. \tag{2.106}$$

Assuming that conditions (2.103) are satisfied, we wish to obtain a numerically stable algorithm. Let us consider the orthonormal family of Jacobi polynomials ($n \geq k$, k fixed),

$$\hat{P}_{n-k}^{\left(k-\frac{1}{2},k+\frac{1}{2}\right)}(\cos\theta) = \frac{(-1)^k}{\sqrt{\pi}}\left\{\frac{(n-k)!}{(n+k)!}\right\}^{\frac{1}{2}}\left(\frac{1}{\sin\theta}\frac{d}{d\theta}\right)^k\left[\frac{\cos\left(n+\frac{1}{2}\right)\theta}{\cos\frac{1}{2}\theta}\right]. \tag{2.107}$$

The coefficients A_n^m admit the representation

$$A_n^m = \sum_{j=m}^{\infty} W_j^m \alpha_{nm}^j, \quad n = 0, 1, 2, \ldots, m-1 \tag{2.108}$$

where the coefficients α_{nm}^j ($j \geq m$) are solutions of the equations

$$\sum_{n=k}^{m-1} \alpha_{nm}^j \left\{\frac{(n+k)!}{(n-k)!}\right\}^{\frac{1}{2}} \hat{P}_{n-k}^{\left(k-\frac{1}{2},k+\frac{1}{2}\right)}(\cos\theta) = \left\{\frac{(j+k)!}{(j-k)!}\right\}^{\frac{1}{2}} \hat{P}_{j-k}^{\left(k-\frac{1}{2},k+\frac{1}{2}\right)}(\cos\theta), \tag{2.109}$$

for $k = 0, 1, 2, ..., m-1$. For every fixed j, the matrix of the system of Equations (2.109) is upper triangular, so the solution can be easily obtained by a recursive procedure.

Now differentiate (2.99) m times to obtain an equivalent system of linear algebraic equations. Accepting the representation (2.108) for the coefficients A_n^m, the final system is

$$\hat{x}_s^m (1 - \mu_s) - \sum_{n=m}^{\infty} \hat{x}_n^m (\varepsilon_n - \mu_n) W_{sn}^m (\theta_0) = \hat{g}_s^m + \sum_{n=m}^{\infty} \left(\hat{f}_n^m - \hat{g}_n^m \right) W_{sn}^m (\theta_0), \quad (2.110)$$

where $s = m, m+1, m+2, ...$ and

$$\left\{ \hat{x}_n^m, \hat{f}_n^m, \hat{g}_n^m \right\} = \left(n + \frac{1}{2} \right)^m \left\{ x_n^m, f_n^m, g_n^m \right\}, \quad (2.111)$$

$$W_{sn}^m (\theta_0) = U_{sn}^m (\theta_0) - \sum_{n=k}^{m-1} \left(j + \frac{1}{2} \right)^m \alpha_{jm}^n U_{sj}^m (\theta_0), \quad (2.112)$$

and

$$U_{sj}^m (\theta_0) = \frac{1}{\pi} \left[\frac{\sin (s-j) \theta_0}{s-j} + (-1)^m \frac{\sin (s+j+1) \theta_0}{s+j+1} \right] \quad (2.113)$$

with the understanding

$$\left[\frac{\sin n\theta_0}{n} \right]_{n=0} = \theta_0.$$

Thus, the initial dual series Equations (2.86) and (2.87), with associated Legendre function kernels, are transformed to the equivalent system of linear algebraic Equations (2.110); it is a second-kind equation that is a completely continuous perturbation of the identity operator in l_2. However it is significantly more stable than (2.101) and (2.103), albeit at the cost of rather more complicated coefficients. Another stable form may be derived as follows. Using the relationship (2.107), we represent (2.99) in equivalent form

$$\sum_{n=m}^{\infty} x_n^m \hat{P}_n^{\left(-\frac{1}{2}, \frac{1}{2}\right)} (\cos \theta) - \sum_{n=0}^{m-1} C_n^m \hat{P}_n^{\left(-\frac{1}{2}, \frac{1}{2}\right)} (\cos \theta) = \begin{cases} F_1(\theta), & \theta \in (0, \theta_0) \\ F_2(\theta), & \theta \in (\theta_0, \pi) \end{cases}, \quad (2.114)$$

where

$$F_1(\theta) = \sum_{n=m}^{\infty} \left(x_n^m \varepsilon_n + f_n^m\right) \hat{P}_n^{\left(-\frac{1}{2},\frac{1}{2}\right)}(\cos\theta),$$

$$F_2(\theta) = \sum_{n=m}^{\infty} \left(x_n^m \mu_n + g_n^m\right) \hat{P}_n^{\left(-\frac{1}{2},\frac{1}{2}\right)}(\cos\theta) + \sum_{n=0}^{m-1} A_n^m \hat{P}_n^{\left(-\frac{1}{2},\frac{1}{2}\right)}(\cos\theta).$$

For these orthonormal Jacobi polynomials the following differentiation formula holds when $k \leq n$, [58],

$$\frac{d^k}{dx^k}\hat{P}_n^{\left(-\frac{1}{2},\frac{1}{2}\right)}(x) = \left[(n+k)!/(n-k)!\right]^{\frac{1}{2}} \hat{P}_{n-k}^{\left(k-\frac{1}{2},k+\frac{1}{2}\right)}(x); \tag{2.115}$$

the k-fold derivative vanishes when $k > n$. Introduce the new unknowns and coefficients

$$\left\{y_n^m, F_n^m, G_n^m\right\} = \left\{\frac{(n+m)!}{(n-m)!}\right\}^{\frac{1}{2}} \left\{x_n^m, f_n^m, g_n^m\right\}. \tag{2.116}$$

It follows from (2.90), (2.93), and(2.111) that

$$\left\{y_n^m, F_n^m, G_n^m\right\}_{n=m}^{\infty} \in l_2 = l_2(0). \tag{2.117}$$

Assuming that condition (2.93) is valid, we may differentiate the Equation (2.114) m times term-by-term with respect to $x = \cos\theta$. Keeping in mind the relationship (2.105), we find (setting $x_0 = \cos\theta_0$),

$$\sum_{n=m}^{\infty} y_n^m \hat{P}_{n-m}^{\left(m-\frac{1}{2},m+\frac{1}{2}\right)}(x) = \begin{Bmatrix} F_1(x), & x \in (-1, x_0) \\ F_2(x), & x \in (x_0, 1) \end{Bmatrix} \tag{2.118}$$

where

$$F_1(x) = \sum_{n=m}^{\infty} \left(y_n^m \mu_n + F_n^m\right) \hat{P}_{n-m}^{\left(m-\frac{1}{2},m+\frac{1}{2}\right)}(x),$$

$$F_2(x) = \sum_{n=m}^{\infty} \left(y_n^m \varepsilon_n + G_n^m\right) \hat{P}_{n-m}^{\left(m-\frac{1}{2},m+\frac{1}{2}\right)}(x).$$

The polynomials $\hat{P}_s^{\left(m-\frac{1}{2},m+\frac{1}{2}\right)}$ are orthonormal on $[-1,1]$ with weight function $w(x) = (1-x)^{m-\frac{1}{2}}(1+x)^{m+\frac{1}{2}}$; multiplying (2.118) by $w\hat{P}_{s-m}^{\left(m-\frac{1}{2},m+\frac{1}{2}\right)}$ and integrating term-by-term over $[-1,1]$, we obtain the infinite system of linear algebraic equations

$$\begin{aligned}(1-\varepsilon_s)\, y_s^m &+ \sum_{n=m}^{\infty} y_n^m (\varepsilon_n - \mu_n)\, \hat{Q}_{s-m,n-m}^{\left(m-\frac{1}{2},m+\frac{1}{2}\right)}(x_0) \\ &= G_n^m + \sum_{n=m}^{\infty} \left(F_n^m - G_n^m\right) \hat{Q}_{s-m,n-m}^{\left(m-\frac{1}{2},m+\frac{1}{2}\right)}(x_0),\end{aligned} \tag{2.119}$$

where $s = m+1, m+2, ...,$ and the usual normalised incomplete inner product has been employed.

Comparing (2.111) with (2.116) we have

$$\left\{\hat{x}_n^m, \hat{f}_n^m, \hat{g}_n^m\right\} = k_n^m \left\{y_n^m, F_n^m, G_n^m\right\} \tag{2.120}$$

where

$$k_n^m = \left(n+\frac{1}{2}\right)^m \left[\frac{(n-m)!}{(n+m)!}\right]^{\frac{1}{2}}. \tag{2.121}$$

Observe that $k_n^m \to 1$ as $n \to \infty$. In addition, it can be shown that the following relationship holds:

$$\hat{Q}_{s-m,r-m}^{\left(m-\frac{1}{2},m+\frac{1}{2}\right)}(\cos\theta_0) = k_r^m \left(k_n^m\right)^{-1} W_{sr}^m(\theta_0). \tag{2.122}$$

Formula (2.122) can be used for calculations of $\hat{Q}_{s-m,r-m}^{\left(m-\frac{1}{2},m+\frac{1}{2}\right)}(\cos\theta_0)$, employing (2.112). The systems (2.110) and (2.119) are practically identical, differing only in the normalisation (2.120).

In summary, we have shown how to regularise the special class of dual series Equations (2.86) and (2.87) containing associated Legendre functions as kernels. The simplest approach essentially integrated the series equations to obtain dual series equations with Legendre polynomial kernels, together with constants of integration that are uniquely determined by some differentiability conditions. This produced (2.100), (2.101), and (2.103). The simplicity in calculating the matrix elements of this system is attractive: however, as already noted, it is unstable for large m and leads to significant errors in the calculation of the coefficients A_n^m. But provided m is not large, this system is quite suitable for numerical calculation. In order to rectify this instability, the modified system (2.110) was derived, and its normalised variant (2.119). Both these systems are stable, but the algorithm for calculation of the matrix coefficients is rather more complicated.

2.4 Symmetric triple series equations involving Jacobi polynomials

Triple series equations present an obvious extension and generalisation of dual series equations. In this section we consider *symmetric* triple series equations, the kernels of which are Jacobi polynomials $P_n^{(\alpha,\beta)}$. Without a significant loss of generality, we restrict attention to kernels of most use in subsequent chapters, the *ultraspherical polynomial* $P_n^{(\alpha,\alpha)}$; the parameter η that occurred in Section 2.1 will be fixed to be $\frac{1}{2}$. Moreover, the interval $[-1,1]$ is subdivided into three subintervals on which the corresponding functional equations are enforced, so that the middle subinterval is symmetric about 0. Thus the term *symmetric equations* highlights two different aspects: equality of the parameters α and β, and a symmetric subdivision of the full interval of definition $[-1, 1]$. Nonsymmetric subdivisions will be deferred to Section 2.7.

Retaining all the notation introduced in Section 2.1 we consider equations of two types, Type A and Type B, being, respectively, the sets of triple equations

$$\sum_{n=0}^{\infty} \{x_n(1-q_n)-g_n\} P_n^{(\alpha,\alpha)}(t) = 0, \; t \in (-1,-t_0), \tag{2.123}$$

$$\sum_{n=0}^{\infty} \lambda_n(\alpha,\alpha;\tfrac{1}{2}) \{x_n(1-r_n)-f_n\} P_n^{(\alpha,\alpha)}(t) = 0, \; t \in (-t_0,t_0), \tag{2.124}$$

$$\sum_{n=0}^{\infty} \{x_n(1-q_n)-g_n\} P_n^{(\alpha,\alpha)}(t) = 0, \; t \in (t_0,1), \tag{2.125}$$

and

$$\sum_{n=0}^{\infty} \lambda_n(\alpha,\alpha;\tfrac{1}{2}) \{x_n(1-r_n)-f_n\} P_n^{(\alpha,\alpha)}(t) = 0, \; t \in (-1,-t_0), \tag{2.126}$$

$$\sum_{n=0}^{\infty} \{x_n(1-q_n)-g_n\} P_n^{(\alpha,\alpha)}(t) = 0, \; t \in (-t_0,t_0), \tag{2.127}$$

$$\sum_{n=0}^{\infty} \lambda_n(\alpha,\alpha;\tfrac{1}{2}) \{x_n(1-r_n)-f_n\} P_n^{(\alpha,\alpha)}(t) = 0, \; t \in (t_0,1), \tag{2.128}$$

where we recall from definition (2.3), the coefficient

$$\lambda_n(\alpha,\alpha;\tfrac{1}{2}) = n+\alpha+\tfrac{1}{2}.$$

The solution $\{x_n\}_{n=0}^{\infty}$ is sought in the class l_2; in addition, we assume that $\{f_n\}_{n=0}^{\infty}, \{g_n\}_{n=0}^{\infty} \in l_2$.

2.4.1 Type A triple series equations

Using the symmetry property of Jacobi polynomials (see Appendix, (B.26), with $\beta = \alpha$),

$$P_n^{(\alpha,\alpha)}(-t) = (-1)^n P_n^{(\alpha,\alpha)}(t)$$

we may transform the Equations (2.123)–(2.125) to two sets of dual series equations, for the odd ($l = 1$) and even ($l = 0$) unknown coefficients, respectively; the interval of definition of the dual equations is halved. The coefficients satisfy (for $l = 0, 1$)

$$\sum_{n=0}^{\infty} \{x_{2n+l}(1 - q_{2n+l}) - g_{2n+l}\} P_{2n+l}^{(\alpha,\alpha)}(z) = 0, \; z \in (-1, -z_0), \quad (2.129)$$

$$\sum_{n=0}^{\infty} \lambda_{2n+l}(\alpha, \alpha; \tfrac{1}{2}) \{x_{2n+l}(1 - r_{2n+l}) - f_{2n+l}\} P_{2n+l}^{(\alpha,\alpha)}(z) = 0, \; z \in (-z_0, 0), \quad (2.130)$$

In itself, this transformation does not construct an effective solution of equations of Type A. The key step is to connect the ultraspherical polynomials with Jacobi polynomials [58]:

$$P_{2n+l}^{(\alpha,\alpha)}(z) = \frac{\Gamma(n+1)}{\Gamma(2n+1+l)} \frac{\Gamma(2n+\alpha+1+l)}{\Gamma(n+\alpha+1)} z^l P_n^{(\alpha, l-\frac{1}{2})}(2z^2 - 1). \quad (2.131)$$

This transforms the dual Equations (2.129) and (2.130), which are defined on $[-1, 0]$, to another set of dual equations that are defined on the complete interval $[-1, 1]$. Setting $u = 2z^2 - 1$ and $u_0 = 2z_0^2 - 1$, we obtain

$$\sum_{n=0}^{\infty} \left(n + \frac{\alpha + l}{2} + \frac{1}{4}\right) \{x_{2n+l}^*(1 - r_{2n+l}) - f_{2n+l}^*\} P_n^{(\alpha, l-\frac{1}{2})}(u) = 0, \quad u \in (-1, u_0) \quad (2.132)$$

$$\sum_{n=0}^{\infty} \{x_{2n+l}^*(1 - q_{2n+l}) - g_{2n+l}^*\} P_n^{(\alpha, l-\frac{1}{2})}(u) = 0, \; u \in (u_0, 1) \quad (2.133)$$

where the rescaled coefficients are

$$\{x_{2n+l}^*, f_{2n+l}^*, g_{2n+l}^*\} = \frac{\Gamma(n+1)\Gamma(2n+\alpha+1+l)}{\Gamma(2n+1+l)\Gamma(n+\alpha+1)} \{x_{2n+l}, f_{2n+l}, g_{2n+l}\}. \quad (2.134)$$

In order to apply the method developed in Section 2.1, rewrite the dual equations as

$$\sum_{n=0}^{\infty} \{\Lambda_n^l (1 - r_{2n+l}) X_{2n+l} - F_{2n+l}\} \hat{P}_n^{(\alpha-\frac{1}{2}, l)}(u) = 0, \; u \in (-1, u_0) \quad (2.135)$$

$$\sum_{n=0}^{\infty} \{(1 - q_{2n+l}) X_{2n+l} - G_{2n+l}\} \hat{P}_n^{(\alpha-\frac{1}{2},l)}(u) = 0, \; u \in (u_0, 1) \tag{2.136}$$

where

$$X_{2n+l} = \frac{\Gamma(n+\alpha+1)}{\Gamma\left(n+\alpha+\frac{1}{2}\right)} \left\{h_n^{(\alpha-\frac{1}{2},l)}\right\}^{\frac{1}{2}} x_{2n+l}^*, \tag{2.137}$$

$$G_{2n+l} = \frac{\Gamma(n+\alpha+1)}{\Gamma\left(n+\alpha+\frac{1}{2}\right)} \left\{h_n^{(\alpha-\frac{1}{2},l)}\right\}^{\frac{1}{2}} g_{2n+l}^*, \tag{2.138}$$

$$F_{2n+l} = \left(n + \frac{\alpha+l}{2} + \frac{1}{4}\right) \frac{\Gamma\left(n+l+\frac{1}{2}\right)}{\Gamma(n+l+1)} \left\{h_n^{(\alpha-\frac{1}{2},l)}\right\}^{\frac{1}{2}} f_{2n+l}^*, \tag{2.139}$$

and

$$\Lambda_n^l = \left(n + \frac{\alpha+l}{2} + \frac{1}{4}\right) \frac{\Gamma\left(n+l+\frac{1}{2}\right)}{\Gamma(n+l+1)} \frac{\Gamma\left(n+\alpha+\frac{1}{2}\right)}{\Gamma(n+\alpha+1)}. \tag{2.140}$$

From Field's formula for the ratio of Gamma functions (see Appendix, (B.7)), we deduce

$$\Lambda_n^l = 1 + O\left(n^{-2}\right), \quad \text{as } n \to \infty, \tag{2.141}$$

and introduce the *asymptotically small* parameter ε_n^l defined by

$$\varepsilon_n^l = 1 - \Lambda_n^l = O\left(n^{-2}\right). \tag{2.142}$$

After some rearrangement (2.135), (2.136) become

$$\sum_{n=0}^{\infty} X_{2n+l} \hat{P}_n^{(\alpha-\frac{1}{2},l)}(u) = \left\{ \begin{array}{ll} F_1(u), & u \in (-1, u_0) \\ F_2(u), & u \in (u_0, 1) \end{array} \right\} \tag{2.143}$$

where

$$\begin{aligned} F_1(u) &= \sum_{n=0}^{\infty} \left\{\left[r_{2n+l} + \varepsilon_n^l (1 - r_{2n+l})\right] X_{2n+l} + F_{2n+l}\right\} \hat{P}_n^{(\alpha-\frac{1}{2},l)}(u), \\ F_2(u) &= \sum_{n=0}^{\infty} \{q_{2n+l} X_{2n+l} + G_{2n+l}\} \hat{P}_n^{(\alpha-\frac{1}{2},l)}(u). \end{aligned}$$

As usual, multiply both sides of Equation (2.143) by the factor

$$(1-u)^{\alpha-\frac{1}{2}} (1+u)^l \hat{P}_n^{(\alpha-\frac{1}{2},l)}(u)$$

and integrate over $[-1,1]$. The result is an infinite system of linear algebraic equations, the matrix operator of which is a completely continuous perturbation of the identity (in l_2):

$$\left\{1-\left[r_{2s+l}+\varepsilon_s^l\left(1-r_{2s+l}\right)\right]\right\} X_{2s+l}-\sum_{n=0}^{\infty} X_{2n+l}\left\{q_{2n+l}-\left[r_{2n+l}+\varepsilon_n^l\left(1-r_{2n+l}\right)\right]\right\} \hat{Q}_{ns}^{\left(\alpha-\frac{1}{2},l\right)}\left(u_0\right)$$
$$=F_{2s+l}+\sum_{n=0}^{\infty}\left(G_{2n+l}-F_{2n+l}\right) \hat{Q}_{ns}^{\left(\alpha-\frac{1}{2},l\right)}\left(u_0\right), \quad (2.144)$$

where $s=0,1,2,\ldots$ .This regularised system is valid for both even ($l=0$) or odd ($l=1$) coefficients.

A remark is in order. When $\alpha=\frac{1}{2}$, the kernels essentially reduce to the trigonometric functions $\sin n\vartheta$ and $\varepsilon_n^l \equiv 0$ for all n. However, the procedure above is applicable only when $\alpha>-\frac{1}{2}$. To circumvent the difficulty encountered when $\alpha=-\frac{1}{2}$, (corresponding to the kernels $\cos n\theta$) we may use those devices applied to obtain solution of similar equations in previous sections (based on Rodrigues' formula, etc.).

2.4.2 Type B triple series equations

A similar argument to that employed in the last section transforms the triple series (2.126)–(2.128) to the analogue of (2.132) and (2.133). Omitting the preliminary steps of this deduction, we obtain (with the same notation)

$$\sum_{n=0}^{\infty}\left\{x_{2n+l}^*\left(1-q_{2n+l}\right)-g_{2n+l}^*\right\} P_n^{\left(\alpha,l-\frac{1}{2}\right)}(u)=0, \quad u \in\left(-1,u_0\right) \quad (2.145)$$

$$\sum_{n=0}^{\infty}\left(n+\frac{\alpha+l}{2}+\frac{1}{4}\right)\left\{x_{2n+l}^*\left(1-r_{2n+l}\right)-f_{2n+l}^*\right\} P_n^{\left(\alpha,l-\frac{1}{2}\right)}(u)=0, \quad u \in\left(u_0,1\right) \quad (2.146)$$

The odd case ($l=1$) of the dual pair (2.145), (2.146) is solvable by means of the general theory developed in Section 2.1, when $\alpha>-1$. We obtain the regularised system

$$\left(1-q_{2s+1}\right) y_{2s+1}+\sum_{n=0}^{\infty}\left\{q_{2n+1}-\left[r_{2n+1}+\mu_n\left(1-r_{2n+1}\right)\right]\right\} y_{2n+1} \hat{Q}_{ns}^{\left(\alpha+\frac{1}{2},0\right)}\left(u_0\right)$$
$$=\hat{g}_{2s+1}+\sum_{n=0}^{\infty}\left(\hat{f}_{2n+1}-\hat{g}_{2n+1}\right) \hat{Q}_{ns}^{\left(\alpha+\frac{1}{2},0\right)}\left(u_0\right), \quad (2.147)$$

where $s = 0, 1, 2, ...$, and

$$y_{2n+1} = \frac{\Gamma\left(n+\frac{3}{2}\right)}{\Gamma\left(n+1\right)} \left\{ h_n^{(\alpha+\frac{1}{2},0)} \right\}^{\frac{1}{2}} x^*_{2n+1}, \tag{2.148}$$

$$\hat{g}_{2n+1} = \frac{\Gamma\left(n+\frac{3}{2}\right)}{\Gamma\left(n+1\right)} \left\{ h_n^{(\alpha+\frac{1}{2},0)} \right\}^{\frac{1}{2}} g^*_{2n+1},$$

$$\hat{f}_{2n+1} = \frac{\Gamma\left(n+\alpha+1\right)}{\Gamma\left(n+\alpha+\frac{3}{2}\right)} \left(n+\frac{\alpha}{2}+\frac{3}{4}\right) \left\{ h_n^{(\alpha+\frac{1}{2},0)} \right\}^{\frac{1}{2}} f^*_{2n+1},$$

and

$$\mu_n = 1 - \left(n+\frac{\alpha}{2}+\frac{3}{4}\right) \frac{\Gamma\left(n+1\right)\Gamma\left(n+\alpha+1\right)}{\Gamma\left(n+\frac{3}{2}\right)\Gamma\left(n+\alpha+\frac{3}{2}\right)}. \tag{2.149}$$

The parameter μ_n is asymptotically small: $\mu_n = O\left(n^{-2}\right)$ as $n \to \infty$.

In the even case ($l = 0$) the parameters fall outside the range of applicability of the method described in Section 2.1. This necessitates the application of another method that was used in the analysis of Equations (2.39) and (2.40), which can be considered as a particular case of the more general Equations (2.145) and (2.146) with values $\alpha = -\frac{1}{2}$, $l = 0$. Although the solution can be obtained in this more general case, we omit the details, and confine attention to a specific example that will be treated in Section 4.4.

2.5 Relationships between series and integral equations

This purpose of this section is to explain the relationship between some classes of series and integral equations, and to show how the scope of the Abel integral equation method may be expanded to establish such connections. Dual integral equations will be considered in their own right in the next section. The results of this section are based upon those obtained by W. E. Williams [76], [77]; A. A. Ashour [3]; and J. S. Lowndes [37].

Let m be a fixed nonnegative integer. We consider two basic kinds of dual series equations. The kernel of the first employs associated Legendre functions

$$\sum_{n=m}^{\infty} a_n^m P_n^m\left(\cos\theta\right) = F_m\left(\theta\right),\ \theta \in \left(0,\theta_0\right), \tag{2.150}$$

$$\sum_{n=m}^{\infty} \left(2n+1\right) a_n^m P_n^m\left(\cos\theta\right) = G_m\left(\theta\right),\ \theta \in \left(\theta_0,\pi\right), \tag{2.151}$$

whilst the second employs trigonometric kernels,

$$\sum_{n=1}^{\infty} b_n \sin n\varphi = f(\varphi), \; \varphi \in (0, \varphi_0), \tag{2.152}$$

$$\sum_{n=1}^{\infty} n b_n \sin n\varphi = g(\varphi), \; \varphi \in (\varphi_0, \pi). \tag{2.153}$$

In addition, we consider two types of dual integral equations. The kernel of the first is a Bessel function of integer order m,

$$\int_0^{\infty} A_m(\lambda) J_m(\lambda\rho)\, d\lambda = E_m(\rho), \; 0 \le \rho < a \tag{2.154}$$

$$\int_0^{\infty} \lambda A_m(\lambda) J_m(\lambda\rho)\, d\lambda = H_m(\rho), \; \rho > a \tag{2.155}$$

whilst the second has a trigonometric kernel,

$$\int_0^{\infty} B(\mu) \sin(\mu x)\, d\mu = e(x), \; 0 \le x < b \tag{2.156}$$

$$\int_0^{\infty} \mu B(\mu) \sin(\mu x)\, d\mu = h(x), \; x > b \tag{2.157}$$

The functions $F_m, G_m, f, g, E_m, H_m, e$ and h occurring on the right-hand sides of (2.150)–(2.157) are assumed to be known; the equations are to be solved for the unknown coefficients a_n^m, b_n and functions A_m, B, respectively. Let us extend the domain of definition of the functions occurring in Equations (2.151) and (2.155) in the following way. Let

$$\sum_{n=m}^{\infty} (2n+1)\, a_n^m P_n^m(\cos\theta) = \begin{Bmatrix} C_m(\theta), & \theta \in (0, \theta_0) \\ G_m(\theta), & \theta \in (\theta_0, \pi) \end{Bmatrix} \tag{2.158}$$

and

$$\int_0^{\infty} \lambda A_m(\lambda) J_m(\lambda\rho)\, d\lambda = \begin{Bmatrix} L_m(\rho), & 0 \le \rho < a \\ H_m(\rho), & \rho > a \end{Bmatrix}. \tag{2.159}$$

The relationship between the coefficients a_n^m and $C_m(\theta)$, or between the coefficients A_m and $L_m(\rho)$, is found using the orthogonality of associated Legendre's functions P_n^m on $[0, \pi]$, or by using the Fourier-Bessel transform as appropriate:

$$a_n^m = \frac{1}{2}\frac{(n-m)!}{(n+m)!} \int_0^{\theta_0} d\theta \sin\theta C_m(\theta) P_n^m(\cos\theta) + \frac{1}{2}\frac{(n-m)!}{(n+m)!} \int_{\theta_0}^{\pi} d\theta \sin\theta G_m(\theta) P_n^m(\cos\theta), \tag{2.160}$$

$$A_m(\lambda) = \int_0^a rL_m(r)\, J_m(\lambda r)\, dr + \int_a^\infty rH_m(r)\, J_m(\lambda r)\, dr. \tag{2.161}$$

Now substitute these expressions for a_n^m or A_m in (2.150) and (2.154). This leads to two first-kind Fredholm integral equations involving the unknown functions C_m and L_m:

$$\int_0^{\theta_0} d\vartheta \sin\vartheta C_m(\vartheta)\, K_1(\vartheta,\theta) = 2F_m(\theta) - G_m^*(\theta),\ \theta\in(0,\theta_0), \tag{2.162}$$

$$\int_0^a dr.rL_m(r)K_2(r,\rho) = E(\rho) - H^*(\rho),\ 0\le\rho<a, \tag{2.163}$$

where

$$G_m^*(\theta) = \frac{1}{\pi}\cot^m\frac{\theta}{2}\int_0^\theta \frac{\tan^{2m}\frac{1}{2}\varphi d\varphi}{(\cos\varphi-\cos\theta)^{\frac{1}{2}}}\int_{\theta_0}^\pi \frac{G_m(\vartheta)\cot^m\frac{1}{2}\vartheta}{(\cos\varphi-\cos\vartheta)^{\frac{1}{2}}}d\vartheta, \tag{2.164}$$

$$H_m^*(\rho) = \frac{2}{\pi}\rho^{-m}\int_0^\rho dz\frac{z^{2m}}{(\rho^2-z^2)^{\frac{1}{2}}}\int_a^\infty dr\frac{rH(r)}{(r^2-z^2)^{\frac{1}{2}}}, \tag{2.165}$$

and the kernels of these integral equations are

$$K_1(\vartheta,\theta) = \sum_{n=m}^\infty \frac{(n-m)!}{(n+m)!}P_n^m(\cos\vartheta)\, P_n^m(\cos\theta), \tag{2.166}$$

$$K_2(r,\rho) = \int_0^\infty J_m(\lambda r)\, J_m(\lambda\rho)\, d\lambda. \tag{2.167}$$

These kernels admit the representation

$$K_1(\vartheta,\theta) = \frac{1}{\pi}\cot^m\frac{\theta}{2}\cot^m\frac{\vartheta}{2}\int_0^{\min(\theta,\vartheta)} \frac{\tan^{2m}\frac{1}{2}\varphi d\varphi}{(\cos\varphi-\cos\theta)^{\frac{1}{2}}(\cos\varphi-\cos\vartheta)^{\frac{1}{2}}}, \tag{2.168}$$

$$K_2(r,\rho) = \frac{2}{\pi}r^{-m}\rho^{-m}\int_0^{\min(r,\rho)} \frac{z^{2m}dz}{(r^2-z^2)^{\frac{1}{2}}(\rho^2-z^2)^{\frac{1}{2}}}. \tag{2.169}$$

With the change of variables

$$z = \tan\frac{\varphi}{2},\ r = \tan\frac{\vartheta}{2},\ \rho = \tan\frac{\theta}{2},$$

it can be shown that

$$K_2\left(\tan\frac{\vartheta}{2},\tan\frac{\theta}{2}\right)=2\cos\frac{\theta}{2}\cos\frac{\vartheta}{2}K_1(\vartheta,\theta), \tag{2.170}$$

so establishing a relationship between Equations (2.162) and (2.163); they are identical provided

$$L_m\left(\tan\frac{\vartheta}{2}\right) = \cos^3\frac{\vartheta}{2}C_m(\vartheta), \tag{2.171}$$

$$E_m\left(\tan\frac{\vartheta}{2}\right) = \cos\frac{\vartheta}{2}F_m(\vartheta), \tag{2.172}$$

$$H\left(\tan\frac{\vartheta}{2}\right) = \cos^3\frac{\vartheta}{2}G_m(\vartheta), \tag{2.173}$$

$$H^*\left(\tan\frac{\vartheta}{2}\right) = \cos\frac{\vartheta}{2}G_m^*(\vartheta). \tag{2.174}$$

Thus, we have demonstrated a one-to-one correspondence between the dual series Equations (2.150) and (2.151) and the dual integral Equations (2.154) and (2.155), and their solutions. If the condition (2.172) holds, we find

$$\sum_{n=m}^{\infty}a_n^m P_n^m(\cos\theta)=\sec\frac{\theta}{2}\int_0^{\infty}A_m(\lambda)J_m\left(\lambda\tan\frac{\theta}{2}\right)d\lambda. \tag{2.175}$$

In a similar way, if the condition (2.173) holds, we find

$$\sum_{n=m}^{\infty}(2n+1)a_n^m P_n^m(\cos\theta)=\sec^3\frac{\theta}{2}\int_0^{\infty}\lambda A_m(\lambda)J_m\left(\lambda\tan\frac{\theta}{2}\right)d\lambda. \tag{2.176}$$

We may now determine the relationship between solutions of these equations. Multiply both parts of equations (2.176) by the factor $\sin\theta P_k^m(\cos\theta)$ and integrate over $[0,\pi]$, to find

$$a_n^m = 2\frac{(n-m)!}{(n+m)!}\times$$
$$\int_0^{\infty}\left\{\int_0^{\infty}\lambda A_m(\lambda)J_m(\lambda u)\,d\lambda\right\}\frac{u}{\sqrt{1+u^2}}P_n^m\left(\frac{1-u^2}{1+u^2}\right)du. \tag{2.177}$$

On the other hand, using the Hankel transform, multiply both parts of (2.176) by $\cos\frac{1}{2}\theta\tan\frac{1}{2}\theta J_m\left(\mu\tan\frac{1}{2}\theta\right)$, and integrate with respect to $\rho=\tan\frac{1}{2}\theta$ over $(0,\infty)$. This gives the relation

$$A_m(\lambda)=\frac{1}{2\sqrt{2}}\int_{-1}^{1}\left\{\sum_{n=m}^{\infty}(2n+1)a_n^m P_n^m(x)\right\}J_m\left(\lambda\sqrt{\frac{1-x}{1+x}}\right)\frac{dx}{\sqrt{1+x}}. \tag{2.178}$$

Thus the solution of dual- (or multiple-) series equations has its counterpart in the solution of the corresponding dual- (or multiple-) integral equations, and vice versa.

Let us now demonstrate that the same is true for the pairs of Equations (2.152) and (2.153) and (2.156) and (2.157). These equations are reducible to first-kind Fredholm integral equations of the form

$$\int_0^{\varphi_0} C(\beta) K_3(\beta,\varphi)\, d\varphi = \frac{\pi}{2}\left[f(\varphi) - g^*(\varphi)\right], \ \varphi \in (0,\varphi_0), \tag{2.179}$$

and

$$\int_0^{b} l(y) K_4(x,y)\, dy = \frac{\pi}{2}\left[e(x) - h^*(x)\right], \ 0 \le x < b, \tag{2.180}$$

where

$$h^*(x) = \int_0^x dt \frac{t}{(x^2-t^2)^{\frac{1}{2}}} \int_b^\infty dy \frac{h(y)}{(y^2-t^2)^{\frac{1}{2}}}, \tag{2.181}$$

$$g^*(\varphi) = \frac{1}{2}\cos\frac{\varphi}{2} \int_0^\varphi \frac{d\alpha \sin\alpha}{\cos^2 \frac{1}{2}\alpha \,(\cos\alpha - \cos\varphi)^{\frac{1}{2}}} \int_{\varphi_0}^\pi \frac{d\beta\, g(\beta)\cos\frac{1}{2}\beta}{(\cos\alpha - \cos\beta)^{\frac{1}{2}}}, \tag{2.182}$$

and the kernels of the integral equations are, respectively

$$K_3(\beta,\varphi) = \sum_{n=1}^\infty \frac{\sin n\beta \sin n\varphi}{n} = \frac{1}{2}\ln\left|\frac{\tan\frac{1}{2}\varphi + \tan\frac{1}{2}\beta}{\tan\frac{1}{2}\varphi - \tan\frac{1}{2}\beta}\right|, \tag{2.183}$$

$$K_4(x,y) = \int_0^\infty \frac{\sin\mu x \sin\mu y}{\mu} d\mu = \frac{1}{2}\ln\left|\frac{x+y}{x-y}\right|. \tag{2.184}$$

The kernels $K_3(\beta,\varphi)$, $K_4(x,y)$ have a representation of the same form as (2.168) and (2.169). A more general representation for this type of kernel is derived later in this section. The relationship between the integral Equations (2.179) and (2.180) is established by observing that under the substitution $x = \tan\frac{1}{2}\varphi$, $y = \tan\frac{1}{2}\beta$,

$$K_4(x,y) = K_3(\beta,\varphi).$$

Thus, the integral equations are equivalent with the identification

$$l\left(\tan\frac{\beta}{2}\right) = 2\cos^2\frac{\beta}{2} C(\beta), \tag{2.185}$$

$$e\left(\tan\frac{\varphi}{2}\right) = f(\varphi), \tag{2.186}$$

$$h^*\left(\tan\frac{\varphi}{2}\right) = g^*(\varphi), \tag{2.187}$$

$$h\left(\tan\frac{\beta}{2}\right) = 2\cos^2\frac{\beta}{2} g(\beta). \tag{2.188}$$

Thus, if the following relation is valid

$$\sum_{k=1}^{\infty} b_n \sin n\varphi = \int_0^{\infty} B(\mu) \sin\left(\mu \tan\frac{\varphi}{2}\right) d\mu, \tag{2.189}$$

then so too is the relation

$$\sum_{k=1}^{\infty} n b_n \sin n\varphi = \sec^2\frac{\varphi}{2} \int_0^{\infty} \mu B(\mu) \sin\left(\mu \tan\frac{\varphi}{2}\right) d\mu. \tag{2.190}$$

Thus, the unknowns $\{b_n\}_{n=1}^{\infty}$ and B are connected by

$$b_n = \frac{2}{\pi n} \int_0^{\pi} d\varphi \sin n\varphi \sec^2\frac{\varphi}{2} \int_0^{\infty} \mu B(\mu) \sin\left(\mu \tan\frac{\varphi}{2}\right) d\mu. \tag{2.191}$$

The relationship stated above between some specific series and integral equations is not special and exists under more general conditions, which we now explore. The kernels of the series equations considered above are essentially Jacobi polynomials with symmetrical indices (see Formulae (1.153) and (2.91)):

$$\sin n\varphi \propto P_{n-1}^{\left(\frac{1}{2},\frac{1}{2}\right)}(\cos\varphi), \; P_n^m(\cos\theta) \propto P_{n-m}^{(m,m)}(\cos\theta).$$

On the other hand, since $\sin\nu x \propto J_{\frac{1}{2}}(\nu x)$, the corresponding integral Equations (2.154)–(2.157) involve the Bessel functions of order equal to $\frac{1}{2}$ or an integer m. We extend our considerations to series equations with ultraspherical polynomial kernels $P_n^{(\alpha,\alpha)}$, having arbitrary index α, and relate these to integral equations with Bessel function kernels of the same order α. So fixing α, let us examine the extended class of dual equations

$$\sum_{n=0}^{\infty} a_n P_n^{(\alpha,\alpha)}(x) = F(x), \; x \in (-1, x_0), \tag{2.192}$$

$$\sum_{n=0}^{\infty} \lambda_n(\alpha,\alpha;\eta) a_n P_n^{(\alpha,\alpha)}(x) = G(x), \; x \in (x_0, 1), \tag{2.193}$$

and

$$\int_0^{\infty} \lambda^{2\eta} A(\lambda) J_\alpha(\lambda\rho) d\lambda = g(\rho), \; 0 \le \rho < 1, \tag{2.194}$$

$$\int_0^{\infty} A(\lambda) J_\alpha(\lambda\rho) d\lambda = f(\rho), \; \rho > 1, \tag{2.195}$$

where the parameter η satisfies $0 \le \eta \le \frac{1}{2}$, and the value $\lambda_n(\alpha,\alpha;\eta)$ defined by (2.3) has the property

$$\lambda_n(\alpha,\alpha;\eta) = \frac{\Gamma(n+\alpha+1+\eta)}{\Gamma(n+\alpha+1-\eta)} = n^{2\eta}\left(1 + O(n^{-1})\right) \text{ as } n \to \infty. \tag{2.196}$$

Paralleling the argument previously employed, let us extend the domain of definition of the functions occurring in Equations (2.193) and (2.194), so that

$$\sum_{n=0}^{\infty} \lambda_n(\alpha,\alpha;\eta)\, a_n P_n^{(\alpha,\alpha)}(x) = \begin{Bmatrix} \hat{G}(x), & x\in(-1,x_0) \\ G(x), & x\in(x_0,1) \end{Bmatrix} \tag{2.197}$$

and

$$\int_0^\infty \lambda^{2\eta} A(\lambda)\, J_\alpha(\lambda\rho)\, d\lambda = \begin{Bmatrix} g(\rho), & 0\le\rho<1 \\ \hat{g}(\rho), & \rho>1 \end{Bmatrix}, \tag{2.198}$$

where $\hat{G}$ and $\hat{g}$ are unknown functions to be determined. Using the orthogonality property of Jacobi polynomials $P_n^{(\alpha,\alpha)}$ (with respect to the weight function $(1-x^2)^\alpha$) on $[-1,1]$, and using the Fourier-Hankel transform, one finds the relationships between a_n and $\hat{G}$, or between A and $\hat{g}$, respectively, to be

$$a_n = \Lambda_n \int_{-1}^{x_0} (1-y^2)^\alpha\, \hat{G}(y)\, P_n^{(\alpha,\alpha)}(y)\, dy +$$

$$\Lambda_n \int_{x_0}^{1} (1-y^2)^\alpha\, G(y)\, P_n^{(\alpha,\alpha)}(y)\, dy \tag{2.199}$$

and

$$A(\lambda) = \lambda^{1-2\eta}\left\{\int_1^\infty r\hat{g}(r)\, J_\alpha(\lambda r)\, dr + \int_0^1 rg(r)\, J_\alpha(\lambda r)\, dr\right\}, \tag{2.200}$$

where

$$\Lambda_n = 2^{-2\alpha-1}(2n+2\alpha+1)\frac{\Gamma(n+\alpha+1-\eta)}{\Gamma(n+\alpha+1+\eta)}\frac{\Gamma(n+2\alpha+1)\,\Gamma(n+1)}{\Gamma^2(n+\alpha+1)}.$$

Substitute these expressions into (2.192) and (2.195), respectively, to obtain first-kind Fredholm equations for the unknown functions $\hat{G}$ and $\hat{g}$:

$$\int_{-1}^{x_0} \hat{G}(y)\,(1-y^2)^\alpha\, K_1^{(\eta)}(x,y)\, dy = \hat{F}(x), \quad x\in(-1,x_0), \tag{2.201}$$

$$\int_1^\infty \hat{g}(r)\, rK_2^{(\eta)}(\rho,r)\, dr = \hat{f}(\rho), \quad \rho\in(1,\infty) \tag{2.202}$$

where the functions $\hat{F}$ and $\hat{f}$ are explicitly calculated from

$$\hat{F}(x) = F(x) - \int_{x_0}^{1} G(y)\,(1-y^2)^\alpha\, K_1^{(\eta)}(x,y)\, dy, \tag{2.203}$$

$$\hat{f}(\rho) = f(\rho) - \int_0^1 g(r)\, r K_2^{(\eta)}(\rho, r)\, dr, \tag{2.204}$$

and the kernels of these integral equations are

$$K_1^{(\eta)}(x, y) = \sum_{n=0}^{\infty} \Lambda_n P_n^{(\alpha,\alpha)}(x)\, P_n^{(\alpha,\alpha)}(y) \tag{2.205}$$

and

$$K_2^{(\eta)}(\rho, r) = \int_0^{\infty} \lambda^{1-2\eta} J_\alpha(\lambda\rho)\, J_\alpha(\lambda r)\, d\lambda. \tag{2.206}$$

Now we transform these kernels using the Abel integral representations for Jacobi polynomials $P_n^{(\alpha,\alpha)}$ (1.171) and Bessel functions J_α (1.180):

$$P_n^{(\alpha,\alpha)}(y) = \frac{(1+y)^{-\alpha}\,\Gamma(n+\alpha+1)}{\Gamma(\eta)\,\Gamma(n+\alpha+1-\eta)} \int_{-1}^{y} \frac{(1+u)^{\alpha-\eta} P_n^{(\alpha+\eta,\alpha-\eta)}(u)}{(y-u)^{1-\eta}} du, \tag{2.207}$$

$$J_\alpha(\lambda r) = \frac{\lambda^\eta r^\alpha}{2^{\eta-1}\Gamma(\eta)} \int_r^{\infty} \frac{v^{-\alpha-\eta+1} J_{\alpha+\eta}(\lambda v)}{(v^2 - r^2)^{1-\eta}} dv. \tag{2.208}$$

We transform the kernel $K_1^{(\eta)}$ by substituting (2.207) into (2.205) and inverting the order of summation and integration to find

$$K_1^{(\eta)}(x, y) = \frac{2^{-2\alpha-1}}{\Gamma(\eta)} (1+y)^{-\alpha} \int_{-1}^{y} du \frac{(1+u)^{\alpha-\eta}}{(y-u)^{1-\eta}} k_1^{(\eta)}(x, u), \tag{2.209}$$

where

$$k_1^{(\eta)}(x, u) = \sum_{n=0}^{\infty} \frac{(2n+2\alpha+1)\,\Gamma(n+1)\,\Gamma(n+2\alpha+1)}{\Gamma(n+\alpha+1)\,\Gamma(n+\alpha+1+\eta)} P_n^{(\alpha,\alpha)}(x)\, P_n^{(\alpha,\alpha)}(u). \tag{2.210}$$

The sum of the series in (2.210) is a discontinuous function; when $-1 \le u < x$, its value is [55]

$$k_1^{(\eta)}(x, u) = 2^{2\alpha+1} \{\Gamma(\eta)\}^{-1} (x-u)^{\eta-1} (1-u)^{-\alpha-\eta} (1+x)^{-\alpha}, \tag{2.211}$$

and when $x < u \le 1$, its value is zero. It follows that $K_1^{(\eta)}$ has the representation

$$K_1^{(\eta)}(x, y) = \frac{(1+x)^{-\alpha}(1+y)^{-\alpha}}{\Gamma^2(\eta)} \int_{-1}^{\min(x,y)} du \frac{(1-u)^{-\alpha-\eta}(1+u)^{\alpha-\eta}}{(x-u)^{1-\eta}(y-u)^{1-\eta}}. \tag{2.212}$$

We transform the kernel $K_2^{(\eta)}$ by substituting (2.208) into (2.206) and interchanging the order of integration. The result is

$$K_2^{(\eta)}(\rho,r) = \frac{r^\alpha}{2^{\eta-1}\Gamma(\eta)}\int_r^\infty dv \frac{v^{-\alpha-\eta+1}}{(v^2-r^2)^{1-\eta}}\int_0^\infty \lambda^{1-\eta}J_\alpha(\lambda\rho)\,J_{\alpha+\eta}(\lambda v)\,d\lambda. \tag{2.213}$$

The inner integral in (2.213) is the discontinuous Weber-Schafheitlin integral [19], [55]; when $0 \le \rho < v$, its value is

$$\int_0^\infty \lambda^{1-\eta}J_\alpha(\lambda\rho)\,J_{\alpha+\eta}(\lambda v)\,d\lambda = \rho^{\alpha 2^{1-\eta}}\{\Gamma(\eta)\}^{-1}(v^2-\rho^2)^{\eta-1}v^{-\alpha-\eta}, \tag{2.214}$$

and when $\rho > v$, its value is zero. Thus, the kernel $K_2^{(\eta)}$ can be expressed as

$$K_2^{(\eta)}(\rho,r) = \frac{\rho^\alpha r^\alpha}{2^{2\eta-2}\Gamma^2(\eta)}\int_{\max(\rho,r)}^\infty \frac{v^{-2\alpha-2\eta+1}}{(v^2-\rho^2)^{1-\eta}(v^2-r^2)^{1-\eta}}dv. \tag{2.215}$$

The relationship between $K_1^{(\eta)}$ and $K_2^{(\eta)}$ can now be stated. Using the substitutions

$$v = (1-u)^{\frac{1}{2}}(1+u)^{-\frac{1}{2}}, \rho = (1-x)^{\frac{1}{2}}(1+x)^{-\frac{1}{2}}, r = (1-y)^{\frac{1}{2}}(1+y)^{-\frac{1}{2}},$$

we obtain

$$K_2^{(\eta)}\left((1-x)^{\frac{1}{2}}(1+x)^{-\frac{1}{2}},(1-y)^{\frac{1}{2}}(1+y)^{-\frac{1}{2}}\right)$$
$$= (1-x)^{\frac{\alpha}{2}}(1+x)^{\frac{\alpha}{2}+1-\eta}(1-y)^{\frac{\alpha}{2}}(1+y)^{\frac{\alpha}{2}+1-\eta}K_1^{(\eta)}(x,y). \tag{2.216}$$

The relationship between the pairs of Equations (2.192) and (2.193) and (2.194) and (2.195) and their solutions a_n and A is now easily obtained, and the details are left to the reader.

Before concluding this section, we draw the reader's attention to one remarkable consequence of the kernel representations (2.212) and (2.215): we can find the analytic solution to both integral Equations (2.201) and (2.202). If we substitute the kernel representation (2.212) into (2.201), it takes the form

$$\int_{-1}^{x_0} dy\,(1-y)^\alpha\,\hat{G}(y)\int_{-1}^{\min(x,y)}\frac{(1-u)^{-\alpha-\eta}(1+u)^{\alpha-\eta}}{(x-u)^{1-\eta}(y-u)^{1-\eta}}du$$
$$= \Gamma^2(\eta)(1+x)^\alpha\,\hat{F}(x),\ x\in(-1,x_0). \tag{2.217}$$

We split the interval of integration for the external integral; symbolically, this operation may be represented as

$$\int_{-1}^{x_0} = \int_{-1}^{x} + \int_{x}^{x_0}. \tag{2.218}$$

Considering the first integral on the right-hand side of (2.218), the upper limit of the inner integral in (2.217) is $\min(x,y) = y(< x)$; for the second integral on the right-hand side of (2.218), the upper limit of the inner integral in (2.217) is $\min(x,y) = x(< y)$. Thus, the integral Equation (2.217) becomes

$$\int_{-1}^{x} dy\,(1-y)^{\alpha}\,\hat{G}(y)\int_{-1}^{y}\frac{(1-u)^{-\alpha-\eta}(1+u)^{\alpha-\eta}}{(x-u)^{1-\eta}(y-u)^{1-\eta}}du +$$
$$\int_{x}^{x_0} dy\,(1-y)^{\alpha}\,\hat{G}(y)\int_{-1}^{x}\frac{(1-u)^{-\alpha-\eta}(1+u)^{\alpha-\eta}}{(x-u)^{1-\eta}(y-u)^{1-\eta}}du$$
$$= \Gamma^2(\eta)(1+x)^{\alpha}\hat{F}(x), \qquad x\in(-1,x_0). \tag{2.219}$$

Transform the first term of the left-hand side of this equation using Dirichlet's extended Formula (1.135); invert the order of integration in the second term. These operations lead to

$$\int_{-1}^{x} du\frac{(1-u)^{-\alpha-\eta}(1+u)^{\alpha-\eta}}{(x-u)^{1-\eta}}\int_{u}^{x_0} dy\frac{(1-y)^{\alpha}\hat{G}(y)}{(y-u)^{1-\eta}}$$
$$= \Gamma^2(\eta)(1+x)^{\alpha}\hat{F}(x),\ x\in(-1,x_0). \tag{2.220}$$

Equation (2.220) may be recognised as Abel's integral equation

$$\int_{-1}^{x}\frac{G_1(u)\,du}{(x-u)^{1-\eta}} = \Gamma^2(\eta)(1+x)^{\alpha}\hat{F}(x),\ x\in(-1,x_0), \tag{2.221}$$

where the (as yet unknown) function G_1 is given by

$$G_1(u) = (1-u)^{-\alpha-\eta}(1+u)^{\alpha-\eta}\int_{u}^{x_0} dy\frac{(1-y)^{\alpha}\hat{G}(y)}{(y-u)^{1-\eta}}. \tag{2.222}$$

From the inverse Formula (1.131), we deduce

$$G_1(u) = \Gamma^2(\eta)\frac{\sin(\eta\pi)}{\pi}\frac{d}{du}\int_{-1}^{u}\frac{(1+x)^{\alpha}\hat{F}(x)}{(x-u)^{\eta}}dx. \tag{2.223}$$

Recognising that (2.222) is also an Abel integral equation, the inversion Formula (1.133) leads to the final and explicit form of the analytic solution to (2.201):

$$\hat{G}(y) = -\frac{\sin^2(\eta\pi)}{\pi^2}\Gamma^2(\eta)(1-y)^{-\alpha}\times$$
$$\frac{d}{dy}\int_{y}^{x_0} du\frac{(1-u)^{\alpha+\eta}(1+u)^{\alpha-\eta}}{(x-y)^{\eta}}\frac{d}{du}\int_{-1}^{u} dx\frac{(1+x)^{\alpha}\hat{F}(x)}{(u-x)^{\eta}}. \tag{2.224}$$

The solution of Equation (2.202) can be obtained in a similar way, and the reader may wish to verify that

$$\hat{g}(r) = -2^{2\eta-2}\Gamma^2(\eta)\frac{\sin^2(\eta\pi)}{\pi^2}r^{-\alpha-1} \times$$
$$\frac{d}{dr}\int_1^r dv \frac{v^{2\alpha+2\eta}}{(r^2-v^2)^\eta}\frac{d}{dv}\int_v^\infty d\rho \frac{\rho^{-\alpha+1}\hat{f}(\rho)}{(\rho^2-v^2)^\eta}. \quad (2.225)$$

2.6 Dual integral equations involving Bessel functions

In this section we demonstrate how to apply Abel's integral transform to obtain the solution of dual integral equations whose kernels are Bessel functions of fixed order α. We shall treat two kinds of dual integral equations, the pair

$$\int_0^\infty \lambda^{2\eta} A(\lambda) J_\alpha(\lambda\rho) d\lambda = g(\rho), \quad 0 \le \rho < 1, \quad (2.226)$$

$$\int_0^\infty A(\lambda) J_\alpha(\lambda\rho) d\lambda = f(\rho), \quad \rho > 1, \quad (2.227)$$

and the complementary pair, in which the subintervals of definition have been interchanged,

$$\int_0^\infty A(\lambda) J_\alpha(\lambda\rho) d\lambda = f(\rho), \quad 0 \le \rho < 1, \quad (2.228)$$

$$\int_0^\infty \lambda^{2\eta} A(\lambda) J_\alpha(\lambda\rho) d\lambda = g(\rho), \quad \rho > 1, \quad (2.229)$$

where A is the unknown function to be determined. The parameter η satisfies $0 < \eta \le \frac{1}{2}$, and g, f are given functions, which possess Fourier-Bessel integral expansions

$$g(\rho) = \int_0^\infty \lambda^{2\eta} G(\lambda) J_\alpha(\lambda\rho) d\lambda, \quad (2.230)$$

$$f(\rho) = \int_0^\infty F(\lambda) J_\alpha(\lambda\rho) d\lambda. \quad (2.231)$$

Denote by $L_2(\mu)$ the space of functions B defined on $[0,\infty)$ satisfying

$$\int_0^\infty \lambda^\mu \left|B(\lambda)\right|^2 d\lambda < \infty.$$

We shall find the solution A of these dual integral equations in the functional class $L_2(2\eta-1)$, assuming that the functions F, G belong to the same class as well:

$$A, F, G \in L_2(2\eta - 1).$$

As we have previously remarked, the condition imposed on the solution class is a reflection of the boundedness of the energy condition (Section 1.3).

Using the Formula (1.181), we integrate Equation (2.226) and obtain the dual equations

$$\int_0^\infty \lambda^{-1+2\eta}\left\{A(\lambda)-G(\lambda)\right\}J_{\alpha+1}(\lambda\rho)d\lambda = 0, \quad 0 \le \rho < 1 \quad (2.233)$$

$$\int_0^\infty \{A(\lambda)-F(\lambda)\}J_\alpha(\lambda\rho)d\lambda = 0, \quad \rho > 1. \quad (2.234)$$

Now substitute for the Bessel functions occurring in these equations, the Abel integral representations derived from (1.178) and (1.180),

$$J_{\alpha+1}(\lambda\rho) = \frac{\lambda^{1-\eta}\rho^{-\alpha-1}}{2^{-\eta}\Gamma\left(1-\eta\right)}\int_0^\rho \frac{v^{\alpha+\eta+1}J_{\alpha+\eta}\left(\lambda v\right)}{\left(\rho^2-v^2\right)^\eta}dv, \quad (2.235)$$

$$J_\alpha(\lambda\rho) = \frac{\lambda^\eta\rho^\alpha}{2^{\eta-1}\Gamma\left(\eta\right)}\int_\rho^\infty \frac{v^{-\alpha-\eta+1}J_{\alpha+\eta}\left(\lambda v\right)}{\left(v^2-\rho^2\right)^{1-\eta}}dv. \quad (2.236)$$

Interchanging the order of integration, one obtains the following pair of homogeneous Abel integral equations:

$$\int_0^\rho \frac{v^{\alpha+\eta+1}}{\left(\rho^2-v^2\right)^\eta}\left\{\int_0^\infty \lambda^\eta\left\{A(\lambda)-G(\lambda)\right\}J_{\alpha+\eta}\left(\lambda v\right)d\lambda\right\}dv = 0, \quad 0 \le \rho < 1 \quad (2.237)$$

$$\int_\rho^\infty \frac{v^{-\alpha-\eta+1}}{\left(v^2-\rho^2\right)^{1-\eta}}\left\{\int_0^\infty \lambda^\eta\left\{A(\lambda)-F(\lambda)\right\}J_{\alpha+\eta}\left(\lambda v\right)d\lambda\right\}dv = 0, \quad \rho > 1. \quad (2.238)$$

These equations possess unique zero solutions; the expressions in brackets therefore vanish, and we deduce a piecewise continuous representation of the sort that has repeatedly appeared in this book:

$$\int_0^\infty \lambda^\eta A(\lambda) J_{\alpha+\eta}(\lambda\rho)d\lambda = \begin{cases} \int_0^\infty \lambda^\eta G(\lambda) J_{\alpha+\eta}(\lambda\rho)d\lambda, & 0 \le \rho < 1 \\ \int_0^\infty \lambda^\eta F(\lambda) J_{\alpha+\eta}(\lambda\rho)d\lambda, & \rho > 1. \end{cases} \tag{2.239}$$

Let us use the Hankel transform to reach the final form of solution of these equations. Multiply both sides of (2.239) by the factor $\rho J_{\alpha+\eta}(\mu\rho)$ and integrate over $(0,\infty)$ to obtain the closed form solution

$$A(\mu) = \mu^{1-\eta} \int_0^1 d\rho.\rho J_{\alpha+\eta}(\mu\rho) \int_0^\infty \lambda^\eta G(\lambda) J_{\alpha+\eta}(\lambda\rho)d\lambda + \mu^{1-\eta} \int_1^\infty d\rho.\rho J_{\alpha+\eta}(\mu\rho) \int_0^\infty \lambda^\eta F(\lambda) J_{\alpha+\eta}(\lambda\rho)d\lambda. \tag{2.240}$$

Notice that this solution is valid provided $\alpha > -\frac{1}{2}$.

The dual Equations (2.228) and (2.229) are solved in a similar way; the solution is

$$A(\mu) = \mu^{1-\eta} \int_0^1 d\rho.\rho J_{\alpha-\eta}(\mu\rho) \int_0^\infty \lambda^\eta F(\lambda) J_{\alpha-\eta}(\lambda\rho)d\lambda + \mu^{1-\eta} \int_1^\infty d\rho.\rho J_{\alpha-\eta}(\mu\rho) \int_0^\infty \lambda^\eta G(\lambda) J_{\alpha-\eta}(\lambda\rho)d\lambda. \tag{2.241}$$

Thus, both pairs of dual integral equations possess a closed-form analytical solution.

More complicated dual integral equations may be transformed to second-kind Fredholm integral equations, provided some suitable and asymptotically small parameters can be identified. For example, we may treat the dual equations

$$\int_0^\infty \lambda^{2\eta} A(\lambda) \{1 + h(\lambda)\} J_\alpha(\lambda\rho)\, d\lambda = g(\rho), \quad 0 \le \rho < 1 \tag{2.242}$$

$$\int_0^\infty A(\lambda) \{1 + p(\lambda)\} J_\alpha(\lambda\rho)\, d\lambda = f(\rho), \quad \rho > 1 \tag{2.243}$$

where the functions h, p satisfy

$$\lim_{\lambda\to\infty} h(\lambda) = \lim_{\lambda\to\infty} p(\lambda) = 0.$$

These conditions ensure that the integral operator in the equation is compact (completely continuous) in $L_2(2\eta-1)$ (see Appendix C.2). In addition, the expansions (2.230), (2.231) for f, g must hold. Following the same steps used to obtain solution of (2.226) and (2.227), we obtain the second-kind Fredholm integral equation

$$\{1+p(\mu)\} A(\mu) + \mu^{1-\eta} \int_0^\infty \lambda^\eta A(\lambda) \{h(\lambda) - p(\lambda)\} K_{\alpha+\eta}(\lambda,\mu)\, d\lambda$$
$$= F(\mu) + \mu^{1-\eta} \int_0^\infty \lambda^\eta \{G(\lambda) - F(\lambda)\} K_{\alpha+\eta}(\lambda,\mu)\, d\lambda \quad (2.244)$$

where the kernel is

$$K_{\alpha+\eta}(\lambda,\mu) = \int_0^1 \rho J_{\alpha+\eta}(\lambda\rho) J_{\alpha+\eta}(\mu\rho) d\rho. \quad (2.245)$$

This second kind equation enjoys the same advantages identified for the second kind matrix systems obtained for dual series equations.

2.7 Nonsymmetrical triple series equations

In Section 2.4 we described an effective algorithm for the solution of symmetric triple series equations. How may one solve such equations in the more general case when the subdivision of the complete interval $[-1, 1]$ of definition is not symmetric? The answer has its basis in results that were derived in Section 2.5. Moreover, the solution of triple integral equations, involving Bessel functions, can be derived from the same results.

First, we consider some particular (but frequently occurring in practice) equations involving associated Legendre functions $P_n^m(\cos\theta)$ or Bessel functions $J_m(v\rho)$. Subsequently, we will extend the method to equations involving Jacobi polynomials $P_n^{(\alpha,\alpha)}$ or Bessel functions $J_\alpha(\lambda x)$ of arbitrary order as well.

Let m be a fixed non-negative integer, and α, β be fixed so that $0 < \alpha < \beta < \pi$. Consider the following two sets of triple series equations.

$$\begin{cases} \sum_{n=m}^{\infty} A_n^m (2n+1) P_n^m (\cos\theta) = F_1^m(\theta), & \theta \in (0,\alpha) \\ \sum_{n=m}^{\infty} A_n^m P_n^m (\cos\theta) = F_2^m(\theta), & \theta \in (\alpha,\beta) \\ \sum_{n=m}^{\infty} A_n^m (2n+1) P_n^m (\cos\theta) = F_3^m(\theta), & \theta \in (\beta,\pi) \end{cases} \tag{2.246}$$

and

$$\begin{cases} \sum_{n=m}^{\infty} C_n^m P_n^m (\cos\theta) = G_1^m(\theta), & \theta \in (0,\alpha) \\ \sum_{n=m}^{\infty} C_n^m (2n+1) P_n^m (\cos\theta) = G_2^m(\theta), & \theta \in (\alpha,\beta) \\ \sum_{n=m}^{\infty} C_n^m P_n^m (\cos\theta) = G_3^m(\theta), & \theta \in (\beta,\pi) \end{cases} \tag{2.247}$$

The solution $\{A_n^m, C_n^m\}_{n=m}^{\infty}$ of these triple series equations is sought in the functional class $l_2(2m)$. We consider only the first set (2.246), because the analysis of the equations (2.247) is similar.

On the basis of the relations (2.175) and (2.176) one may show that the triple Equations (2.246) are equivalent to the following triple integral equations, with Bessel function kernels,

$$\begin{cases} \int_0^\infty \lambda . A_m(\lambda) J_m(\lambda\rho)\, d\lambda = \left(1+\rho^2\right)^{-\frac{3}{2}} F_1^m(2\arctan\rho), & 0 \le \rho < \rho_0 \\ \int_0^\infty A_m(\lambda) J_m(\lambda\rho)\, d\lambda = \left(1+\rho^2\right)^{-\frac{1}{2}} F_2^m(2\arctan\rho), & \rho_0 < \rho < \rho_1 \\ \int_0^\infty \lambda . A_m(\lambda) J_m(\lambda\rho)\, d\lambda = \left(1+\rho^2\right)^{-\frac{3}{2}} F_3^m(2\arctan\rho), & \rho_1 < \rho \end{cases} \tag{2.248}$$

where $\rho_0 = \tan\frac{1}{2}\alpha$, $\rho_1 = \tan\frac{1}{2}\beta$, and $\rho = \tan\frac{1}{2}\theta$. The relation between the coefficients A_n^m and the function A_m is given by (2.177), (2.178). The transform $\rho = \tan\frac{1}{2}\theta$ may be geometrically visualised as a stereographic projection of the sphere onto a plane. The "symmetrisation" of equations (2.248), producing a symmetric partition of the domain of definition for each functional equation of the set, is realised by an "inversion in a circle." Introduce the new variable

$$r = (\rho_0\rho_1)^{-\frac{1}{2}} \rho = (\rho_0\rho_1)^{-\frac{1}{2}} \tan\frac{1}{2}\theta, \tag{2.249}$$

so that

$$\theta = \theta(r) = 2\arctan\left[(\rho_0\rho_1)^{\frac{1}{2}} r\right],$$

and transform Equations (2.248) to

$$\begin{cases} \int_0^\infty \mu . A_1(\mu) J_m(\mu r)\, d\mu = \rho_0\rho_1 (1+\rho_0\rho_1 r^2)^{-\frac{3}{2}} F_1^m\{\theta(r)\}, & 0 \le r < r_0 \\ \int_0^\infty A_1(\mu) J_m(\mu r)\, d\mu = (\rho_0\rho_1)^{\frac{1}{2}} \left(1+\rho_0\rho_1 r^2\right)^{-\frac{1}{2}} F_2^m\{\theta(r)\}, & r_0 < r < r_1 \\ \int_0^\infty \mu . A_1(\mu) J_m(\mu r)\, d\mu = \rho_0\rho_1 \left(1+\rho_0\rho_1 r^2\right)^{-\frac{3}{2}} F_3^m\{\theta(r)\}, & r_1 < r \end{cases} \tag{2.250}$$

where $\mu = (\rho_0\rho_1)^{\frac{1}{2}} \lambda$, $r_0 = (\rho_0/\rho_1)^{\frac{1}{2}} = r_1^{-1}$, and $A_1(\mu) = A(\lambda)$.

After the final change of variables

$$r = \tan\frac{1}{2}\vartheta = (\rho_0\rho_1)^{-\frac{1}{2}} \tan\frac{1}{2}\theta, \tag{2.251}$$

so that

$$\theta = \theta(\vartheta) = 2\arctan\left[(\rho_0\rho_1)^{\frac{1}{2}} \tan\frac{1}{2}\vartheta\right],$$

(which may be visualised geometrically as *reconstruction* of the spherical surface from its stereographic projection in the plane), we obtain the following symmetric triple series equations, involving the associated Legendre functions P_n^m as kernels:

$$\begin{cases} \sum_{n=m}^\infty (2n+1)\, B_n^m P_n^m(\cos\vartheta) = \rho_0\rho_1 u(\vartheta)^{-\frac{3}{2}} F_1^m(\theta(\vartheta)), & \vartheta \in (0, \vartheta_0) \\ \sum_{n=m}^\infty B_n^m P_n^m(\cos\vartheta) = (\rho_0\rho_1)^{\frac{1}{2}}\, u(\vartheta)^{-\frac{1}{2}} F_2^m(\theta(\vartheta)), & \vartheta \in (\vartheta_0, \pi-\vartheta_0) \\ \sum_{n=m}^\infty (2n+1)\, B_n^m P_n^m(\cos\vartheta) = \rho_0\rho_1 u(\vartheta)^{-\frac{3}{2}} F_3^m(\theta(\vartheta)), & \vartheta \in (\pi-\vartheta_0, \pi) \end{cases} \tag{2.252}$$

where $u(\vartheta) = (\cos^2 \frac{1}{2}\vartheta + \rho_0\rho_1 \sin^2 \frac{1}{2}\vartheta)$, $\vartheta_0 = 2\arctan(\rho_0/\rho_1)^{\frac{1}{2}}$, and so $\cos\vartheta_0 = (\rho_1-\rho_0)/(\rho_1+\rho_0)$. Note that in deriving (2.252) we used a relationship comparable to (2.176):

$$\sum_{s=m}^\infty (2s+1)\, B_s^m P_s^m(\cos\vartheta) = \sec^3\frac{\vartheta}{2} \int_0^\infty \mu A_1(\mu)\, J_m\left(\mu \tan\frac{\vartheta}{2}\right) d\mu \tag{2.253}$$

Using (2.253) and (2.175), the relationship between the coefficients A_n^m and B_n^m is

$$A_n^m = \frac{1}{\sqrt{2}} \frac{(n-m)!}{(n+m)!} \int_{-1}^1 \frac{dx\, P_n^m\{z(x)\}}{\sqrt{1+\rho_0\rho_1 + (1-\rho_0\rho_1)\, x}} \sum_{s=m}^\infty (2s+1) B_s^m P_s^m(x) \tag{2.254}$$

where

$$z(x) = \frac{1 - \rho_0\rho_1 + (1 + \rho_0\rho_1)\,x}{1 + \rho_0\rho_1 + (1 - \rho_0\rho_1)\,x}.$$

It is obvious that if $\rho_1 = \rho_0^{-1}$, then $A_n^m \equiv B_n^m$, and the Equations (2.252) will be identical with (2.246).

The relation between Jacobi polynomials $P_{n-m}^{(m,m)}$ and associated Legendre functions P_n^m,

$$P_n^m(z) = 2^{-m}\left(1 - z^2\right)^{\frac{m}{2}} \frac{\Gamma(n+m+1)}{\Gamma(n+1)} P_{n-m}^{(m,m)}(z)$$

enables us to convert (2.252) to the type of symmetric triple series equations solved in Section 2.4. Thus, the linear-fractional transform

$$\cos\theta = z = \frac{1 - \rho_0\rho_1 + (1 + \rho_0\rho_1)\,x}{1 + \rho_0\rho_1 + (1 - \rho_0\rho_1)\,x}, \tag{2.255}$$

with $x = \cos\vartheta$, "symmetrises" the initial Equations (2.246) and converts them to the symmetric triple series Equations (2.252) with a new set of unknown coefficients $\{B_s\}_{s=m}^{\infty}$, for which the relationship with the original set of unknowns $\{A_n^m\}_{n=m}^{\infty}$ is given by the Formula (2.254).

From this derivation, one further result should be noted: the triple integral equations involving the Bessel functions $J_m(\lambda\rho)$, with arbitrary fragmentation of the complete range of the variable ρ, may be transformed to a set of triple series equations involving $P_n^m(\cos\theta)$, with symmetrical fragmentation of the corresponding interval.

Triple series equations involving the trigonometric functions ($\sin n\varphi$ or $\cos n\varphi$) can be solved in an analogous manner. Although other methods have previously been reported in the literature, the attractive approach suggested here is based on [77], [3], and [37].

Consider the triple series equations

$$\begin{cases} \sum\limits_{n=1}^{\infty} n a_n \sin n\varphi = f_1(\varphi), & \varphi \in (0, \varphi_0) \\ \sum\limits_{n=1}^{\infty} a_n \sin n\varphi = f_2(\varphi), & \varphi \in (\varphi_0, \varphi_1) \\ \sum\limits_{n=1}^{\infty} n a_n \sin n\varphi = f_3(\varphi), & \varphi \in (\varphi_1, \pi) \end{cases} \tag{2.256}$$

where f_1, f_2, and f_3 are given functions, and ϕ_0, ϕ_1 are fixed so that $0 < \phi_0 < \phi_1 < \pi$. From (2.189) and (2.190) it can be shown that Equations

(2.256) are equivalent to the following triple integral equations:

$$\begin{cases} \int_0^\infty \mu A(\mu)\sin(\mu x)\,d\mu = \left(1+x^2\right)^{-1} f_1(2\arctan x), & x < x_0 \\ \int_0^\infty A(\mu)\sin(\mu x)\,d\mu = f_2(2\arctan x), & x_0 < x < x_1 \\ \int_0^\infty \mu A(\mu)\sin(\mu x)\,d\mu = \left(1+x^2\right)^{-1} f_3(2\arctan x), & x_1 < x \end{cases} \tag{2.257}$$

where $x = \tan\frac{1}{2}\varphi$, $x_0 = \tan\frac{1}{2}\varphi_0$, and $x_1 = \tan\frac{1}{2}\varphi_1$. The unknown coefficients $\{a_n\}_{n=1}^{\infty}$ and function A are related by

$$a_n = \frac{8}{\pi n}\int_0^\infty \left[\int_0^\infty \mu A(\mu)\sin(\mu u)\,d\mu\right]\frac{u}{1+u^2}U_{u-1}\left(\frac{1-u^2}{1+u^2}\right)du \tag{2.258}$$

and

$$A(\mu) = \frac{1}{\pi}\int_{-1}^{1}\left\{\sum_{n=1}^{\infty} n a_n U_{n-1}(z)\right\}\sqrt{\frac{1-z}{1+z}}\sin\left(\mu\sqrt{\frac{1-z}{1+z}}\right)dz \tag{2.259}$$

where $U_n(\cos\varphi) = \sin(n+1)\varphi/\sin\varphi$ is the Chebyshev polynomial of the second kind. Applying the change of variables connected with inversion in a circle

$$y = (x_0 x_1)^{-\frac{1}{2}} x, \qquad v = (x_0 x_1)^{\frac{1}{2}}\mu,$$

so that

$$\varphi = \varphi(y) = 2\arctan\left[(x_0 x_1)^{\frac{1}{2}} y\right],$$

the triple Equations (2.257) become the symmetric triple integral equations

$$\begin{cases} \int_0^\infty vA_1(v)\sin(vy)dv = x_0x_1\left(1+x_0x_1y^2\right)^{-1} f_1\{\varphi(y)\}, & y < y_0 \\ \int_0^\infty A_1(v)\sin(vy)dv = (x_0x_1)^{\frac{1}{2}} f_2\{\varphi(y)\}, & y_0 < y < y_1 \\ \int_0^\infty vA_1(v)\sin(vy)dv = x_0x_1\left(1+x_0x_1y^2\right)^{-1} f_3\{\varphi(y)\}, & y_1 < y \end{cases} \tag{2.260}$$

where $y_0 = \left(\tan\frac{1}{2}\varphi_0\cot\frac{1}{2}\varphi_1\right)^{\frac{1}{2}}$, $y_1 = y_0^{-1}$ and $A_1(v) = A(\mu)$.

Using the transform $\vartheta = 2\arctan y$, so that

$$\varphi = \varphi(\vartheta) = 2\arctan\left[(x_0x_1)^{\frac{1}{2}}\tan\frac{1}{2}\vartheta\right],$$

and the relationships

$$\sum_{n=1}^{\infty} nb_n \sin n\vartheta = \sec^2 \frac{\vartheta}{2} \int_0^{\infty} vA_1(v) \sin(vy)\, dv, \tag{2.261}$$

$$\sum_{n=1}^{\infty} b_n \sin n\vartheta = \int_0^{\infty} A_1(v) \sin(vy)\, dv, \tag{2.262}$$

one may reduce Equations (2.260) to the following symmetric series equations with trigonometric kernels, to be solved for the new set of unknowns $\{b_n\}_{n=1}^{\infty}$:

$$\left\{ \begin{array}{ll} \sum_{n=1}^{\infty} nb_n \sin n\vartheta = x_0 x_1 \left(\cos^2 \frac{1}{2}\vartheta + x_0 x_1 \sin^2 \frac{1}{2}\vartheta\right)^{-1} f_1\{\varphi(\vartheta)\}, & \\ \qquad\qquad\qquad\qquad \vartheta \in (0, \vartheta_0) & \\ \sum_{n=1}^{\infty} b_n \sin n\vartheta = (x_0 x_1)^{\frac{1}{2}} f_2\{\varphi(\vartheta)\}, & \vartheta \in (\vartheta_0, \pi - \vartheta_0) \\ \sum_{n=1}^{\infty} nb_n \sin n\vartheta = x_0 x_1 \left(\cos^2 \frac{1}{2}\vartheta + x_0 x_1 \sin^2 \frac{1}{2}\vartheta\right)^{-1} f_3\{\varphi(\vartheta)\}, & \\ \qquad\qquad\qquad\qquad \vartheta \in (\vartheta_0, \pi) & \end{array} \right. \tag{2.263}$$

where $\vartheta_0 = 2\arctan\left(\tan \frac{1}{2}\varphi_0 \cot \frac{1}{2}\varphi_1\right)$. From (2.261) and (2.258) we obtain the relationship between the two sets of coefficients:

$$a_n = \frac{4}{\pi n} \int_{-1}^{1} dz \frac{\sqrt{1-z^2}}{1 + x_0 x_1 + (1 - x_0 x_1)z} \times$$
$$U_{n-1}\left\{\frac{1 - x_0 x_1 + (1 + x_0 x_1)z}{1 + x_0 x_1 + (1 - x_0 x_1)z}\right\} \sum_{s=1}^{\infty} sb_s U_{s-1}(z). \tag{2.264}$$

Due to the symmetrical subdivision of the complete interval $[0, \pi]$, the system of Equations (2.263) may be solved by the method developed in Section 2.4, by reducing it to two decoupled dual series equations.

2.8 Coupled series equations

Coupled systems arise in several contexts including elasticity. A recent example is the crack analysis of Martin [39]. Although coupled systems will be briefly encountered in Section 7.5, some general consideration of them is included for completeness. Thus, we consider coupled series equations of the following type,

$$\sum_{n=0}^{\infty} \lambda_n(\alpha,\beta;\eta)\{a(1-r_n)x_n + b(1-s_n)y_n\} P_n^{(\alpha,\beta)}(x) = F_1(x), \tag{2.265}$$

$$\sum_{n=0}^{\infty} \lambda_n(\gamma,\delta;\varepsilon)\{c(1-t_n)x_n + d(1-u_n)y_n\} P_n^{(\gamma,\delta)}(x) = F_2(x), \tag{2.266}$$

$$\sum_{n=0}^{\infty} (1-p_n)x_n P_n^{(\alpha,\beta)}(x) = G_1(t), \tag{2.267}$$

$$\sum_{n=0}^{\infty} (1-q_n)y_n P_n^{(\gamma,\delta)}(x) = G_2(x), \tag{2.268}$$

where the first pair holds for $x \in (-1, x_0)$, and the second pair holds for $x \in (x_0, 1)$. The unknowns x_n, y_n are to be found; the parameters $\alpha, \beta, \eta, \gamma, \delta, \varepsilon$ obey the constraint conditions of Section 2.1 and $\eta, \varepsilon \in (0,1)$. The sequence terms $r_n, s_n, t_n, u_n, p_n, q_n$ vanish as $n \to \infty$. The right-hand sides of these equations have Fourier-Jacobi expansions

$$F_1(x) = \sum_{n=0}^{\infty} \lambda_n(\alpha,\beta;\eta) f_n^1 P_n^{(\alpha,\beta)}(x), \tag{2.269}$$

$$F_2(x) = \sum_{n=0}^{\infty} \lambda_n(\gamma,\delta;\varepsilon) f_n^2 P_n^{(\gamma,\delta)}(x), \tag{2.270}$$

$$G_1(x) = \sum_{n=0}^{\infty} g_n^1 P_n^{(\alpha,\beta)}(x), \tag{2.271}$$

$$G_2(x) = \sum_{n=0}^{\infty} g_n^2 P_n^{(\gamma,\delta)}(x). \tag{2.272}$$

The following regularisation procedure is justified by the same sort of arguments as employed in Section 2.1, and so we omit any discussion of this aspect, and present the formal technique. The systems are nontrivially coupled provided $bc \neq 0$. Without loss of generality, we may suppose that $p_n = q_n = 0$. Multiply (2.265) by $(1+x)^\beta$ and integrate, then use the integral representation of Abel type (1.172) to obtain

$$\sum_{n=0}^{\infty} c_n \left\{a(1-r_n)x_n + b(1-s_n)y_n - f_n^1\right\} P_n^{(\alpha-\eta,\beta+\eta)}(x) = 0, \quad x \in (-1, x_0), \tag{2.273}$$

where

$$c_n = \frac{\Gamma(\alpha+n+1)}{\Gamma(\alpha+n+1-\eta)};$$

similarly, multiply (2.265) by $(1+x)^{\delta}$ and integrate, then use the integral representation (1.172) to obtain

$$\sum_{n=0}^{\infty} d_n \left\{c(1-t_n)x_n + d(1-u_n)y_n - f_n^2\right\} P_n^{(\gamma-\varepsilon,\delta+\varepsilon)}(x) = 0,$$
$$x \in (-1, x_0), \quad (2.274)$$

where

$$d_n = \frac{\Gamma(\gamma+n+1)}{\Gamma(\gamma+n+1-\varepsilon)}.$$

On the other hand, using the integral representation (1.171) for $P_n^{(\alpha,\beta)}$ and $P_n^{(\gamma,\delta)}$, we obtain

$$\sum_{n=0}^{\infty} \left(x_n - g_n^1\right) c_n P_n^{(\alpha-\eta,\beta+\eta)}(x) = 0, \; x \in (x_0, 1), \quad (2.275)$$

$$\sum_{n=0}^{\infty} \left(y_n - g_n^2\right) d_n P_n^{(\gamma-\varepsilon,\delta+\varepsilon)}(x) = 0, \; x \in (x_0, 1). \quad (2.276)$$

Rearrange (2.273) and (2.275) in the form

$$\sum_{n=0}^{\infty} a x_n c_n P_n^{(\alpha-\eta,\beta+\eta)}(x) = \begin{Bmatrix} H_1(x), & x \in (-1, x_0) \\ H_2(x), & x \in (x_0, 1) \end{Bmatrix}, \quad (2.277)$$

where

$$H_1(x) = \sum_{n=0}^{\infty} c_n \left\{a r_n x_n - b(1-s_n) y_n + a f_n^1\right\} P_n^{(\alpha-\eta,\beta+\eta)}(x), \quad (2.278)$$

$$H_2(x) = \sum_{n=0}^{\infty} c_n a g_n^1 P_n^{(\alpha-\eta,\beta+\eta)}(x); \quad (2.279)$$

similarly, rearrange (2.274) and (2.276) in the form

$$\sum_{n=0}^{\infty} d_n d y_n P_n^{(\gamma-\varepsilon,\delta+\varepsilon)}(x) = \begin{Bmatrix} H_3(x), & x \in (-1, x_0) \\ H_4(x), & x \in (x_0, 1) \end{Bmatrix} \quad (2.280)$$

where

$$H_3(x) = \sum_{n=0}^{\infty} d_n \left\{ d u_n y_n - c(1-t_n)\, x_n + f_n^2 \right\} P_n^{(\gamma-\varepsilon,\delta+\varepsilon)}(x), \tag{2.281}$$

$$H_4(x) = \sum_{n=0}^{\infty} d_n \left\{ d g_n^2 \right\} P_n^{(\gamma-\varepsilon,\delta+\varepsilon)}(x). \tag{2.282}$$

A standard orthogonality argument produces the coupled i.s.l.a.e.

$$\begin{aligned} a\,\mathrm{diag}(1-r_n)Ix + aK_1x + bK_2y &= f, \\ d\,\mathrm{diag}(1-u_n)Ix + cK_3x + dK_4y &= g, \end{aligned}$$

where $x = \{x_n\}_{n=1}^{\infty}, y = \{y_n\}_{n=1}^{\infty}$, $\mathrm{diag}(1-r_n)$ and $\mathrm{diag}(1-u_n)$ denote diagonal operators formed from the sequences $\{r_n\}_{n=1}^{\infty}, \{u_n\}_{n=1}^{\infty}$, I denotes the identity operator, and K_1, K_2, K_3, K_4 denote compact operators whose matrix entries are calculated in terms of unnormalised incomplete scalar products $Q_{nm}^{(\gamma-\varepsilon,\delta+\varepsilon)}(x_0)$ and $Q_{nm}^{(\alpha-\eta,\beta+\eta)}(x_0)$; also f, g are explicitly known. Provided $ad \neq 0$, the system is a Fredholm system of second kind; numerically, when the truncation method is used, it has the same advantages as previously noted for uncoupled systems.

2.9 A class of integro-series equations

The approach developed in the previous sections provides a unified treatment for both series and integral equations. It, therefore, provides perhaps the most suitable foundation for investigating a certain class of functional equations, the so-called *integro-series* equations (I.S.E.). This novel class arises from mixed boundary value problems in potential theory or diffraction for structures composed of plane or curvilinear conducting surfaces. Let us briefly describe the type of equations in this class, but defer further description of solution techniques until Section 8.5, where a specific problem of this type concerning a spherical cap and a circular disc, will be encountered.

In operator notation, the integro-series equations take the form

$$L_{11}(u)\{A(\mu)\} + L_{12}\{v(u)\}\{B_n\} = F_1(u), \quad a \le u \le c, \tag{2.283}$$

$$L_{22}(u)\{B_n\} + L_{21}\{u(v)\}\{A(\mu)\} = F_2(v), \quad \alpha \le v \le \gamma, \tag{2.284}$$

where

$$L_{i1}\{A(\mu)\} = \int_a^c A(\mu) \left\{ \begin{array}{l} K_{i1}^{(1)}(u,\mu), \quad a \le u < b \\ K_{i1}^{(2)}(u,\mu), \quad b < u \le c \end{array} \right\} d\mu, \quad i = 1,2 \tag{2.285}$$

and

$$L_{i2}(v)\{B_n\} = \sum_{n=0}^{\infty} B_n \left\{ \begin{array}{l} K_{i2}^{(1)}(v,n), \ \alpha \le v < \beta \\ K_{i2}^{(2)}(v,n), \ \beta < v \le \gamma \end{array} \right\}, \qquad i = 1, 2. \quad (2.286)$$

The solution of the I.S.E. is sought in the standard functional space:

$$\{B_n\}_{n=0}^{\infty} \in l_2, \text{ and } A \in L_2(a,c). \qquad (2.287)$$

The main technical difficulty encountered in solving these equations is the expansion of the kernels defined in (2.285) and (2.286), in terms of eigenfunctions of the Laplace operator in some other coordinate system. Using the relations connecting different coordinate systems, in which the considered shells are described intrinsically as parts of coordinate surfaces $u = u(v)$, $v = v(u)$, these re-expansions take the form

$$K_{i1}^{(1,2)}(u(v),\mu) = \sum_{n=0}^{\infty} C_n^{(1,2)}(\mu)\, K_{i2}^{(1,2)}(v,n) \qquad (2.288)$$

and

$$K_{i2}^{(1,2)}(v(u),u) = \int_a^c D_u^{(1,2)}(\mu)\, K_{i1}^{(1,2)}(u,\mu)\, d\mu \qquad (2.289)$$

The substitution of (2.288) and (2.289) into (2.283)–(2.286), and application of Abel's integral equation method leads ultimately to an I.S.E. of the second kind, which is a perturbation of the identity by a completely continuous operator, in the Cartesian product of functional spaces $l_2 \times L_2(a,c)$. The method is valid for arbitrary location of shells that make no contact or intersection.

When the (imaginary) continuation of that coordinate surface that describes the open shell intersects the real surface of another shell, some technical difficulties may appear. These difficulties are not insurmountable and can be overcome by a correct representation of the desired solution.

3

Electrostatic Potential Theory for Open Spherical Shells

Spherical geometry provides the simplest and most attractive setting for three-dimensional potential theory. The electrostatic potential surrounding a closed conducting sphere on which the surface potential is specified is easily calculated in terms of spherical harmonics; it has an especially simple form if the surface is an equipotential surface. When apertures are introduced, some of this simplicity is retained provided the surface is punctured in a rotationally symmetric fashion.

A single circular aperture, characterised by the angle θ_1 it subtends at the centre of the spherical structure, is the topologically simplest such structure, though rather different forms of the shell appear as θ_1 varies, from the nearly enclosed spherical cavity ($\theta_1 \to 0$) through an open spherical cap ($0 < \theta_1 < \pi$) to a slightly curved circular disc ($\theta_1 \to \pi$). Closed-form solutions that can be obtained for this family of shells by solving an appropriate set of dual series equations, are presented in Section 3.1.

Closed-form solutions do not exist for more complicated shell structures, such as the axisymmetric spherical barrel (in which the spherical surface is punctured by two equal circular holes) or the complementary surface, a pair of spherical caps. Perhaps the best criterion by which to judge a solution is its accuracy and effectiveness for numerical calculation. The potential problem for the barrel (or caps) may be formulated as triple series equations; the regularisation and conversion to a second-kind Fredholm matrix system provides an excellent basis for both approximate analytical estimates as well as precise numerical calculation because the norm of the compact operator occurring the resulting system is small (rather less than unity). Thus, the impact of edges and the influence of the cavity on the potential distribution can be assessed with relative ease. Some examples of the potential distribution around these structures are given in Section 3.2, together with capacitance estimates for the condensor formed from an oppositely charged pair of caps. Section 3.3 extends the triple series approach to a barrel with unequal holes (but located axisymmetrically), and to its complementary surface, a pair of unequally sized spherical caps.

Section 3.4 considers pairs of spherical caps which lie on different but touching spheres. The classical tool of inversion (in an appropriate sphere) produces planar structures. The potential distribution may be described by dual integral equations; these may be regularised to produce a system that is well suited to effective numerical calculation.

A variant of the barrel structures already considered provides a model for a type of electronic lens; this is discussed in some detail in Section 3.5.

The final two sections (3.6 and 3.7) provide a contrast to the previous sections. The magnetostatic potential surrounding superconducting surfaces gives rise to mixed boundary value problems, but Neumann (rather than Dirichlet) boundary conditions are enforced on the spherical surface. However, the resulting series equations are amenable to the standard approach developed in this chapter for spherical geometry, and the magnetic field is determined inside a spherical shell.

3.1 The open conducting spherical shell

The spatial distribution of the electrostatic potential surrounding a charged spherical cap has been investigated by many authors [41], [11], and [25]. As mentioned in the introduction, it provides one of the simplest three-dimensional mixed boundary value problems for Laplace's equation. In this section, we reformulate this well-known problem in terms of dual equations involving Jacobi polynomials. The techniques described in Chapter 2 provide a standard method for the deduction of the solution; furthermore, they provide a rational basis from which more complicated problems may be tackled.

Let $U^0 = U^0(\theta,\varphi)$ be the electrostatic potential that is assumed to be known on the spherical cap, of radius a and subtending an angle θ_0 at the spherical centre (see Figure 3.1). The only requirement on the function U^0 is that it has a Fourier-Legendre series expansion:

$$U^0(\theta,\varphi) = \sum_{m=0}^{\infty} (2-\delta_m^0)\cos m(\varphi-\varphi_0) \sum_{n=m}^{\infty} \alpha_n^m P_n^m(\cos\theta), \tag{3.1}$$

where α_n^m are known (Fourier) coefficients.

We seek a potential $U(r,\theta,\varphi)$, that satisfies the Laplace equation, is continuous across the closed spherical surface $r=a$, and decays at infinity

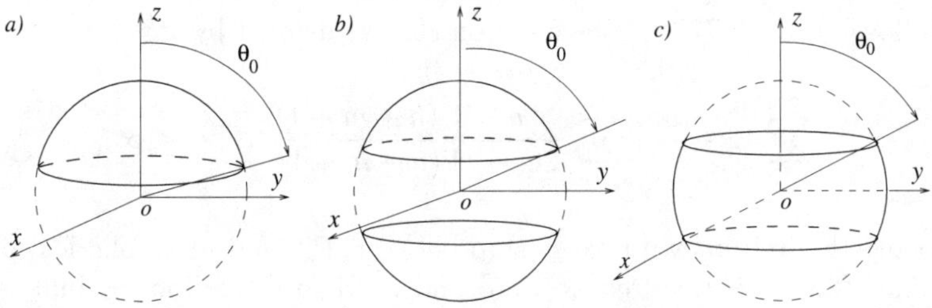

FIGURE 3.1. Spherical shell geometry: (a) the spherical cap, (b) a symmetrical pair of spherical caps, and (c) a symmetrical spherical barrel.

according to $U(r,\theta,\varphi) = O\left(r^{-1}\right)$ as $r \to \infty$. Thus, for suitable A_n^m (to be determined), U has the form

$$\sum_{m=0}^{\infty} \left(2-\delta_m^0\right) \cos m\left(\varphi - \varphi_0\right) \sum_{n=m}^{\infty} A_n^m P_n^m(\cos\theta) \left\{ \begin{array}{c} (r/a)^n, 0 \le r < a \\ (r/a)^{-n-1}, r > a \end{array} \right\}. \tag{3.2}$$

It is clear that, with no loss of generality, we may assume $\varphi_0 = 0$. The mixed boundary conditions to be enforced on the spherical surface $r = a$ are (for $\varphi \in (0, 2\pi)$)

$$U(a,\theta,\varphi) = U^0(\theta,\varphi), \qquad \theta \in (0,\theta_0), \tag{3.3}$$

$$\left[\frac{\partial}{\partial r} U(r,\theta,\varphi)\right]_{r=a-0}^{r=a+0} = 0, \qquad \theta \in (\theta_0, \pi). \tag{3.4}$$

The latter condition (3.4) reflects the continuity of the normal derivative of the potential function across the aperture. Due to the completeness and orthogonality of the set of trigonometric functions $\{\cos m\varphi\}_{m=0}^{\infty}$ on $(0, 2\pi)$, the solution for each index m may be considered independently. Enforcing the mixed boundary conditions leads to the dual series equations

$$\sum_{n=m}^{\infty} A_n^m P_n^m(\cos\theta) = \sum_{n=m}^{\infty} \alpha_n^m P_n^m(\cos\theta), \theta \in (0,\theta_0), \tag{3.5}$$

$$\sum_{n=m}^{\infty} (2n+1) A_n^m P_n^m(\cos\theta) = 0, \qquad \theta \in (\theta_0,\pi). \tag{3.6}$$

Let us determine the solution class for the coefficients A_n^m, guided by the boundedness condition for energy integral (Section 1.3). The integration region is most conveniently chosen as the sphere of radius a, so that

$$W = \int_0^{2\pi} d\varphi \int_0^a dr . r^2 \int_0^{\pi} d\theta \sin\theta \left|\operatorname{grad} U\right|^2 < \infty. \tag{3.7}$$

It follows from (3.7) that the solution class is defined by

$$\sum_{m=0}^{\infty}\left(2-\delta_m^0\right)\sum_{n=m}^{\infty}\frac{n}{2n+1}\frac{\Gamma(n+m+1)}{\Gamma(n-m+1)}\left|A_n^m\right|^2<\infty. \tag{3.8}$$

Using the well-known relationship between P_n^m and ultraspherical polynomials (2.91) we may reduce (3.5) and (3.6) to the following dual series equations, involving Jacobi polynomials as kernels (setting $n-m=s$):

$$\sum_{s=0}^{\infty}X_{s+m}^m(s+m+\frac{1}{2})P_s^{(m,m)}(z)=0, \qquad z\in(-1,z_0) \tag{3.9}$$

$$\sum_{s=0}^{\infty}X_{s+m}^m P_s^{(m,m)}(z)=\sum_{s=0}^{\infty}\beta_{s+m}^m P_s^{(m,m)}(z), \qquad z\in(z_0,1) \tag{3.10}$$

where $z=\cos\theta$, $z_0=\cos\theta_0$ and

$$\left\{\begin{array}{c}X_{s+m}^m\\ \beta_{s+m}^m\end{array}\right\}=\frac{\Gamma(s+2m+1)}{\Gamma(s+m+1)}\left\{\begin{array}{c}A_{s+m}^m\\ \alpha_{s+m}^m\end{array}\right\}. \tag{3.11}$$

Equations of this kind are readily solved by the techniques outlined in Chapter 2. Equations (3.9) and (3.10) may be recognised as equations of the form (2.4), (2.5) with the identification $\alpha=\beta=m$, $f_n=q_n=r_n=0$, and g_n replaced by β_{n+m}; furthermore, $\eta=\frac{1}{2}$, and $\lambda_s\left(m,m;\frac{1}{2}\right)=s+m+\frac{1}{2}$.

From (2.23), we deduce that the analytical solution is

$$Y_{p+m}^m=\sum_{s=0}^{\infty}\hat{\beta}_{s+m}^m\hat{Q}_{sp}^{(m-\frac{1}{2},m+\frac{1}{2})}(z_0) \tag{3.12}$$

where $p=0,1,2,\ldots$, and

$$\left\{Y_{p+m}^m,\hat{\beta}_{p+m}^m\right\}=\frac{\Gamma(p+m+1)}{\Gamma(p+m+\frac{1}{2})}\left[h_p^{(m-\frac{1}{2},m+\frac{1}{2})}\right]^{\frac{1}{2}}\left\{X_{p+m}^m,\beta_{p+m}^m\right\}. \tag{3.13}$$

The normalised incomplete scalar product $\hat{Q}_{sp}^{(\alpha,\beta)}(z_0)$ is defined by (2.24).

In conclusion, it should be noted that the solution belongs to the required class (3.8); this can be proved using the properties (B.171) and (B.172) of the function $\hat{Q}_{sp}^{(\alpha,\beta)}$ (see Appendix B.6).

When the cap is an equipotential surface, only the index 0 coefficient is nonzero, and the summation (3.12) comprises a single term; the solution simplifies to that already obtained in Section 1.4.

3.2 A symmetrical pair of open spherical caps and the spherical barrel

The most striking feature of the problem considered in the previous section is that an analytical, closed form of the electrostatic potential was obtained; this solution is obviously independent of the method used. Such a simple and satisfactory solution cannot be expected for more complicated conductors, such as a pair of charged spherical caps, or a charged spherical shell with two holes (the so-called spherical *barrel*). These structures present very particular cases of *two-body* problems in physics, in which the goal is to calculate effectively the mutual impact of two bodies. In a general situation, the method of successive approximations is used. However, this is effective in only a few situations, for example, objects with dimensions very much smaller than their separation. Such situations are somewhat exceptional. However, if the bodies are identical there is a high degree of symmetry in their mutual impact, so that there is some hope of describing the dominant part of their interaction analytically, even when they are very closely coupled.

In this section we consider two examples of this highly symmetric situation; we produce semi-analytic solutions for the electrostatic potential around a pair of symmetrically located, charged spherical caps and around a spherical barrel. The approach is completely based on the effective procedure of solving triple series equations involving Legendre polynomials, described in Section 2.4.

The geometry is shown in Figures 3.1b and 3.1c. Two spherical caps occupy the region

$$r = a, \quad \theta \in (0, \theta_0) \cup (\pi - \theta_0, \pi),$$

whilst the barrel occupies the complementary portion of the spherical surface defined by

$$r = a, \quad \theta \in (\theta_0, \pi - \theta_0).$$

The conditions that the potential satisfies are similar to those for a single spherical cap (see Section 3.1), except that the given potential is now assumed to be constant over each conductor surface. We seek the rotationally symmetric potential U in the form

$$U = U(r, \theta) = \sum_{n=0}^{\infty} x_n P_n(\cos\theta) \left\{ \begin{array}{l} (r/a)^n, 0 \le r < a \\ (r/a)^{-n-1}, r > a \end{array} \right\} \tag{3.14}$$

where the unknown coefficients $\{x_n\}_{n=0}^{\infty}$ satisfy (cf. (3.8))

$$W = 4\pi a \sum_{n=1}^{\infty} \frac{n}{2n+1} |x_n|^2 < \infty, \tag{3.15}$$

so that $\{x_n\}_{n=0}^{\infty}$ lies in the solution class $l_2 \equiv l_2(0)$.

First let us consider the pair of charged caps, the upper (in the region $z > 0$) and lower being maintained at potentials 1 and ± 1, respectively. Enforcement of the mixed boundary conditions leads to the symmetric triple series equations

$$\begin{cases} \sum_{n=0}^{\infty} x_n P_n(z) = (-1)^l, & z \in (-1, -z_0) \\ \sum_{n=0}^{\infty} \left(n + \frac{1}{2}\right) x_n P_n(z) = 0, & z \in (-z_0, z_0) \\ \sum_{n=0}^{\infty} x_n P_n(z) = 1, & z \in (z_0, 1) \end{cases} \tag{3.16}$$

where $z = \cos\theta$, $z_0 = \cos\theta_0$, and the index l takes the values 0 or 1.

We may use the approach described in Section 2.4, to reduce (3.16) to the following dual series equations involving the Jacobi polynomials $P_n^{(0,l-\frac{1}{2})}$

$$\begin{cases} \sum_{n=0}^{\infty} \left(n + \frac{1}{2}l + \frac{1}{4}\right) x_{2n+l} P_n^{(0,l-\frac{1}{2})}(u) = 0, & u \in (-1, u_0), \\ \sum_{n=0}^{\infty} x_{2n+l} P_n^{(0,l-\frac{1}{2})}(u) = (-1)^l \left\{\frac{1}{2}(1+u)\right\}^{-\frac{l}{2}}, & u \in (u_0, 1), \end{cases} \tag{3.17}$$

where $u = 2z^2 - 1$ and $u_0 = 2z_0^2 - 1 = \cos 2\theta_0$. These equations are now transformed in the standard way to the following infinite systems of linear algebraic equations (i.s.l.a.e.) of the second kind. Denoting

$$b_{2n+l} = \frac{\Gamma(n+1)}{\Gamma\left(n+\frac{1}{2}\right)} \left\{h_n^{(-\frac{1}{2},l)}\right\}^{-\frac{1}{2}} x_{2n+l} \tag{3.18}$$

and

$$\varepsilon_n^l = 1 - \left(n + \frac{l}{2} + \frac{1}{4}\right) \frac{\Gamma\left(n+l+\frac{1}{2}\right)\Gamma\left(n+\frac{1}{2}\right)}{\Gamma(n+l+1)\Gamma(n+1)}, \tag{3.19}$$

so that $\varepsilon_n^l = O\left(n^{-2}\right)$ as $n \to \infty$, the system for the even ($l = 0$) coefficients is

$$\left(1 - \varepsilon_m^0\right) b_{2m} + \sum_{n=0}^{\infty} b_{2n} \varepsilon_n^0 \hat{Q}_{nm}^{(-\frac{1}{2},0)}(u_0) = \frac{2^{\frac{3}{4}}}{\sqrt{\pi}} \hat{Q}_{0m}^{(-\frac{1}{2},0)}(u_0) \tag{3.20}$$

where $m = 0, 1, 2, \ldots$; the system for the odd ($l = 1$) coefficients is

$$\left(1 - \varepsilon_m^1\right) b_{2m+1} + \sum_{n=0}^{\infty} b_{2n+1} \varepsilon_n^1 \hat{Q}_{nm}^{(-\frac{1}{2},1)}(u_0)$$

$$= -2 \left\{\frac{1 - u_0}{\pi\left(m+\frac{1}{2}\right)(m+1)}\right\}^{\frac{1}{2}} \hat{P}_m^{(\frac{1}{2},0)}(u_0), \tag{3.21}$$

where $m = 0, 1, 2, \ldots$; recall that the incomplete scalar products $\hat{Q}_{nm}^{(\alpha,\beta)}$ are defined by Formula (2.24).

It is convenient to rearrange these second-kind systems by replacing the angle parameter u_0 (or θ_0) by $u_1 = -u_0$ $\left(\text{or } \theta_1 = \frac{\pi}{2} - \theta_0\right)$, and using Equation (B.170) (see Appendix) to transform the incomplete scalar products. This leads to the following equivalent i.s.l.a.e., in which the index 0 equations have been separated out. Let

$$c_{2m+l} = (-1)^m \, b_{2m+l} \text{ where } l = 0 \text{ or } 1. \tag{3.22}$$

The even index system is

$$\left\{1 - \varepsilon_0^0 \hat{Q}_{00}^{\left(0,-\frac{1}{2}\right)}(u_1)\right\} c_0 = \frac{2^{\frac{3}{4}}}{\sqrt{\pi}} \left\{1 - \hat{Q}_{00}^{\left(0,-\frac{1}{2}\right)}(u_1)\right\} + \sum_{n=1}^{\infty} c_{2n} \varepsilon_n^0 \hat{Q}_{n0}^{\left(0,-\frac{1}{2}\right)}(u_1), \tag{3.23}$$

and

$$c_{2m} - \sum_{n=1}^{\infty} c_{2n} \varepsilon_n^0 \hat{Q}_{nm}^{\left(0,-\frac{1}{2}\right)}(u_1) = c_0 \varepsilon_0^0 \hat{Q}_{0m}^{\left(0,-\frac{1}{2}\right)}(u_1) - \frac{2^{\frac{3}{4}}}{\sqrt{\pi}} \hat{Q}_{0m}^{\left(0,-\frac{1}{2}\right)}(u_1). \tag{3.24}$$

for $m = 1, 2, \ldots$. The odd index system comprises

$$\left\{1 - \varepsilon_0^1 \hat{Q}_{00}^{\left(1,-\frac{1}{2}\right)}(u_1)\right\} c_1 = -\frac{2\sqrt{2}}{\sqrt{\pi}} (1 + u_1)^{\frac{1}{2}} \hat{P}_0^{\left(1,-\frac{1}{2}\right)}(u_1) + \sum_{n=1}^{\infty} c_{2n+1} \varepsilon_n^1 \hat{Q}_{n0}^{\left(1,-\frac{1}{2}\right)}(u_1), \tag{3.25}$$

and

$$c_{2m+1} - \sum_{n=1}^{\infty} c_{2n+1} \varepsilon_n^1 \hat{Q}_{nm}^{\left(1,-\frac{1}{2}\right)}(u_1) = c_1 \varepsilon_0^1 \hat{Q}_{0m}^{\left(1,-\frac{1}{2}\right)}(u_1) - 2 \left\{\frac{1 + u_1}{\pi \left(m + \frac{1}{2}\right)(m+1)}\right\}^{\frac{1}{2}} \hat{P}_m^{\left(0,-\frac{1}{2}\right)}(u_1). \tag{3.26}$$

for $m = 1, 2, \ldots$. Formulae (3.23) and (3.25) provide the values of c_0 and c_1 for replacement in (3.24) and (3.26) producing systems for $\{c_{2n}\}_{n=1}^{\infty}$ and $\{c_{2n+1}\}_{n=1}^{\infty}$. Bounds, which are uniform in the parameter u_1, on the norms p and q of the completely continuous operators of the systems (3.24) and (3.26) are

$$\begin{aligned} q &\leq \max \left|\varepsilon_n^1\right| = \left|\varepsilon_1^1\right| = 1 - \tfrac{5\pi}{16} \simeq 0.018 \ll 1, \\ p &\leq \max \left|\varepsilon_n^1\right| = \left|\varepsilon_1^1\right| = \left|1 - \tfrac{21\pi}{64}\right| \simeq 0.031 \ll 1. \end{aligned} \tag{3.27}$$

(This estimate follows from the observation that the matrix operator with components $\hat{Q}_{nm}^{(l,-\frac{1}{2})}$ is a projection operator of norm at most unity.) Thus, the systems (3.24) and (3.26) can be solved very effectively by the method of successive approximations for any value of the parameter u_1 (or θ_1). Approximate analytical expressions for capacitance given at the end of this section are derived in this way.

Let us now turn attention to the charged spherical barrel. Assume that this doubly-connected conductor is charged to unit potential, i.e.,

$$U(a,\theta)=1, \qquad \theta\in(\theta_0,\pi-\theta_0). \tag{3.28}$$

Following a similar procedure to the above produces the dual series equations

$$\begin{cases} \sum_{n=0}^{\infty}\left(n+\frac{1}{4}\right)b_n P_n^{(-\frac{1}{2},0)}(u)=0, & u\in(-1,u_1) \\ \sum_{n=0}^{\infty} b_n P_n^{(-\frac{1}{2},0)}(u)=1, & u\in(u_1,1) \end{cases} \tag{3.29}$$

where $b_n=(-1)^n x_{2n}$ and $u_1=-u_0$ $(\theta_1=\frac{\pi}{2}-\theta_1)$. A preliminary integration is necessary to transform these equations to the standard form considered in Section 2.1.

$$\sum_{n=1}^{\infty}\frac{n+\frac{1}{4}}{n}b_n P_{n-1}^{(\frac{1}{2},1)}(u) = \frac{\sqrt{2}-(1-u)^{\frac{1}{2}}}{(1-u)^{\frac{1}{2}}(1+u)}b_0, \quad u\in(-1,u_1) \tag{3.30}$$

$$\sum_{n=1}^{\infty}\frac{b_n}{n}P_{n-1}^{(\frac{1}{2},1)}(u) = \frac{4}{1+u}(1-b_0), \qquad u\in(u_1,1) \tag{3.31}$$

The final format of the solution is deduced from the results of Section 2.1; omitting details, it is

$$d_s-\sum_{n=1}^{\infty}d_n\mu_n\left\{\hat{Q}_{n-1,m-1}^{(1,\frac{1}{2})}(u_1)+\frac{2\sqrt{2}}{\alpha(u_1)}Q_n(u_1)Q_m(u_1)\right\} = \frac{2\sqrt{2}}{\alpha(u_1)}Q_s(u_1), \tag{3.32}$$

where $s=1,2,\ldots$; the coefficient b_0 is computed from the formula

$$b_0=(\alpha(u_1))^{-1}\left\{1+\sum_{n=1}^{\infty}d_n\mu_n Q_n(u_1)\right\} \tag{3.33}$$

Furthermore,

$$d_n=\left(n+\frac{1}{4}\right)h_{n-1}^{(0,\frac{3}{2})}\left\{h_{n-1}^{(1,\frac{1}{2})}\right\}^{-\frac{1}{2}}\frac{\Gamma(n+1)}{\Gamma\left(n+\frac{3}{2}\right)}b_n,$$

$$\mu_n = 1 - \frac{n\left(n+\frac{1}{2}\right)}{n+\frac{1}{4}}\left[\frac{\Gamma\left(n+\frac{1}{2}\right)}{\Gamma\left(n+1\right)}\right]^2 = O\left(n^{-2}\right) \text{ as } n \to \infty, \tag{3.34}$$

$$\alpha\left(u_1\right) = 1 - \frac{1}{\pi}\left(\frac{1+u_1}{2}\right)^{\frac{1}{2}} - \frac{1}{2\pi}\ln\left[\frac{1-\sqrt{\left(1+u_1\right)/2}}{1+\sqrt{\left(1+u_1\right)/2}}\right],$$

and

$$Q_n\left(u_1\right) = \frac{1}{\sqrt{\pi}}\left(\frac{1+u_1}{2}\right)^{\frac{3}{2}}\frac{\hat{P}_{n-1}^{\left(0,\frac{3}{2}\right)}\left(u_1\right)}{\sqrt{n\left(n+\frac{l}{2}\right)}}.$$

The norm of the compact operator H associated with the system (3.32) has the bound

$$\|H\| \le \max\left|\mu_n\right| = \mu_1 = 1 - \frac{3\pi}{10} \simeq 0.057 \ll 1; \tag{3.35}$$

this estimate is uniform in the parameter u_1. Hence, the solution of the system (3.32) is effectively computed by means of successive approximations for any value of the parameter u_1.

We shall now calculate capacitances of these structures. The capacitance C is related to the total charge q on a conductor at potential U by

$$q = CU.$$

Thus, at unit potential, the capacitance C numerically coincides with value of the charge q. Charge is determined by integration of the surface charge density σ on the conductor surface; it is proportional to the jump in the normal component of the electric field $\overrightarrow{E} = \operatorname{grad} U$ on the conductor surface

$$\sigma\left(\theta\right) = \frac{1}{4\pi}\left\{E_r\left(a+0,\theta\right) - E_r\left(a-0,\theta\right)\right\}.$$

(This follows from Equation (1.2).) The concrete expression for σ is

$$\sigma\left(\theta\right) = \frac{1}{4\pi a}\sum_{n=0}^{\infty}\left(2n+1\right)x_n P_n\left(\cos\theta\right). \tag{3.36}$$

3.2.1 *Approximate analytical formulae for capacitance*

Let us first consider two caps at equal potential ($l = 0$). The charge $q_{1,1}$ on each spherical cap is determined by integration of the function $\sigma\left(\theta\right)$ over the appropriate portion of the spherical surface $r = a$:

$$q_{1,1} = \frac{1}{2}ax_0 = 2^{-\frac{7}{4}}\sqrt{\pi}ab_0. \tag{3.37}$$

θ_1	0°	10°	20°	30°	40°
$a^{-1}q_{1,1}^{(0)}$	0.5	0.49401	0.47600	0.44583	0.40326
$a^{-1}q_{1,1}^{(1)}$	0.5	0.49399	0.47583	0.44525	0.40220
θ_1	50°	60°	70°	80°	90°
$a^{-1}q_{1,1}^{(0)}$	0.34807	0.28004	0.19912	0.10554	0
$a^{-1}q_{1,1}^{(1)}$	0.34654	0.27835	0.19778	0.10498	0

TABLE 3.1. Approximate capacitances of the charged cap pair.

From the trivial approximation ($c_{2n}^0 = 0$) one readily obtains from (3.24) the approximation for c_0 :

$$c_0^{(0)} \approx \frac{2^{\frac{3}{4}}}{\sqrt{\pi}} \frac{1 - \hat{Q}_{00}^{(0,-\frac{1}{2})}(u_1)}{1 - \varepsilon_0^0 \hat{Q}_{00}^{(0,-\frac{1}{2})}(u_1)} = \frac{2^{\frac{3}{4}}}{\sqrt{\pi}} \cdot \frac{\cos\theta_1}{1 - \left(1 - \frac{\pi}{4}\right)(1 - \cos\theta_1)}. \tag{3.38}$$

Substituting (3.38) in (3.37) produces the approximation

$$q_{1,1} \approx \frac{1}{2} a \frac{\cos\theta_1}{1 - \left(1 - \frac{\pi}{4}\right)(1 - \cos\theta_1)}. \tag{3.39}$$

The simplest approximation for the capacitance of this pair of conductors is thus

$$C_{1,1}^{(0)} = 2q_{1,1}^{(0)} = a \frac{\cos\theta_1}{1 - \left(1 - \frac{\pi}{4}\right)(1 - \cos\theta_1)}. \tag{3.40}$$

It is worth noting that the same problem was solved in [42]. Despite obtaining a Fredholm integral equation of the second kind (which in itself does not guarantee solution effectiveness), further analytical investigation was impossible because the solution was highly dependent on the cap dimensions; only numerical results were obtained. Let us make some comparison of results (those of [42] are given in parentheses): when $\theta_1 = 60°$, $a^{-1}q_{1,1}^{(0)} = 0.280\,(0.278)$; when $\theta_1 = 30°$, $a^{-1}q_{1,1}^{(0)} = 0.445\,(0.445)$. Formula (3.40) is thus appealing in its simplicity and relatively good accuracy, demonstrating the advantages of the method presented here.

The first successive approximation provides a more accurate estimate of capacitance (or charge), and an approximate analytical expression for the potential distribution; we obtain the following approximation for the coefficients:

$$b_{2m}^{(1)} = \left(b_0 \varepsilon_0^0 - \frac{2^{\frac{3}{4}}}{\sqrt{\pi}} \right) (-1)^m \hat{Q}_{0m}^{(0,-\frac{1}{2})}(u_1), \qquad (m = 1, 2, \ldots). \tag{3.41}$$

In this approximation the charge is

$$q_{1,1}^{(1)} = \frac{1}{2} a \frac{\cos\theta_1 - \beta(\theta_1)}{1 - \left(1 - \frac{\pi}{4}\right)\{1 - \cos\theta_1 + \beta(\theta_1)\}} \tag{3.42}$$

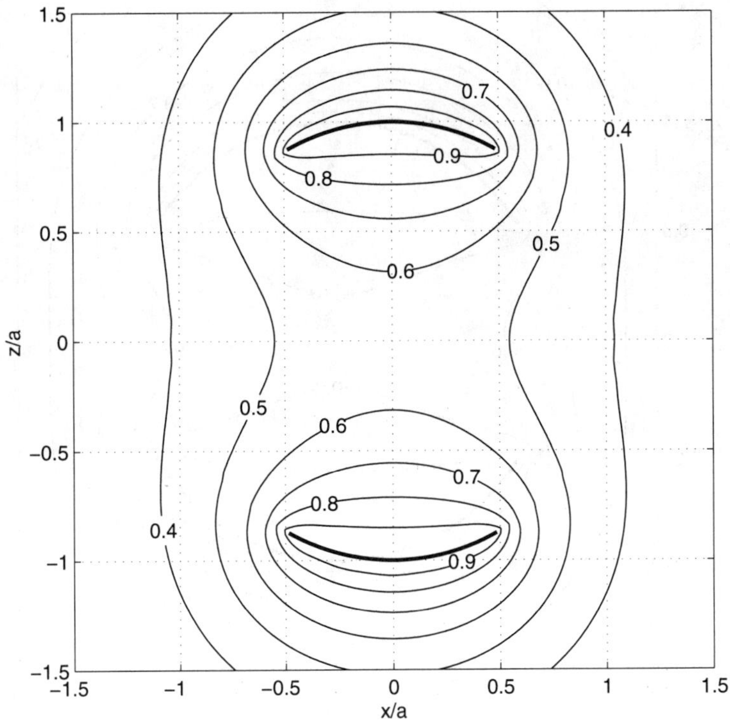

FIGURE 3.2. Electrostatic potential near a pair of symmetrical spherical caps charged to unit potential with subtended angle $\theta_0 = 30^o$. Truncation number $N_{tr} = 11$.

where

$$\beta(\theta_1) = \sum_{n=1}^{\infty} \varepsilon_n^0 \hat{Q}_{n0}^{\left(0,-\frac{1}{2}\right)}(u_1)\, \hat{Q}_{0n}^{\left(0,-\frac{1}{2}\right)}(u_1). \tag{3.43}$$

An approximation for the function β with relative error not exceeding 3.10^{-4} is

$$\beta(\theta_1) \approx \sum_{n=1}^{5} \Delta\varepsilon_n^0 \left[P_{2n-1}(\cos\theta_1) - P_{2n+1}(\cos\theta_1)\right]^2 / (4n+1) +$$

$$\frac{1}{8}\left\{-2\cos\theta_1(1-\cos\theta_1)\ln 2 + \frac{1}{2}\sin^2\theta_1 - \frac{1}{2}(1-\cos\theta_1)^2 \ln(1-\cos\theta_1)\right\}$$

$$+\frac{1}{8}\left\{2\cos\theta_1 \ln 2 - \cos\theta_1(1-\cos\theta_1) - \cos\theta_1(1+\cos\theta_1)\ln(1+\cos\theta_1)\right\}$$

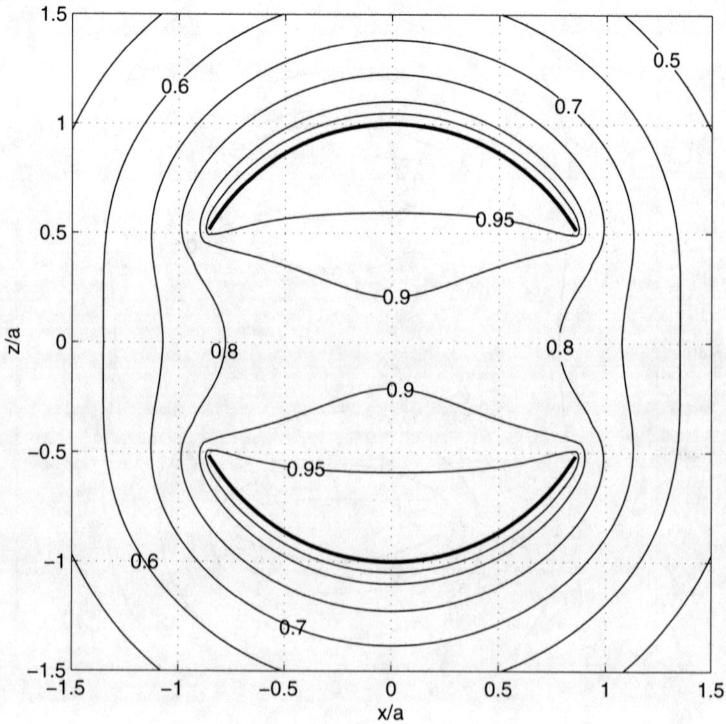

FIGURE 3.3. Electrostatic potential near a symmetrical pair of spherical caps charged to unit potential with subtended angle $\theta_0 = 60^o$. Truncation number $N_{tr} = 11$.

where

$$\Delta\varepsilon_n^0 = \varepsilon_n^0 - \frac{1}{16n\,(2n+1)}. \tag{3.44}$$

Formulae (3.40) and (3.42) were used to calculate $q_{1,1}^{(0)}$ and $q_{1,1}^{(1)}$, respectively. The values of $q_{1,1}^{(1)}$ agree perfectly with data in [42]. Some computed results are presented in Table 3.1.

The spatial distribution of the potential U computed from (3.14), after solution of (3.24) is displayed in Figures 3.2 and 3.3 for the pair of caps, at unit potential with angle parameter $\theta_0 = 30°$ and $\theta_0 = 60°$, respectively. It is evident that mutual coupling of the electrostatic fields produced by the smaller pair of charged caps is small. The larger pair exhibits much stronger coupling; the resultant field appears not as the composition of two individual fields, but as a single electrostatic field surrounding the entire structure.

Furthermore, these figures illustrate that well-separated small caps might be readily analysed by a method of successive approximations, utilising the

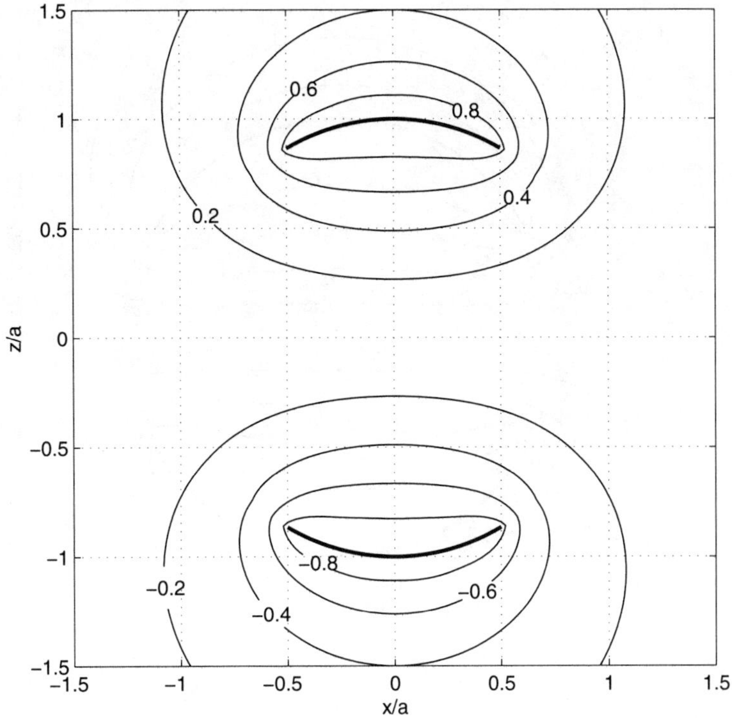

FIGURE 3.4. Electrostatic potential near a spherical condensor with subtended angle $\theta_0 = 30^o$. Truncation number $N_{tr} = 11$.

known potential of a single isolated charged cap. However, such an approach will fail for larger caps (Figure 3.3); the choice of method applied is critical in producing an efficient mathematical tool for analytical treatment of the problem.

When oppositely charged ($l = 1$), the caps form a capacitor or condensor. The charge on the lower cap is

$$q_{-1,1} = \frac{1}{2}a\sum_{n=0}^{\infty} x_{2n+1}\left[P_{2n}(0) - P_{2n+2}(0)\right] =$$

$$\frac{1}{2}a\sum_{n=0}^{\infty}(-1)^n c_{2n+1}\frac{\Gamma\left(n+\frac{1}{2}\right)}{\Gamma(n+1)}\left[h_n^{(-\frac{1}{2},1)}\right]^{-\frac{1}{2}}\left[P_{2n}(0) - P_{2n+2}(0)\right]. \quad (3.45)$$

The first approximation in solving Equations (3.25) and (3.26) produces

$$c_1^{(1)} = -2^{\frac{3}{4}}\left(\frac{3}{\pi}\right)^{\frac{1}{2}}\frac{\cos\theta_1}{1+(\frac{3\pi}{8}-1)(1-\frac{3}{2}\cos\theta_1+\frac{1}{2}\cos^3\theta_1)},$$

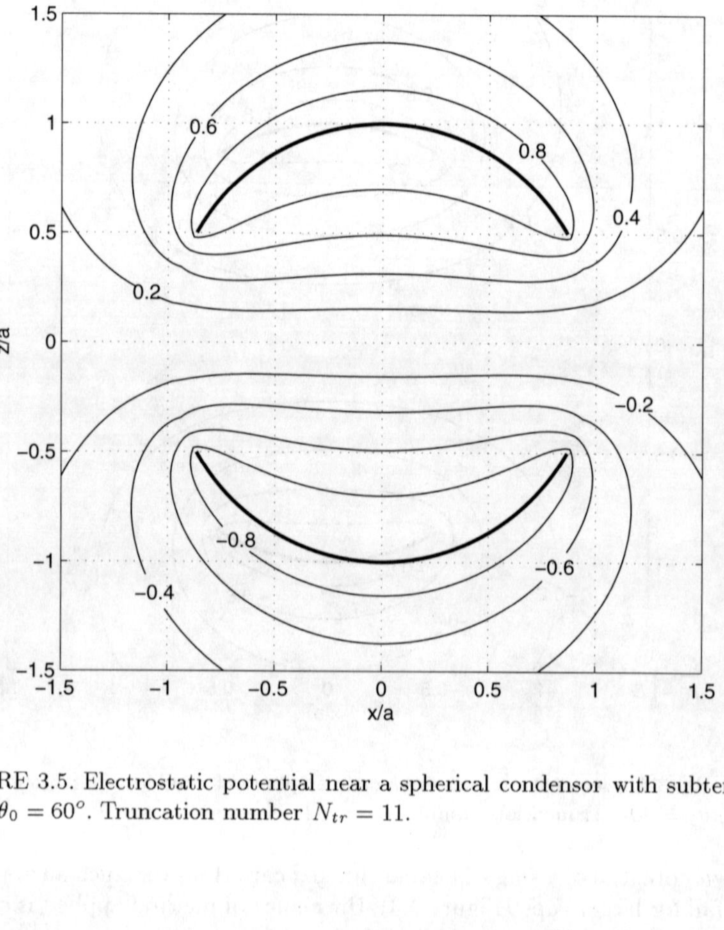

FIGURE 3.5. Electrostatic potential near a spherical condensor with subtended angle $\theta_0 = 60^o$. Truncation number $N_{tr} = 11$.

and

$$c_{2n+1}^{(1)} = c_1 \varepsilon_0^1 \hat{Q}_{0n}^{\left(1,-\frac{1}{2}\right)}(u_1) - 2\left\{\frac{1+u_1}{\pi\left(n+\frac{1}{2}\right)(n+1)}\right\}^{\frac{1}{2}} \hat{P}_n^{\left(0,\frac{1}{2}\right)}(u_1). \quad (3.46)$$

Substitution of these values in the formula (3.45) yields an approximate analytical expression for $q_{-1,1}$:

$$\begin{aligned} q_{-1,1}^{(1)} = &-\frac{1}{2}a\sum_{n=1}^{\infty}\left[P_{2n}(0) - P_{2n+2}(0)\right]^2 P_{2n+1}(\cos\theta_1) \\ &-\frac{9}{8}a\cos\theta_1\left\{1 - \left(\frac{3\pi}{8}-1\right)\left(1 - \frac{3}{2}\cos\theta_1 + \frac{1}{2}\cos^3\theta_1\right)\right\}^{-1} \\ &\times\left\{1 - \frac{2}{3}\left(\frac{3\pi}{8}-1\right)\sum_{n=1}^{\infty}\left[P_{2n}(0) - P_{2n+2}(0)\right]^2 V_n(\cos\theta_1)\right\} \quad (3.47)\end{aligned}$$

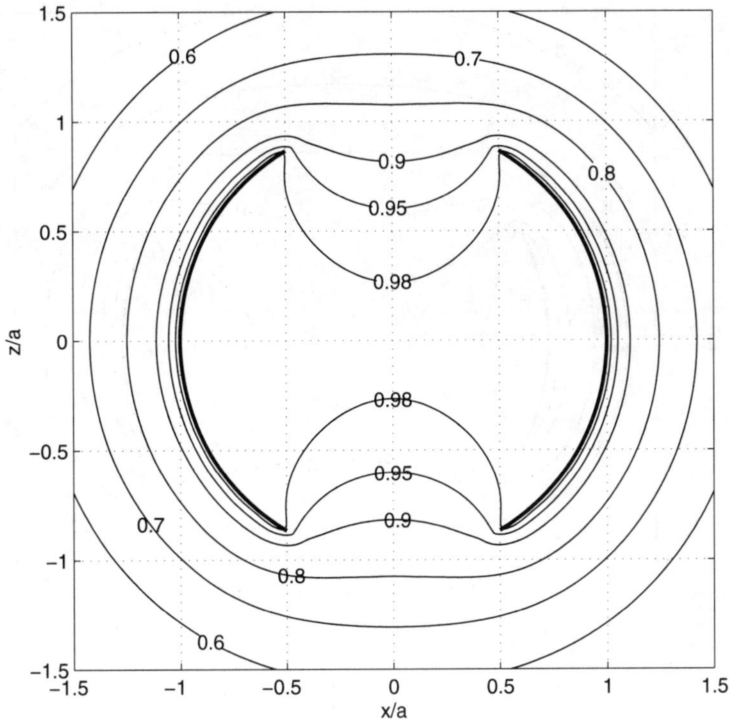

FIGURE 3.6. Electrostatic potential near a spherical barrel charged to unit potential with aperture subtending angle $\theta_0 = 30^o$. Truncation number $N_{tr} = 11$.

where

$$\begin{aligned} V_n(\cos\theta_1) = & -\sin^2\theta_1 P_{2n+1}(\cos\theta_1) \\ & -2(4n+3)^{-1}\cos\theta_1\left[P_{2n+2}(\cos\theta_1) - P_{2n}(\cos\theta_1)\right] \\ & +2(4n+3)^{-1}(4n+5)^{-1}\left[P_{2n+3}(\cos\theta_1) - P_{2n+1}(\cos\theta_1)\right] \\ & -2(4n+3)^{-1}(4n+1)^{-1}\left[P_{2n+1}(\cos\theta_1) - P_{2n-1}(\cos\theta_1)\right]. \end{aligned}$$

Some calculated results are reproduced in Table 3.2. A comparison of the tabulated results with those obtained by numerical solution of (3.25) and (3.26) shows that Formula (3.47) is accurate to three significant digits (over the whole range of θ_1). The spatial distribution of the potential around capacitors with angle parameter $\theta_0 = 30°$ and $60°$ are shown in Figures 3.4 and 3.5, respectively. This was computed from (3.14) after solution of (3.26).

Finally, we calculate the capacitance of the spherical barrel. The charge q_1, and hence the capacitance of the doubly-connected spherical barrel

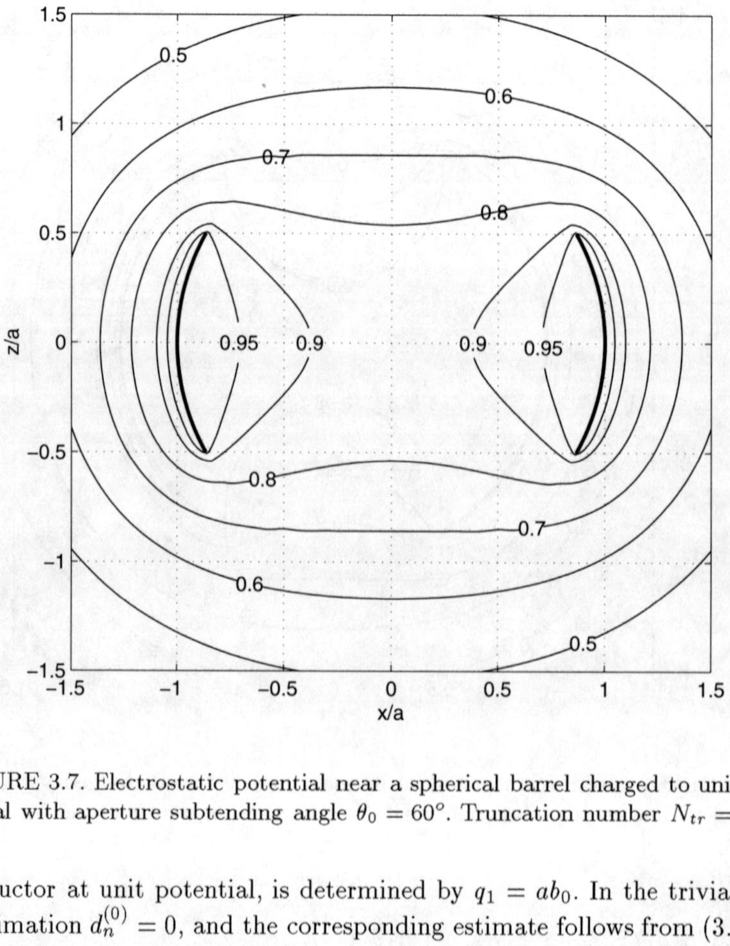

FIGURE 3.7. Electrostatic potential near a spherical barrel charged to unit potential with aperture subtending angle $\theta_0 = 60^o$. Truncation number $N_{tr} = 11$.

conductor at unit potential, is determined by $q_1 = ab_0$. In the trivial approximation $d_n^{(0)} = 0$, and the corresponding estimate follows from (3.32) :

$$q_1^{(0)} = c_1^{(0)} = a\left(1 - \frac{1}{\pi}\cos\theta_1 - \frac{1}{2\pi}\ln\left[\frac{1-\cos\theta_1}{1+\cos\theta_1}\right]\right)^{-1}. \tag{3.48}$$

In the limiting case of free space $(\theta_1 = 0)$, Formula (3.48) produces the expected result that $q_1^{(0)} = 0$. For the other limiting case of a closed spherical shell $\left(\theta_1 = \frac{\pi}{2}\right)$, it produces the expected result $q_1^{(0)} = a$. A thin cylin-

θ_1	10°	20°	30°	40°
$a^{-1}q_{-1,1}^{(1)}$	-1.729	-1.262	-0.967	-0.739
θ_1	50°	60°	70°	80°
$a^{-1}q_{-1,1}^{(1)}$	-0.549	-0.391	-0.241	-0.113

TABLE 3.2. Total charge on the lower cap of the spherical condensor.

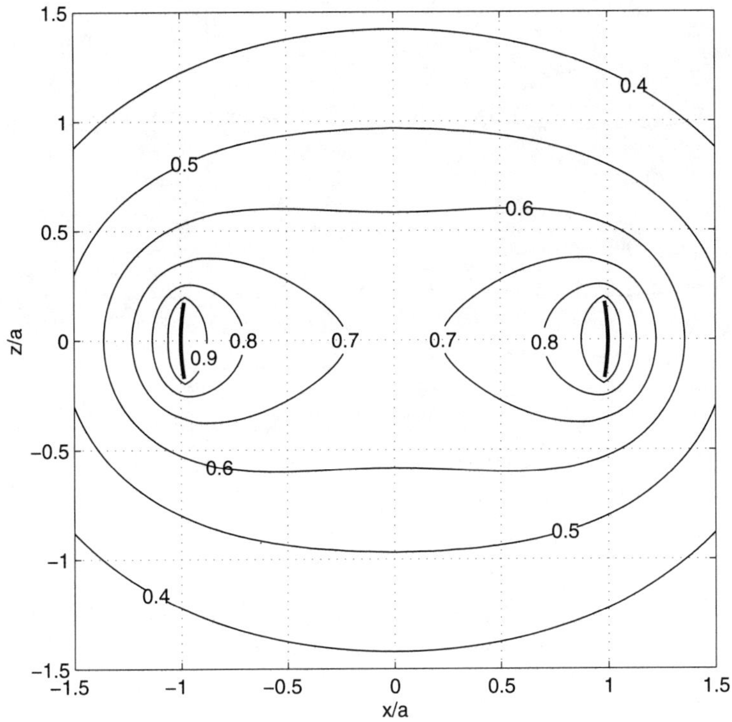

FIGURE 3.8. Electrostatic potential near a spherical barrel or ring charged to unit potential with aperture subtending angle $\theta_0 = 80^\circ$. Truncation number $N_{tr} = 11$.

drical ring ($\theta_1 \ll 1$) has the approximate charge

$$q_1^{(0)} \cong \frac{\pi a}{\pi - 1 + \ln(2/\theta_1)} \cong \frac{\pi a}{0.07 + \ln(16/\theta_1)}$$

where we have employed the approximation $\pi - 1 \cong \ln 8 + 0.07$. This is very close to the known result for the charge on a narrow cylindrical ring [29]. The estimate of q_1 improves with the next approximation. Sample calculations of $q_1^{(0)}$ are reproduced in Table 3.3.

The distribution of the electrostatic potential surrounding three differently shaped barrels ($\theta_0 = 30^\circ, 60^\circ$, and 80°) is displayed in Figures 3.6, 3.7, and 3.8. This was computed from (3.14) after solving (3.32). As might

θ_1	10°	20°	30°	40°	50°	60°	70°	80°
$a^{-1}q_1^{(0)}$	0.683	0.797	0.874	0.927	0.963	0.984	0.995	0.999

TABLE 3.3. Approximate capacitance of the spherical barrel as a function of angle $\theta_1 = \frac{\pi}{2} - \theta_0$.

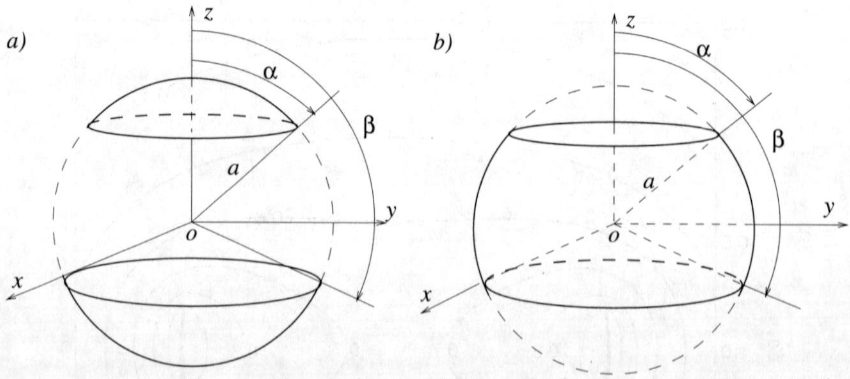

FIGURE 3.9. (a) An asymmetrical pair of spherical caps, (b) an asymmetric spherical barrel.

be expected, the potential is nearly constant inside the largest barrel. When the angle $\theta_0 = 80°$, the spherical barrel becomes a "ring."

3.3 An asymmetrical pair of spherical caps and the asymmetric barrel

In the previous section we considered two spherical caps that subtended equal angles at the origin of the common spherical surface on which they both lie. Retaining axial symmetry of the entire structure, we now allow the caps to subtend different angles, α and $\pi - \beta$, as shown in Figure 3.9(a). When charged to constant potential, the standard formulation of this boundary value problem for Laplace's equation produces the nonsymmetric triple series equations (which are similar to the symmetric triple equations of the previous section)

$$\begin{cases} \sum_{n=0}^{\infty} a_n P_n(\cos\theta) = 1, & \theta \in (0, \alpha) \\ \sum_{n=0}^{\infty} (2n+1)\, a_n P_n(\cos\theta) = 0, & \theta \in (\alpha, \beta) \\ \sum_{n=0}^{\infty} a_n P_n(\cos\theta) = 1, & \theta \in (\beta, \pi). \end{cases} \tag{3.49}$$

Proceeding as in Section 2.7, we may transform the Equations (3.49) to the equivalent symmetric triple series equations

$$\sum_{n=0}^{\infty} b_n P_n(x) = \left\{\frac{2\rho_0\rho_1}{1+\rho_0\rho_1+(1-\rho_0\rho_1)x}\right\}^{\frac{1}{2}},$$
$$x \in (-1,-x_0)\cup(x_0,1) \tag{3.50}$$

$$\sum_{n=0}^{\infty}(2n+1)b_n P_n(x) = 0, \qquad x \in (-x_0,x_0). \tag{3.51}$$

Here

$$x_0 = \frac{\rho_0-\rho_1}{\rho_0+\rho_1} = \frac{\sin\Delta}{\sin\Delta_0},$$

where the parameter $\Delta_0 = \frac{1}{2}(\alpha+\beta)$ is the angular coordinate of the middle of the slot and $\Delta = \frac{1}{2}(\beta-\alpha)$ is its semi-width.

The right-hand side of (3.50) has the Fourier-Legendre expansion

$$\left\{\frac{2\rho_0\rho_1}{1+\rho_0\rho_1+(1-\rho_0\rho_1)x}\right\}^{\frac{1}{2}} = \sum_{n=0}^{\infty} d_n P_n(x) \tag{3.52}$$

where

$$d_n = \sqrt{\frac{\cos\Delta-\cos\Delta_0}{\cos\Delta+\cos\Delta_0}}\left[1-q(\Delta,\Delta_0)\right]q^n(\Delta,\Delta_0) \tag{3.53}$$

and

$$q(\Delta,\Delta_0) = \frac{\sqrt{\cos^2\Delta-\cos^2\Delta_0}-\cos\Delta}{\cos\Delta_0}.$$

In calculating the coefficients d_n we used the integral

$$\int_{-1}^{1}\frac{P_s(z)}{\sqrt{a+bz}}dz = \frac{1}{(s+\frac{1}{2})\sqrt{a+b}}\left[1-q(a,b)\right]q^s(a,b), \qquad a>b \tag{3.54}$$

which may be obtained from the Dirichlet-Mehler integral representation for the Legendre polynomials P_n and the tabulated definite integral [19]

$$\int_0^{\pi}\frac{\cos(sx)}{a+b\cos x}dx = \frac{\pi}{\sqrt{a^2-b^2}}\left(\frac{\sqrt{a^2-b^2}-a}{b}\right)^s, \qquad a>b.$$

In contrast to the symmetrical case where the final solution requires only even or only odd coefficients (according as the pair of shells are equally or oppositely charged), the solution to the nonsymmetrical structure requires both even ($l = 0$) and odd ($l = 1$) coefficients. The systems are solved separately and the results are combined. Based on results obtained in the previous section, the final form of the solution is (for $l = 0, 1$) :

$$c_l \left[1 - \varepsilon_0^l \hat{Q}_{00}^{(l,-\frac{1}{2})}(u_1) \right] = f_l \left[1 - \hat{Q}_{00}^{(l,-\frac{1}{2})}(u_1) \right] + \sum_{n=1}^{\infty} \left(c_{2n+l}\varepsilon_n^l - f_{2n+l} \right) \hat{Q}_{n0}^{(l,-\frac{1}{2})}(u_1), \quad (3.55)$$

and when $m \geq 1$,

$$c_{2m+l} - \sum_{n=1}^{\infty} \left(c_{2n+l}\varepsilon_n^l - f_{2n+l} \right) \hat{Q}_{nm}^{(l,-\frac{1}{2})}(u_1) = f_{2m+l} + \left(c_l \varepsilon_0^l - f_l \right) \hat{Q}_{0m}^{(l,-\frac{1}{2})}(u_1) \quad (3.56)$$

where

$$u_1 = 1 - 2\frac{\sin^2 \Delta}{\sin^2 \Delta_0},$$

$$\left\{ \begin{array}{c} f_{2n+l} \\ c_{2n+l} \end{array} \right\} = (-1)^n \frac{\Gamma(n+1)}{\Gamma\left(n+\frac{1}{2}\right)} \left\{ h_n^{(-\frac{1}{2},l)} \right\}^{\frac{1}{2}} \left\{ \begin{array}{c} d_{2n+l} \\ b_{2n+l} \end{array} \right\}, \quad n = 0, 1, 2, \ldots,$$

and the rest of the notation coincides with that used in the previous section. Just as for equally sized caps, this problem may be effectively solved by the method of successive approximations because the same estimates of the norm given by (3.27) are valid.

Suppressing some details, the electrostatic potential of the charged nonsymmetrical spherical barrel (displayed in Figure 3.9(b)) also leads to the nonsymmetric triple series equations

$$\begin{aligned} \sum_{n=0}^{\infty} (2n+1) a_n P_n(\cos\theta) &= 0, \quad \theta \in (0,\alpha) \cup (\beta,\pi) \\ \sum_{n=0}^{\infty} a_n P_n(\cos\theta) &= 1, \quad \theta \in (\alpha,\beta), \end{aligned} \quad (3.57)$$

which are converted in the usual way to the dual series equations ($l = 0, 1$)

$$\begin{aligned} \sum_{n=0}^{\infty} \left(b_{2n+l} - d_{2n+l} \right) P_n^{(0,l-\frac{1}{2})}(u) &= 0, \quad u \in (-1, u_0) \\ \sum_{n=0}^{\infty} \left(n + \tfrac{1}{2}l + \tfrac{1}{4} \right) b_{2n+l} P_n^{(0,l-\frac{1}{2})}(u) &= 0, \quad u \in (u_0, 1) \end{aligned} \quad (3.58)$$

where $u_0 = \cos 2\theta_0$. The standard solution process eventually yields a fast converging i.s.l.a.e. of the second kind for the Fourier coefficients. The odd $(l = 1)$ index system is

$$B_m - \sum_{n=0}^{\infty} B_n \tau_n \hat{Q}_{nm}^{\left(\frac{1}{2},0\right)}(u_0) = D_m - \sum_{n=0}^{\infty} D_n \hat{Q}_{nm}^{\left(\frac{1}{2},0\right)}(u_0) \tag{3.59}$$

where

$$\tau_n = 1 - \left(n + \frac{3}{4}\right)\left[\frac{\Gamma(n+1)}{\Gamma\left(n+\frac{3}{2}\right)}\right]^2 = O\left(n^{-2}\right) \text{ as } n \to \infty,$$

and

$$\{B_n, D_n\} = \frac{\Gamma\left(n+\frac{3}{2}\right)}{\Gamma(n+1)}\left\{h_n^{\left(\frac{1}{2},0\right)}\right\}^{\frac{1}{2}}\{b_{2n+1}, d_{2n+1}\};$$

the even $(l = 0)$ index system is

$$G_m - \sum_{n=1}^{\infty} G_n \mu_n \left\{\hat{Q}_{n-1,m-1}^{\left(1,\frac{1}{2}\right)}(u_1) + 2\sqrt{2}\{\gamma(u_1)\}^{-1} Q_n(u_1) Q_m(u_1)\right\}$$
$$= 2\sqrt{2}\{\gamma(u_1)\}^{-1} D_0 Q_m(u_1) +$$
$$\sum_{n=1}^{\infty} D_n \left\{\hat{Q}_{n-1,m-1}^{\left(1,\frac{1}{2}\right)}(u_1) + 2\sqrt{2}\{\gamma(u_1)\}^{-1} Q_n(u_1) Q_m(u_1)\right\} \tag{3.60}$$

where $\gamma(u_1) \equiv \alpha(u_1)$,

$$\gamma(u_1) b_0 = D_0 + \sum_{n=1}^{\infty} (G_n \mu_n + D_n) Q_n(u_1),$$

$$G_m = \frac{m+\frac{1}{4}}{m+\frac{1}{2}} \frac{\Gamma(m+1)}{\Gamma\left(m+\frac{1}{2}\right)} h_{m-1}^{\left(0,\frac{3}{2}\right)} \left\{h_{m-1}^{\left(1,\frac{1}{2}\right)}\right\}^{-\frac{1}{2}} (-1)^m b_{2m},$$

and

$$D_m = (-1)^m d_{2m} \frac{\Gamma\left(m+\frac{1}{2}\right)}{\Gamma(m)} h_{m-1}^{\left(0,\frac{3}{2}\right)} \left\{h_{m-1}^{\left(1,\frac{1}{2}\right)}\right\}^{-\frac{1}{2}}.$$

The remaining notation coincides with that which we used in the solution of the spherical barrel with equal-sized apertures (Section 3.2).

Some remarks about computation of the electrostatic fields are in order. It is not necessary to compute the original Fourier coefficients $\{a_n\}_{n=0}^{\infty}$. Calculations can be done in terms of the secondary coefficients $\{b_n\}_{n=0}^{\infty}$.

For instance, to derive formulae for capacitance and potential distribution along the z-axis, use Formula (2.254) in which we set $m=0$:

$$a_n = \frac{\sqrt{\cos\Delta+\cos\Delta_0}}{2}\int_{-1}^{1}\frac{dx}{\sqrt{\cos\Delta+\cos\Delta_0 x}} \times P_n\left(\frac{\cos\Delta_0+\cos\Delta x}{\cos\Delta+\cos\Delta_0 x}\right)\sum_{s=0}^{\infty}(2s+1)\,b_s P_s(x). \tag{3.61}$$

The total charge accumulated on both caps is $Q=a.a_0$; from (3.54) one finds that

$$\begin{aligned} Q &= \frac{1}{2}a\sqrt{\cos\Delta+\cos\Delta_0}\sum_{s=0}^{\infty}(2s+1)\,b_s\int_{-1}^{1}\frac{P_s(x)\,dx}{\sqrt{\cos\Delta+\cos\Delta_0 x}} \\ &= a\left[1-q(\Delta,\Delta_0)\right]\sum_{s=0}^{\infty}b_s q^s(\Delta,\Delta_0). \end{aligned} \tag{3.62}$$

(Observe that for symmetric caps $\Delta=\frac{\pi}{2}$, $q\left(\Delta,\frac{\pi}{2}\right)=0$, and the expression (3.62) reduces to the previously stated form, namely $Q=a.a_0$.)

The electrostatic potential taken along the z-axis (so that $\cos\theta=\pm1$) is given by

$$U(t,\pm1)=\sum_{n=0}^{\infty}a_n(\pm t)^n, \tag{3.63}$$

where $t=r/a\le 1$. Upon substituting (3.61) and taking account of the series

$$\sum_{n=0}^{\infty}P_n\left(\frac{\cos\Delta_0+\cos\Delta x}{\cos\Delta+\cos\Delta_0 x}\right)(\pm t)^n=\left(1\mp 2t\frac{\cos\Delta_0+\cos\Delta x}{\cos\Delta+\cos\Delta_0 x}+t^2\right)^{-\frac{1}{2}}$$

(derived from the generating function for P_n, see Appendix, (B.59)), and the value of the integral given by (3.54), we obtain the final formula for the distribution of the electrostatic potential along the z-axis in terms of the coefficients b_n:

$$U(t,\pm1)=\frac{1}{1\mp t}\left[1-R(\Delta,\Delta_0;t)\right]\sum_{s=0}^{\infty}b_s R^s(\Delta,\Delta_0;t), \tag{3.64}$$

where

$$R(\Delta,\Delta_0;t)=\frac{(1-t^2)\sqrt{\cos^2\Delta-\cos^2\Delta_0}-(1+t^2)\cos\Delta\pm 2t\cos\Delta_0}{(1+t^2)\cos\Delta_0\mp 2t\cos\Delta}. \tag{3.65}$$

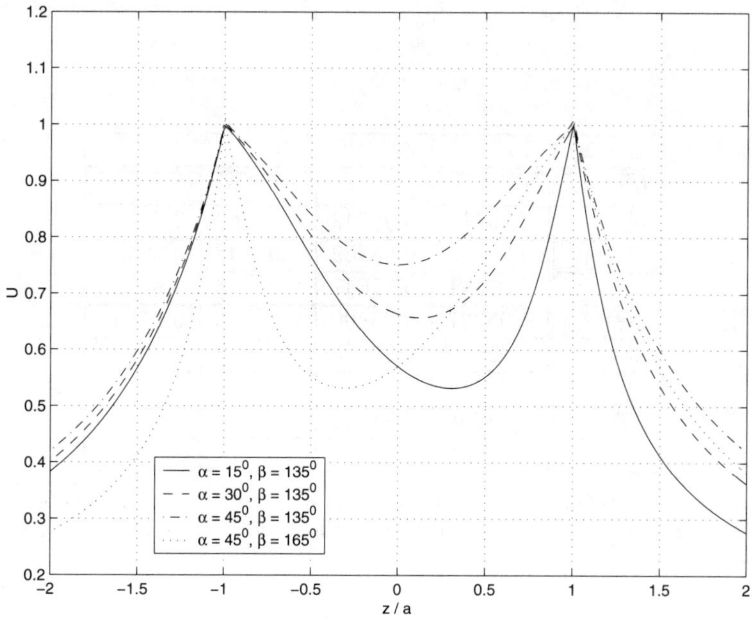

FIGURE 3.10. Electrostatic potential along the z-axis for an asymmetrical pair of spherical caps charged to unit potential and subtending angles α and $\pi - \beta$.

When $r > a$, we use the formula

$$U(\rho, \pm 1) = \sum_{n=0}^{\infty} a_n (\pm \rho)^{-n-1} = t \sum_{n=0}^{\infty} a_n (\pm t)^n, \tag{3.66}$$

where $\rho = t^{-1} = r/a > 1$, so that the expression (3.64) may be employed. Note that with the limiting values $t = 0, 1$ we have

$$R(\Delta, \Delta_0; 0) = q(\Delta, \Delta_0), \qquad R(\Delta, \Delta_0; 1) = \pm 1.$$

Some calculations of the total charge on spherical caps of unequal size are displayed in Table 3.4, and the similar calculations for nonsymmetrical spherical barrels are displayed in Table 3.5. An illustrative example of the electrostatic potential distribution along the z-axis for an asymmetrical pair of spherical caps are shown in Figure 3.10.

3.4 The method of inversion

The method of inversion in a sphere is described in many classical books on electromagnetism (see for example [54], [66]). In three-dimensional potential (electrostatic) problems this method plays, to some extent, the same

$\Delta \diagdown \Delta_0$	75°	60°	45°	30°
0°	1	1	1	1
5°	0.99710	0.99742	0.99788	0.99850
10°	0.98839	0.98959	0.99148	0.99405
15°	0.97381	0.97647	0.98076	0.98684
20°	0.95327	0.95800	0.96568	0.97751
25°	0.92673	0.93409	0.94625	0.96776
30°	0.89413	0.90469	0.92265	—
45°	0.75968	0.78368	—	—
60°	0.57036	—	—	—

TABLE 3.4. Normalised total charge on two nonsymmetrical spherical caps $a^{-1}Q_{1,1}$. Δ_0 is the angular coordinate of the middle of the slot, Δ is its semi-width.

$\Delta \diagdown \Delta_0$	75°	60°	45°	30°
0°	0	0	0	0
5°	0.58555	0.53669	0.45724	0.34964
10°	0.67380	0.62062	0.53415	0.41722
15°	0.73665	0.68163	0.59228	0.47193
20°	0.78640	0.73100	0.64119	0.52084
25°	0.82735	0.77264	0.68415	0.56621
30°	0.86158	0.80843	0.72264	—
45°	0.93426	0.88970	—	—
60°	0.97410	—	—	—

TABLE 3.5. Normalised total charge on a nonsymmetrical spherical barrel $a^{-1}Q_{1,1}$. Δ_0 is the angular coordinate of the middle of the slot, Δ is its semi-width.

role as conformal mapping does in two-dimensional problems. It is mainly used in the calculation of capacitance of closed charged shells. To this end, let us state a very useful theorem first formulated by C. J. Bouwkamp [7] in 1958.

Theorem 10 *Consider an isolated (or solitary) conductor bounded by a surface S. Let $S\prime$ be the image of S under inversion in the sphere of radius a, centred at a given fixed point M. Let U^0 be the free-space potential due to a unit negative charge located at M. Let $U^0 + U^1$ be the total potential induced by this unit charge at M in the presence of S' when it is grounded (i.e., $U^0 + U^1 = 0$ on S'). If V_0 is the value of the induced potential U^1 at M $\left(V_0 = U^1(M)\right)$, then the capacitance C of the conductor S equals $a^2 V_0$.*

We introduce two well-known examples to illustrate the use of this theorem in the simplest cases. The first example calculates the capacitance of a single spherical cap. The second, borrowed from [7], calculates the capacitance of two touching spherical shells.

We have already calculated the capacitance C_{cap} of the spherical cap in Section 1.4: $C_{cap} = a.a_0$, where a is radius of the sphere and a_0 is lowest Fourier coefficient of the expansion of the electrostatic potential in Fourier-Legendre series; thus $a_0 = \pi^{-1}(\theta_0 + \sin\theta_0)$, and $C_{cap} = a\pi^{-1}(\theta_0 + \sin\theta_0)$. Let us demonstrate an alternative way of arriving at this result via inversion.

Consider the spherical cap subtending an angle θ_0 at the origin as shown in Figure 3.11. It occupies the region $0 \le \theta \le \theta_0$ of the spherical surface $r = a$. Before performing an inversion about the centre M located at $r = a, \theta = 0$, we relocate the cap so that it occupies the area $\pi - \theta_0 \le \theta \le \pi$ on the surface $r = a$. Under an inversion in the sphere of centre M and inversion radius $R = 2a$, the spherical cap is transformed to the circular disc shown with centre O'. The capacitance calculation for a spherical cap is transformed to the equivalent calculation of the potential U for the grounded circular disc of radius b in the presence of the unit negative charge, which is placed at the centre of inversion.

Let O' be the origin of a cylindrical polar coordinate system (ρ, z), so that the coordinates of the inversion centre M are $\rho = 0$, $z = 2a$; the inversion procedure described above is given by $\rho = R\tan\frac{1}{2}\theta$, and the radius of the circular disc image is $b = R\tan\frac{1}{2}\theta_0$. The potential function emanating from the negative unit charge is $U^0 = -\left(\rho^2 + z^2\right)^{-\frac{1}{2}}$. By the method of separation of variables, we may seek the axisymmetric electrostatic potential $U \equiv U(\rho, z)$ as the sum $U = U^0 + U^1$, where the induced potential U^1 has the form

$$U^1 = \int_0^\infty f(\nu) J_0(\nu\rho) e^{-\nu|z-a|} d\nu \tag{3.67}$$

and the unknown function f is to be determined. Upon enforcing the mixed boundary conditions one readily obtains the following dual series equations,

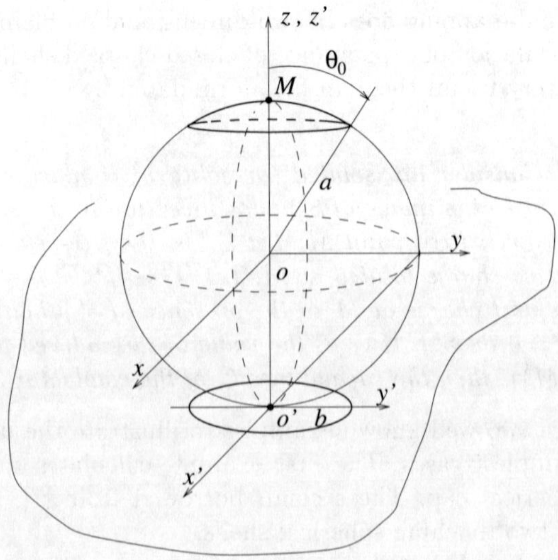

FIGURE 3.11. The spherical cap and its image (the circular disc) under the inversion procedure (see text).

involving Bessel functions:

$$\begin{aligned}\int_0^\infty f(\nu) J_0(\nu\rho)\, d\nu &= \left(\rho^2+4a^2\right)^{-\frac{1}{2}}, \qquad 0 \le \rho < b, \qquad (3.68)\\ \int_0^\infty \nu f(\nu) J_0(\nu\rho)\, d\nu &= 0, \qquad \rho > b.\end{aligned}$$

We may use the results of Section 2.6 to find

$$f(\nu) = \frac{4a}{\pi}\int_0^b \frac{\cos(\nu\rho)}{\rho^2+4a^2} d\rho, \qquad (3.69)$$

$b = 2a\tan\frac{1}{2}\theta_0$. According to Bouwkamp's theorem, the capacitance is

$$C_{cap} = R^2 U^1(M) = 4a^2\int_0^\infty f(\nu)\, e^{-2\nu a} d\nu. \qquad (3.70)$$

Thus, the capacitance equals

$$C_{cap} = \frac{4}{\pi}a^3\int_0^\infty d\nu e^{-2\nu a}\int_0^b \frac{d\rho\cos(\nu\rho)}{\rho^2+4a^2} = \frac{32}{\pi}a^4\int_0^b \frac{d\rho}{\left(\rho^2+4a^2\right)^2},$$

and an elementary calculation leads to

$$C_{cap} = \frac{a}{\pi}\left(\theta_0+\sin\theta_0\right),$$

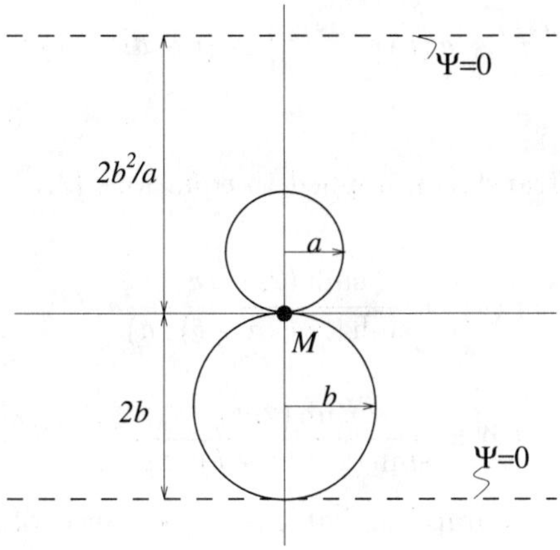

FIGURE 3.12. Touching spheres of radii a, b.

which is in accord with the previous result.

Our second example is the calculation of capacitance of two touching spheres of radii a and b, $a \leq b$ (Figure 3.12). In [7], this problem was treated by the image method. With a view to extending it to open touching spherical shells, we derive a solution by the method of separation of variables. The inversion sphere has centre at the point of contact M and radius $2b$. Let M be the origin of polar cylindrical coordinates. The transformation (given by $\rho = 2a \tan \frac{1}{2}\theta$) transforms the electrostatic problem for two touching spheres that are charged to unit potential to the equivalent electrostatic calculation for two grounded infinite planes, separated by a distance $l = 2b\,(1 + b/a)$, in the presence of a unit negative charge at inversion centre M.

As before we may seek a solution in the form

$$U(\rho, z) = U^0 + U^1, \tag{3.71}$$

where

$$U^0 = -\left(\rho^2 + z^2\right)^{-\frac{1}{2}}, \tag{3.72}$$

$$U^1 = \int_0^\infty \left\{ f(\nu)\, e^{-\nu z} + g(\nu)\, e^{\nu z} \right\} J_0(\nu\rho)\, d\nu, \tag{3.73}$$

with f and g to be determined. Notice $U(\rho, z) \to 0$ as $\rho \to \infty$. The boundary conditions $U(\rho, -2b) = U\left(\rho, 2b^2/a\right) = 0$ (each plane is grounded) imply

$$\int_0^\infty \left\{ f(\nu)\, e^{2\nu b} + g(\nu)\, e^{-2\nu b} \right\} J_0(\nu\rho)\, d\nu = \left(\rho^2 + 4b^2\right)^{-\frac{1}{2}},$$

$$0 < \rho < \infty, \tag{3.74}$$

$$\int_0^\infty \left\{ f(\nu) e^{-2\nu b^2/a} + g(\nu) e^{2\nu b^2/a} \right\} J_0(\nu\rho)\, d\nu = \left(\rho^2 + 4b^2 (b/a)^2\right)^{-\frac{1}{2}},$$
$$0 < \rho < \infty. \quad (3.75)$$

A Bessel integral transform, applied to equations (3.74) and (3.75) shows that

$$f(\nu) = \frac{\sinh(2\nu b^2/a)}{\sinh(2\nu b(a+b)/a)} e^{-2\nu b}, \quad (3.76)$$

$$g(\nu) = \frac{\sinh(2\nu b)}{\sinh(2\nu b(a+b)/a)} e^{-2\nu b^2/a}. \quad (3.77)$$

Bouwkamp's theorem implies that the capacitance of the two touching spheres is

$$C_{a,b} = 4b^2 U^1(M) = 4b^2 \int_0^\infty \{f(\nu) + g(\nu)\}\, d\nu \quad (3.78)$$
$$= -\frac{ab}{a+b}\left\{2\gamma + \psi\left(\frac{a}{a+b}\right) + \psi\left(\frac{b}{a+b}\right)\right\}, \quad (3.79)$$

where γ is Euler's constant, and ψ denotes the logarithmic derivative of the Gamma function Γ (see [1]),

$$\psi(-x) = -\gamma + x^{-1} - x\sum_{n=1}^{\infty} \frac{1}{n(n-x)}.$$

When the spherical radii are equal ($a = b$), $\psi(\frac{1}{2}) = -\gamma - 2\ln 2$, and the capacitance $C_{b,b}$ equals $2b\ln 2$.

Let us extend the last example to consider open spherical caps; various configurations are shown in Figure 3.13. We restrict ourselves to spheres of equal radii a, and shells subtending equal angles θ_0, and concentrate on the last two configurations (c) and (d); the solution to the first two is easily deduced from the last two (using image theory). From the symmetry after inversion, it is sufficient to consider the problem in the upper half-space ($z \geq 0$). Thus, we find the distribution of the electrostatic potential U in R^3, which is due to the unit negative charge located at inversion centre M in presence of two grounded circular discs, separated by a distance $l = 4a$.

As before, the free-space potential emanating from the negative unit charge is $U^0 = -\left(\rho^2 + z^2\right)^{-\frac{1}{2}}$. Subdivide the space into two regions. In region I , $0 < z \leq 2a$, we seek a solution in the form

$$U^I = U^0 + U^{(i)} \quad (3.80)$$

where

$$U^{(i)} = \int_0^\infty f(\nu) J_0(\nu\rho) \cosh(\nu z)\, d\nu; \quad (3.81)$$

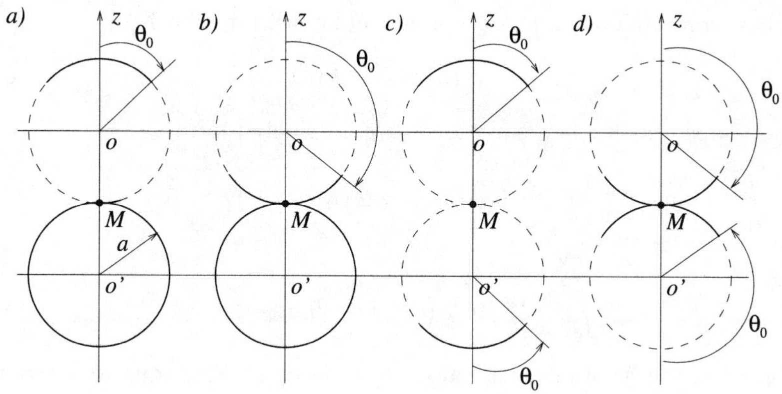

FIGURE 3.13. Various configurations of spherical cap pairs.

in region II , $z > 2a$, we seek a solution in the form

$$U^{II} = U^0 + U^{(e)} \tag{3.82}$$

where

$$U^{(e)} = \int_0^\infty g(\nu) J_0(\nu\rho) e^{-\nu z} d\nu \tag{3.83}$$

and the functions f, g are to be determined. (The form of $U^{(i)}$ and $U^{(e)}$ is a superposition of partial solutions to Laplace's equation, which vanish at infinity.) From the continuity condition

$$U^I(\rho, 2a) = U^{II}(\rho, 2a), 0 \leq \rho < \infty$$

we deduce

$$\cosh(2\nu a) f(\nu) = e^{-2\nu a} g(\nu). \tag{3.84}$$

The mixed boundary conditions applied on the plane $z = 2a$ give

$$U^{(i)}(\rho, 2a) = U^{(e)}(\rho, 2a) = -U^{(0)}(\rho, 2a), \qquad 0 \leq \rho < b, \tag{3.85}$$

$$\frac{\partial U^{(i)}}{\partial z}(\rho, 2a) = \frac{\partial U^{(e)}}{\partial z}(\rho, 2a), \qquad \rho > b, \tag{3.86}$$

where $b = 2a \tan \frac{1}{2}\theta_0$. We therefore obtain the following dual integral equations for the unknown function f :

$$\int_0^\infty f(\nu) \cosh(2\nu a) J_0(\nu\rho) d\nu = (\rho^2 + 4a^2)^{-\frac{1}{2}}, \quad 0 \leq \rho < b, \tag{3.87}$$

$$\int_0^\infty \nu f(\nu) e^{2\nu a} J_0(\nu\rho) d\nu = 0, \qquad \rho > b. \tag{3.88}$$

It is convenient to introduce a new unknown function F by

$$F(\nu) = e^{2\nu a} f(\nu), \tag{3.89}$$

and transform the dual equations to the weighted form

$$\int_0^\infty \left(1 + e^{-4\nu a}\right) F(\nu) J_0(\nu\rho)\, d\nu = 2\left(\rho^2 + 4a^2\right)^{-\frac{1}{2}}, \quad 0 \le \rho < b, \tag{3.90}$$

$$\int_0^\infty \nu F(\nu) J_0(\nu\rho)\, d\nu = 0, \qquad \rho > b. \tag{3.91}$$

Following the Abel integral transform technique, these equations produce

$$\int_0^\infty \nu^{\frac{1}{2}} F(\nu) J_{-\frac{1}{2}}(\nu\rho)\, d\nu$$
$$= \left(\frac{2}{\pi}\right)^{\frac{1}{2}} \frac{4a\rho^{-\frac{1}{2}}}{\rho^2 + 4a^2} - \int_0^\infty \nu^{\frac{1}{2}} F(\nu) e^{-4\nu a} J_{-\frac{1}{2}}(\nu\rho)\, d\nu, \ \rho < b, \tag{3.92}$$

$$\int_0^\infty \nu^{\frac{1}{2}} F(\nu) J_{-\frac{1}{2}}(\nu\rho)\, d\nu = 0, \qquad \rho > b. \tag{3.93}$$

Application of the Bessel-Fourier integral transform to both parts of this equation produces a Fredholm integral equation of the second kind. From a computational point of view, however, the discrete form of solution is preferable. To reduce (3.92) and (3.93) to an i.s.l.a.e., we use the Hankel transform to obtain

$$\mu^{-\frac{1}{2}} F(\mu) = 4a\left(\frac{2}{\pi}\right)^{\frac{1}{2}} \int_0^b \frac{\rho^{\frac{1}{2}} J_{-\frac{1}{2}}(\mu\rho)}{\rho^2 + 4a^2} d\rho$$
$$- \int_0^b \rho J_{-\frac{1}{2}}(\mu\rho) \left\{ \int_0^\infty \nu^{\frac{1}{2}} F(\nu) e^{-4\nu a} J_{-\frac{1}{2}}(\nu\rho)\, d\nu \right\} d\rho \tag{3.94}$$

and then represent unknown function F by a Neumann series

$$F(\mu) = \left(\frac{2}{\pi b \mu}\right)^{\frac{1}{2}} \sum_{n=0}^{\infty} (4n+1)^{\frac{1}{2}} x_n J_{2n+\frac{1}{2}}(\mu b) \tag{3.95}$$

where it can be shown that $\{x_n\}_{n=0}^{\infty} \in l_2$.

Substitute (3.95) into (3.94). Then multiply both sides of (3.94) by $(4m+1)^{\frac{1}{2}} J_{2m+\frac{1}{2}}(\mu b)$, integrate over $[0, \infty)$, and use the well-known integral formula [19],

$$\int_0^\infty t^{-1} J_{\nu+2n+1}(t) J_{\nu+2m+1}(t)\, dt = (4n + 2\nu + 2)^{-1} \delta_{nm}. \tag{3.96}$$

This yields an i.s.l.a.e. of the second kind for the coefficients $\{x_n\}_{n=0}^{\infty}$,

$$x_m + \sum_{n=0}^{\infty} \alpha_{nm} x_n = \beta_m, \tag{3.97}$$

where $m = 0, 1, 2, ...$, and

$$\alpha_{nm} = [(4n+1)(4m+1)]^{\frac{1}{2}} \int_0^{\infty} \nu^{-1} e^{-4a\nu} J_{2n+\frac{1}{2}}(\nu b) J_{2m+\frac{1}{2}}(\nu b)\, d\nu, \tag{3.98}$$

$$\beta_m = 2\tan\frac{\theta_0}{2}(-1)^m (4m+1)^{\frac{1}{2}} \int_0^1 \frac{P_{2m}(t)}{1+t^2\tan^2\frac{1}{2}\theta_0} dt. \tag{3.99}$$

Let us determine the capacitance C of two spherical caps in terms of the Fourier coefficients x_n. As before,

$$C = 4a^2 \int_0^{\infty} f(\nu)\, d\nu = 4a^2 \int_0^{\infty} F(\nu) e^{-2\nu a} d\nu, \tag{3.100}$$

so substituting for F from (3.95), we finally deduce that the capacitance C equals

$$\frac{2a}{\sqrt{\pi}} \sum_{n=0}^{\infty} x_n (4n+1)^{\frac{1}{2}} \frac{\Gamma(2n+1)}{\Gamma\left(2n+\frac{3}{2}\right)} \frac{\tan^{2n}\frac{1}{2}\theta_0}{2^{2n}} \times$$
$${}_2F_1\left(n+\frac{1}{2}, n+1; 2n+\frac{3}{2}; -\tan^2\frac{\theta_0}{2}\right). \tag{3.101}$$

Both Formulae (3.100) and (3.101) are valid for $\theta_0 < \frac{\pi}{2}$. For small caps $(\theta_0 \ll 1)$, one can deduce approximate analytical expressions for capacitance in powers of the small parameter $\varepsilon = \tan\frac{1}{2}\theta_0 \ll 1$. To estimate of their accuracy, we express α_{nm} as a hypergeometric function by direct calculation [14] of the integral in (3.98):

$$\alpha_{nm} = \frac{[(4n+1)(4m+1)]^{\frac{1}{2}}}{2^{4n+4m+2}} \left(\tan\frac{\theta_0}{2}\right)^{2n+2m+1} \frac{\Gamma(2n+2m+1)}{\Gamma\left(2n+\frac{3}{2}\right)\Gamma\left(2m+\frac{3}{2}\right)}$$
$$\times {}_4F_3\left(p, p+\frac{1}{2}, p-\frac{1}{2}, p; 2p, 2n+\frac{3}{2}, 2m+\frac{3}{2}; -\tan^2\frac{\theta_0}{2}\right) \tag{3.102}$$

where $p = n+m+1$. Also we may calculate from (3.99) using the tabulated integral [14], that

$$\beta_m = (4m+1)^{\frac{1}{2}} \frac{\Gamma(m+1)\Gamma\left(m+\frac{1}{2}\right)}{\Gamma\left(2m+\frac{3}{2}\right)} \tan^{2m+1}\frac{\theta_0}{2} \times$$
$${}_2F_1\left(m+\frac{1}{2}, m+1; 2m+\frac{3}{2}; -\tan^2\frac{\theta_0}{2}\right). \tag{3.103}$$

If $\varepsilon = \tan\frac{1}{2}\theta_0 \ll 1$, then

$$\beta_m = (4m+1)^{\frac{1}{2}}\,\varepsilon^{2m+1}\frac{\Gamma\left(m+\frac{1}{2}\right)\Gamma\left(m+1\right)}{\Gamma\left(2m+\frac{3}{2}\right)}\times \left\{1-\frac{\left(m+\frac{1}{2}\right)\left(m+1\right)}{2m+\frac{3}{2}}\varepsilon^2+O\left(\varepsilon^4\right)\right\}. \quad (3.104)$$

We may now apply the method of successive approximations to (3.97):

$$x_m^{(i+1)} = \beta_m - \sum_{n=0}^{\infty}\alpha_{nm}x_n^{(i)}, \quad (3.105)$$

where $i = 0, 1, ...,$ and $x_m^{(0)} = 0$. So

$$x_m^{(1)} = \beta_m,$$

$$x_m^{(2)} = \beta_m - \sum_{n=0}^{\infty}\alpha_{nm}x_n^{(1)} = \beta_m - \sum_{n=0}^{\infty}\alpha_{nm}\beta_m,$$

and so on (for $m = 0, 1, ...$).

From (3.101), it can be readily shown that accuracy of order $O\left(\varepsilon^2\right)$ is obtained for x_0 by neglecting the rest of Fourier coefficients x_n $(n \geq 1)$. Thus, since

$$\begin{aligned} x_0^{(1)} &= 2\varepsilon + O\left(\varepsilon^3\right), \\ x_0^{(2)} &= 2\varepsilon\left(1-\frac{1}{\pi}\varepsilon\right)+O\left(\varepsilon^3\right), \end{aligned} \quad (3.106)$$

an approximate formula for capacitance is

$$C = \frac{4a}{\pi}x_0^{(2)} + O\left(\varepsilon^3\right) \quad (3.107)$$

so that the capacitance of two spherical caps is approximately

$$C = \frac{4\theta_0}{\pi}\left(1-\frac{1}{2\pi}\theta_0\right)+O\left(\theta_0^3\right). \quad (3.108)$$

This formula has a clear physical interpretation. The first term is the sum of the capacitances of two isolated spherical caps. The second quadratic term reflects the interaction or mutual impact of the caps.

The capacitance of the structure shown in Figure 3.13(d) is obtained in a similar way. This approach can be extended to consider spherical shells of differing radii and angle.

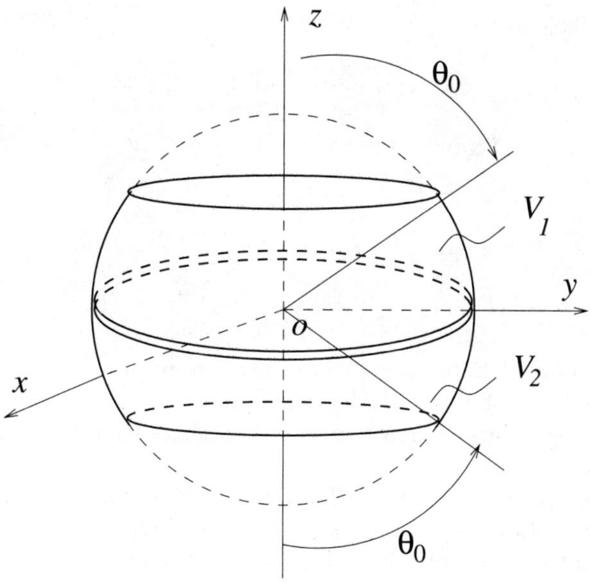

FIGURE 3.14. Spherically-shaped electronic lens.

3.5 Electrostatic fields in a spherical electronic lens

In this section another illustration of methods developed for applications in a spherical geometry context is given. We calculate the electrostatic field of a spherically-shaped electronic lens, shown in Figure 3.14. The spherically-shaped lens is a variant of a widely used electronic lens that comprises two charged, finite hollow cylinders at different potentials V_1 and V_2, aligned along a common axis of rotational symmetry. The upper electrode is the spherical shell segment given by $r = a$, $\theta_0 \leq \theta \leq \frac{\pi}{2} - \delta$; the lower electrode is its mirror image in the xy-plane. The distance between electrodes is negligibly small compared with the electrode dimension ($\delta \approx 0$), so that we model the lens by closely adjoined electrodes, electrically isolated by an infinitesimally thin layer of dielectric.

Let the upper electrode be charged to potential V_1 and the lower one charged to potential V_2. Due to the rotational symmetry of the problem we seek the electrostatic potential $V = V(r, \theta)$ as an expansion in a Fourier-Legendre series (cf. (3.14))

$$V = \sum_{n=0}^{\infty} x_n P_n(\cos\theta) \left\{ \begin{array}{ll} (r/a)^n, & r < a \\ (r/a)^{-n-1}, & r > a \end{array} \right\}. \tag{3.109}$$

Use of the mixed boundary conditions at $r = a$ and of symmetry, produces the following decoupled dual series equations for the even and odd index

Fourier coefficients:

$$\begin{cases} \sum_{n=0}^{\infty}(n+\frac{1}{4})x_{2n}P_{2n}(\cos\theta)=0, & \theta\in(0,\theta_0) \\ \sum_{n=0}^{\infty}x_{2n}P_{2n}(\cos\theta)=\frac{1}{2}(V_1+V_2), & \theta\in(\theta_0,\frac{\pi}{2}) \end{cases}. \tag{3.110}$$

$$\begin{cases} \sum_{n=0}^{\infty}(n+\frac{3}{4})x_{2n+1}P_{2n+1}(\cos\theta)=0, & \theta\in(0,\theta_0) \\ \sum_{n=0}^{\infty}x_{2n+1}P_{2n+1}(\cos\theta)=\frac{1}{2}(V_1-V_2), & \theta\in(\theta_0,\frac{\pi}{2}) \end{cases} \tag{3.111}$$

The first pair of Equations (3.110) have essentially been solved in Section 3.2, and may be identified with Equations (3.29) once we set $b_n = (-1)^n x_{2n}$; the solution given by (3.32) must be multiplied by a factor $\frac{1}{2}(V_1+V_2)$.

The technique developed in Chapter 2 may be followed to reduce the second pair of Equations (3.111) to the following i.s.l.a.e. with a matrix operator that is a completely continuous perturbation of the identity (in l_2). Temporarily, we replace the right-hand side of the second equation in (3.111) by unity, so that in the final solution each Fourier coefficient must be multiplied by a factor $\frac{1}{2}(V_1-V_2)$:

$$(1-p_m)X_{2m+1}+\sum_{n=0}^{\infty}X_{2n+1}p_n\hat{Q}_{nm}^{(0,\frac{1}{2})}(u_1)=A_m, \tag{3.112}$$

where $m = 0, 1, 2, \ldots$, and

$$p_n = 1-\left(n+\frac{3}{4}\right)\left[\frac{\Gamma(n+1)}{\Gamma(n+\frac{3}{2})}\right]^2 = O\left(n^{-2}\right) \text{ as } n\to\infty,$$

$$x_{2n+1}=(-1)^n\frac{\Gamma(n+1)}{\Gamma(n+\frac{3}{2})}\left\{h_n^{(0,\frac{1}{2})}\right\}^{-\frac{1}{2}}X_{2n+1}, \tag{3.113}$$

and

$$A_m=\left(\frac{2}{\pi}\right)^{\frac{1}{2}}\int_{u_1}^{1}(1-u)^{-\frac{1}{2}}(1+u)^{\frac{1}{2}}\hat{P}_m^{(0,\frac{1}{2})}(u)\,du.$$

An approximate analytical formula for the electrostatic potential along the axis of an electronic lens may be deduced. Set $\theta = \left\{ {0 \atop \pi} \right\}$ in (3.109), and let $q = r/a$ $(r < a)$ so that

$$V\left(q,\begin{matrix}0\\ \pi\end{matrix}\right)=\frac{1}{2}(V_1+V_2)\sum_{n=0}^{\infty}x_{2n}q^{2n}\pm\frac{1}{2}(V_1-V_2)\,q\sum_{n=0}^{\infty}x_{2n+1}q^{2n} \tag{3.114}$$

(the plus and minus signs are associated with 0 and π, respectively). Then, using the approximate analytical solution for even and odd Fourier coefficients (see (3.32) and (3.33)),

$$x_0 \simeq (\alpha(u_1))^{-1},$$

$$x_{2n} \simeq (\alpha(u_1))^{-1} \frac{(-1)^n \Gamma\left(n+\frac{1}{2}\right)}{\sqrt{\pi}\;\Gamma(n+1)} \left(\frac{1+u_1}{2}\right)^{\frac{3}{2}} P_{n-1}^{\left(0,\frac{3}{2}\right)}(u_1), \tag{3.115}$$

and

$$x_{2n+1} \simeq \frac{(-1)^n \Gamma\left(n+\frac{3}{2}\right)}{\sqrt{\pi}\;\Gamma(n+1)} \int_{u_1}^{1} (1-u)^{-\frac{1}{2}} (1+u)^{\frac{1}{2}} P_n^{\left(0,\frac{1}{2}\right)}(u)\,du. \tag{3.116}$$

Substituting in the formula (3.114) we obtain

$$V\left(q, \begin{matrix}0\\ \pi\end{matrix}\right) =$$

$$\frac{(V_1+V_2)}{2\alpha(u_1)} \left\{1 + \frac{1}{\sqrt{\pi}}\left(\frac{1+u_1}{2}\right)^{\frac{3}{2}} \sum_{n=1}^{\infty} (-1)^n \frac{\Gamma\left(n+\frac{1}{2}\right)}{\Gamma(n+1)} q^{2n} P_{n-1}^{\left(0,\frac{3}{2}\right)}(u_1)\right\}$$

$$\pm \frac{(V_1-V_2)\,q}{2\sqrt{\pi}} \sum_{n=0}^{\infty} (-1)^n \frac{\Gamma\left(n+\frac{3}{2}\right)}{\Gamma(n+1)} q^{2n} \int_{u_1}^{1} \frac{(1+u)^{\frac{1}{2}}}{(1-u)^{\frac{1}{2}}} P_n^{\left(0,\frac{1}{2}\right)}(u)\,du \tag{3.117}$$

The integral contained in (3.117) is tabulated in [14] so that (if $u_1 = -u_0$)

$$\int_{u_1}^{1} \frac{(1+u)^{\frac{1}{2}}}{(1-u)^{\frac{1}{2}}} P_n^{\left(0,\frac{1}{2}\right)}(u)\,du$$

$$= (-1)^n \int_{-1}^{u_0} \frac{(1-v)^{\frac{1}{2}}}{(1+v)^{\frac{1}{2}}} P_n^{\left(\frac{1}{2},0\right)}(v)\,dv$$

$$= 2(1-u_1)^{\frac{1}{2}}\; {}_3F_2\left(-n-\frac{1}{2}, n+1, \frac{1}{2}; 1, \frac{3}{2}; \frac{1-u_1}{2}\right). \tag{3.118}$$

Since $q < 1$, we may change the order of summation and integration in the last term of (3.117) and so are led to the series also tabulated in [14],

$$S(q,u) = \sum_{n=0}^{\infty} \frac{\Gamma\left(n+\frac{3}{2}\right)}{\Gamma(n+1)} \left(-q^2\right)^n P_n^{\left(0,\frac{1}{2}\right)}(u)$$

$$= \sum_{n=0}^{\infty} \frac{\Gamma\left(n+\frac{3}{2}\right)}{\Gamma(n+1)} \left(q^2\right)^n P_n^{\left(\frac{1}{2},0\right)}(-u)$$

$$= \frac{\sqrt{\pi}}{2}\left(1+q^2\right)^{-\frac{3}{2}}\; {}_2F_1\left(\frac{3}{4}, \frac{5}{4}; 1; \frac{2q^2}{(1+q^2)^2}(1+u)\right). \tag{3.119}$$

This completes the derivation of an approximate formula for the potential distribution along the axis. Note at once that the value of the electrostatic potential at the origin ($z = 0$) is

$$V\left(0, \begin{matrix} 0 \\ \pi \end{matrix}\right) \cong \frac{1}{2}\left(V_1 + V_2\right)\left\{1 - \frac{1}{\pi}\cos\theta_1 - \frac{1}{\pi}\ln\left(\tan\frac{\theta_1}{2}\right)\right\}^{-1} ; \quad (3.120)$$

it is uniformly valid with respect to the parameter $\theta_1 \in \left(0, \frac{\pi}{2}\right)$.

Further approximate analytical expressions which are uniformly valid with respect to the electrode dimensions, are rather complicated except for the limiting case of short electrodes ($|u_1 - 1| \ll 1$ or $\theta_1 \ll 1$). A crude approximation to the electrostatic field for narrow or very short electrodes is

$$V\left(q, \begin{matrix} 0 \\ \pi \end{matrix}\right) \cong \frac{1}{2}\left(V_1 + V_2\right)\left(1 + q^2\right)^{-\frac{1}{2}} \frac{\pi}{\ln\left(2/\theta_1\right)} + O\left(\theta_1\right). \quad (3.121)$$

For general lens parameters, numerical calculations may be simply and satisfactorily performed. If a truncation number N_{tr} of 6 to 8 is used to solve systems (3.112), (3.32), and (3.33), at least four significant digits in the values of Fourier coefficients X_{2n}, X_{2n+1} can be obtained stably.

3.6 *Frozen* magnetic fields inside superconducting shells

In contrast to previous sections, we now consider a physical problem that mathematically reduces to a Neumann problem. The physical situation concerns a spherical thin shell with two symmetrically located circular holes ("doubly-connected" in a topological sense), manufactured from superconducting material with critical temperature T_c. Suppose this material is a superconductor of the first kind so that when $T > T_c$ this material behaves as normal metal, but when $T \leq T_c$, it behaves as a superconductor. Place this shell (at $T > T_c$) in some region of space that is permeated by a homogeneous magnetic field. Cool the shell in order to make the transition to the superconducting state ($T \leq T_c$), and *switch off* the magnetic field. Assuming a perfect (ideal) Meissner effect, the magnetic flux $\Phi = \pi a^2 H_0$ is *frozen* in the shell's cavity. The design of special magnetic field compressors that raises the threshold sensitivity of superconducting magnetic systems exploits this principle.

A mathematical analysis of this phenomenon requires the solution of a mixed boundary-value problem for the magnetostatic potential $U^m(r, \theta)$ with a Neumann boundary condition given on the shell's surface. In addition, the *frozen* magnetic flux must take constant value through any arbitrarily taken cross-section of the shell, including a contour on the surface of the shell.

Considering Laplace's equation, together with the continuity condition for the normal derivative of U^m at $r = a$ and the $O\left(r^{-1}\right)$ behaviour of the potential at infinity $(r \to \infty)$, one may seek a solution in the form

$$U^m(r,\theta) = \frac{\Phi}{\pi a}\sum_{n=1}^{\infty} A_n P_n(\cos\theta)\left\{\begin{array}{ll}(r/a)^n, & r < a\\ -(n/(n+1))(r/a)^{-n-1}, & r > a\end{array}\right\}, \tag{3.122}$$

where $\Phi = \pi a^2 H_0$ is the *frozen* magnetic flux, H_0 is the effective mean value of the magnetic field taken at cross-section $z = 0$, and $\{A_n\}_{n=1}^{\infty}$ are the unknown coefficients to be determined; the finiteness of energy condition (see Section 1.3) requires

$$\sum_{n=1}^{\infty} |A_n|^2 < \infty.$$

Superconducting shells are usually modelled by ideal diamagnetic materials of zero relative permeability; the normal component of magnetic field vanishes at the shell surface. The boundary conditions on the potential are determined by continuity of radial and tangential components of the magnetic field $\vec{H} = -\operatorname{grad} U^m$ on the superconducting portion of the shell (specified by the angular segment $(\theta_0, \pi - \theta_0)$) and aperture, respectively:

$$\begin{aligned} H_r^m(a-0,\theta) &= H_r^m(a+0,\theta) = 0, \quad \theta \in (\theta_0, \pi-\theta_0),\\ H_\theta^m(a-0,\theta) &= H_\theta^m(a+0,\theta), \quad \theta \in (0,\theta_0) \cup (\pi-\theta_0, \pi). \end{aligned}$$

The constancy of the magnetic flux through any arbitrarily taken cross-section of the shell requires that if $\theta \in (\theta_0, \pi - \theta_0)$,

$$2\pi a^2 \int_0^\theta H_r^m(a,\theta)\sin\theta d\theta = \Phi.$$

Applying these conditions to (3.122), we obtain the following triple symmetric equations for the modified Fourier coefficients $x_n = A_n/(n+1)$,

$$\sum_{n=1}^{\infty}(2n+1)x_n P_n^1(\cos\theta) = 0, \quad \theta \in (0,\theta_0) \cup (\pi-\theta_0,\pi) \tag{3.123}$$

$$\sum_{n=1}^{\infty} x_n P_n^1(\cos\theta) = -\frac{1}{2}\operatorname{cosec}\theta, \quad \theta \in (\theta_0, \pi-\theta_0). \tag{3.124}$$

Because of the symmetry, $x_{2n} \equiv 0$ and these triple equations are equivalent to the dual pair

$$\sum_{n=0}^{\infty}\left(n+\frac{3}{4}\right)x_{2n+1}P_{2n+1}^{1}(z) = 0, \ z\in(-1,-z_0) \tag{3.125}$$

$$\sum_{n=0}^{\infty}x_{2n+1}P_{2n+1}^{1}(z) = \frac{-1}{2\sqrt{1-z^2}}, \ z\in(-z_0,0) \tag{3.126}$$

where $z=\cos\theta$, and $z_0=\cos\theta_0$.

As previously done, (see Sections 3.2 and 3.3), we use the substitutions $u=2z^2-1$ and

$$P_{2n+1}^{1}(z)=\sqrt{2}(n+\frac{1}{2})(1-u)^{\frac{1}{2}}P_n^{(1,-\frac{1}{2})}(u) \tag{3.127}$$

in Equations (3.125) and (3.126), and integrate them to obtain dual series equations with Jacobi polynomials $P_n^{(0,\frac{1}{2})}$,

$$\sum_{n=0}^{\infty}x_{2n+1}P_n^{(0,\frac{1}{2})}(u)=2^{-\frac{3}{2}}(1+u)^{-\frac{1}{2}}\ln\left[\frac{1-\sqrt{(1+u)/2}}{1+\sqrt{(1+u)/2}}\right], \ u\in(-1,u_0) \tag{3.128}$$

$$\sum_{n=0}^{\infty}\left(n+\frac{3}{4}\right)x_{2n+1}P_n^{(0,\frac{1}{2})}(u)=2^{\frac{1}{2}}(1+u)^{-\frac{1}{2}}C, \ u\in(u_0,1) \tag{3.129}$$

where $u_0=2z_0^2-1=\cos 2\theta_0$, and C is an integration constant determined by the condition $\sum_{n=1}^{\infty}|A_n|^2<\infty$.

Equations similar to this were solved in Section 3.3; omitting details of its deduction, the final system is

$$X_{2m+1}-\sum_{n=0}^{\infty}X_{2n+1}\tau_n\Pi_{nm}(u_0)=A_m, \tag{3.130}$$

where $m=0,1,2,\ldots$, and

$$X_{2m+1}=2^{\frac{1}{4}}\left[\frac{(m+\frac{1}{2})(m+1)}{m+\frac{3}{4}}\right]^{\frac{1}{2}}\frac{\Gamma(m+\frac{3}{2})}{\Gamma(m+1)}x_{2m+1}. \tag{3.131}$$

Furthermore,

$$\Pi_{nm}(u_0)= \hat{Q}_{nm}^{(-\frac{1}{2},1)}(u_0)-\sqrt{2}\frac{R_n(u_0)R_m(u_0)}{\ln\left[\left(1+((1-u_0)/2)^{\frac{1}{2}}\right)/\left(((1+u_0)/2)^{\frac{1}{2}}\right)\right]},$$

$$R_s(u_0) = \left(\frac{1-u_0}{2\left(s+\frac{1}{2}\right)(s+1)} \right)^{\frac{1}{2}} \hat{P}_s^{\left(\frac{1}{2},0\right)}(u_0), \tag{3.132}$$

$$A_m = -2^{-\frac{3}{2}} \frac{\pi^{\frac{1}{2}} R_m(u_0)}{\ln\left[\left(1+((1-u_0)/2)^{\frac{1}{2}}\right) / \left(((1+u_0)/2)^{\frac{1}{2}}\right)\right]},$$

and

$$\tau_n = 1 - \left(n+\frac{3}{4}\right)\left[\frac{\Gamma(n+1)}{\Gamma\left(n+\frac{3}{2}\right)}\right]^2 = O\left(n^{-2}\right), \text{ as } n \to \infty.$$

In the same way as in Sections 3.2 and 3.3, the system (3.130) has an approximate analytical solution for the Fourier coefficients X_{2n+1} that is uniformly valid with respect to the dimension of the circular holes. In fact, the norm of the completely continuous part H is bounded by the estimate

$$\|H\| \leq \max|\tau_n| = \tau_0 = 1 - \frac{3}{\pi} < 0.046 \ll 1;$$

this is uniformly valid in the parameter u_0. The method of successive approximations may be used to solve (3.130); remarkably, only one step of the iteration process is needed to obtain an approximate analytical solution of high accuracy (3 to 4 correct digits in values of A_n). The result of one iteration is

$$A_{2n+1} \simeq -2^{-\frac{1}{2}} \pi^{\frac{1}{2}} \sin\theta_0 \frac{\Gamma\left(n+\frac{1}{2}\right)}{\Gamma(n+1)} \frac{P_n^{\left(\frac{1}{2},0\right)}(\cos 2\theta_0)}{\ln[1+\sin\theta_0] - \ln[\cos\theta_0]}. \tag{3.133}$$

We may use (3.133) to derive the magnetic field distribution along the shell axis (z-axis). Due to symmetry we need only consider the positive z-axis ($z \geq 0$, $\theta = 0$) and obtain

$$H_r^m(q,0) = -\frac{\Phi}{\pi a^2} \sum_{n=0}^{\infty} (2n+1) A_{2n+1} q^{2n} \tag{3.134}$$

where $q = r/a$. Use the tabulated value of the series [14] to rewrite (3.134) in the form

$$H_r^m(q,0) = \frac{\Phi}{\pi a^2} \cdot \frac{\pi}{2} \cdot \frac{\sin\theta_0}{\ln[1+\sin\theta_0] - \ln[\cos\theta_0]} \times$$
$$\left(1+q^2\right)^{-\frac{3}{2}} {}_2F_1\left(\frac{3}{4}, \frac{5}{4}; 1; \frac{4q^2\cos^2\theta_0}{\left(1+q^2\right)^2}\right). \tag{3.135}$$

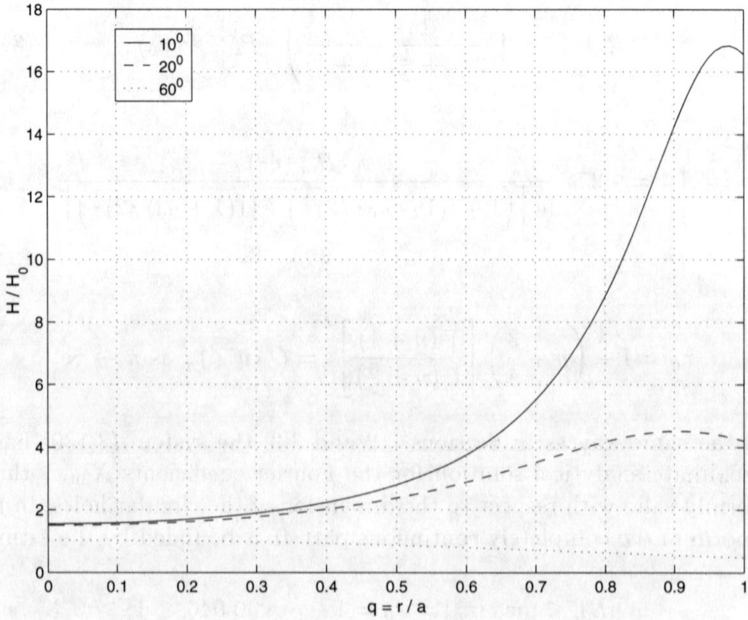

FIGURE 3.15. Frozen magnetic field along z-axis, for various angles θ_0.

The hypergeometric function in (3.135) admits a quadratic transformation to the Legendre function

$$
{}_2F_1\left(\frac{3}{4},\frac{5}{4};1;\frac{4q^2\cos^2\theta_0}{(1+q^2)^2}\right)=\left[\frac{1+q^2}{R(q,\theta_0)}\right]^{\frac{3}{2}} P_{\frac{1}{2}}\left[\frac{1+q^2}{R(q,\theta_0)}\right] \tag{3.136}
$$

where $R(q,\theta_0)=\left(1-2q^2\cos 2\theta_0+q^4\right)^{\frac{1}{2}}$; the Legendre function $P_{\frac{1}{2}}$ is related to the complete elliptic integral of the second kind E by (see Appendix, (B.82))

$$
P_{\frac{1}{2}}\left[\frac{1+q^2}{R(q,\theta_0)}\right]=\frac{2}{\pi}\frac{R_0(q,\theta_0)}{R^{\frac{1}{2}}(q,\theta_0)}E\left[\frac{\sqrt{4q\cos\theta_0}}{R_0(q,\theta_0)}\right], \tag{3.137}
$$

where $R_0(q,\theta_0)=\left(1+2q\cos\theta_0+q^2\right)^{\frac{1}{2}}$.

It can easily be shown that if $\theta_0 \ll 1$ the value of the magnetic field increases in proportion to θ_0^{-2}. Representative calculations of $H_0^{-1}H\,(q,0)$ are plotted in Figure 3.15. Computations based upon Formulae (3.135)–(3.137) and on the numerical solution of System (3.130) were found to be in almost perfect agreement.

In conclusion we remark that the growth of the magnetic field concentration at the apertures is restricted by some threshold value of the magnetic field, the so-called *critical* value, H_c. (This is characteristic for superconductors of the first kind, such as lead, tin, and niobium.) It is interesting that this phenomenon could be used for quite different purposes, such as localised concentration of the magnetic field, or attenuation (i.e., suppression) of the magnetic field in some localised region of space.

If the transition of the shell ($T > T_c$) to the superconducting state ($T \leq T_c$) is induced by a refrigeration process that starts from the equatorial zone of the shell, the initial *frozen* magnetic flux is $\Phi_e = \pi a^2 H_0$. As the superconducting state occupies a larger part of the surface of the shell, the magnitude of the magnetic field increases, attaining its largest value on the aperture planes where the refrigeration process terminates. By contrast, if the refrigeration process starts at the shell rims, the initial *frozen* magnetic flux is $\Phi_r = \pi a^2 \sin^2 \theta_0 . H_0$, and the movement of the superconducting phase to the equatorial zone leads to the attenuation of the mean value of the magnetic field because the *frozen* magnetic flux has a constant value at any cross-section of the shell.

3.7 Screening number of superconducting shells

In this section, we consider another example of a mixed boundary-value problem for Laplace's equation in which Neumann boundary conditions are specified on a spherical shell surface. We consider a superconducting shell, shaped as a thin spherical shell with a single circular hole. It is placed in an external magnetostatic homogeneous field $\overrightarrow{H_0}$, directed at angle α relative to the z-axis (see Figure 3.16), which is the axis of rotational symmetry of the shell.

With no loss of generality, we may suppose that vector $\overrightarrow{H_0}$ lies in a plane xOz, so that its vertical and horizontal components are

$$H_z^0 = H_0 \cos \alpha \equiv H_{\parallel}^0, \quad H_x^0 = H_0 \sin \alpha \equiv H_{\perp}^0. \tag{3.138}$$

The magnetostatic potential function $\Psi^0(r, \theta, \varphi)$ describing this magnetic field $\overrightarrow{H_0} = -\nabla \Psi^0$ in spherical coordinates is

$$\Psi^0(r, \theta, \varphi) = -H_0 . r (\cos \alpha \cos \theta + \sin \alpha \sin \theta \cos \varphi) \tag{3.139}$$

$$= -H_{\parallel}^0 . r \cos \theta - H_{\perp}^0 . r \sin \theta \cos \varphi. \tag{3.140}$$

In the interior region $0 \leq r < a$, the total potential has the form

$$\Psi^{(i)} = H_{\parallel}^0 . a \sum_{n=0}^{\infty} a_n^{(i)} \left(\frac{r}{a}\right)^n P_n(\cos \theta) + H_{\perp}^0 . a \sum_{n=1}^{\infty} b_n^{(i)} \left(\frac{r}{a}\right)^n P_n^1(\cos \theta) \cos \varphi, \tag{3.141}$$

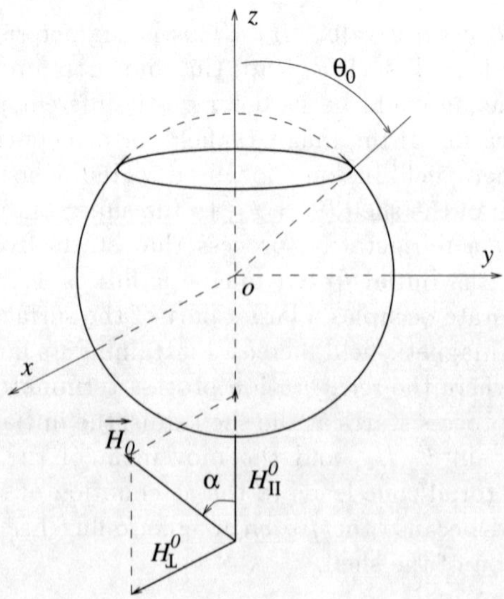

FIGURE 3.16. Spherically-shaped superconducting shell.

whereas in the unbounded region $r > a$, the total potential has the form

$$\Psi^{(e)} = \Psi^0 + H_{||}^0 . a \sum_{n=0}^{\infty} a_n^{(e)} \left(\frac{r}{a}\right)^{-n-1} P_n(\cos\theta)$$
$$+ H_{\perp}^0 . a \sum_{n=1}^{\infty} b_n^{(e)} \left(\frac{r}{a}\right)^{-n-1} P_n^1(\cos\theta)\cos\varphi. \quad (3.142)$$

As mentioned in the previous section, superconducting shells are modelled by ideal diamagnetic materials of zero relative permeability, so that the normal component of magnetic field (in this case, H_r) vanishes at the shell surface. The continuity condition at $r = a$ takes the form

$$H_r^{(i)}(a, \theta, \varphi) = H_r^{(e)}(a, \theta, \varphi), \quad \theta \in (0, \pi), \; \varphi \in (0, 2\pi), \quad (3.143)$$

where the superscripts i and e refer to the interior and exterior regions, respectively. Furthermore, on the screen surface, the normal components satisfy

$$H_r^{(i)}(a, \theta, \varphi) = H_r^{(e)}(a, \theta, \varphi) = 0, \quad \theta \in (\theta_0, \pi), \; \varphi \in (0, 2\pi). \quad (3.144)$$

Also we require continuity on the aperture $(r = a, \theta \in (0, \theta_0), \varphi \in (0, 2\pi))$ for the other magnetic field components:

$$H_\theta^{(i)}(a, \theta, \varphi) = H_\theta^{(e)}(a, \theta, \varphi), \quad (3.145)$$

$$H_\varphi^{(i)}(a, \theta, \varphi) = H_\varphi^{(e)}(a, \theta, \varphi). \quad (3.146)$$

To these conditions are added the finiteness of the energy integral

$$\iiint_V \left|\nabla\Psi^{(i)}\right|^2 dV < \infty, \tag{3.147}$$

which determines the solution class for Fourier coefficients $a_n^{(i,e)}$ and $b_n^{(i,e)}$.

Condition (3.143) implies (for $n = 1, 2, 3, ...$)

$$na_n^{(i)} = -\delta_{1n} - (n+1)\, a_n^{(e)}, \tag{3.148}$$

$$nb_n^{(i)} = -\delta_{1n} - (n+1)\, b_n^{(e)}. \tag{3.149}$$

Enforcing the conditions (3.144)– (3.146) leads to two independent systems of dual series equations for the internal Fourier coefficients,

$$\sum_{n=1}^{\infty} \frac{2n+1}{n+1} a_n^{(i)} P_n^1(\cos\theta) = -\frac{3}{2}\sin\theta, \quad \theta \in (0, \theta_0,) \tag{3.150}$$

$$\sum_{n=1}^{\infty} na_n^{(i)} P_n^1(\cos\theta) = 0, \quad \theta \in (\theta_0, \pi), \tag{3.151}$$

and

$$\sum_{n=1}^{\infty} \frac{2n+1}{n+1} b_n^{(i)} P_n^1(\cos\theta) = -\frac{3}{2}\sin\theta, \quad \theta \in (0, \theta_0), \tag{3.152}$$

$$\sum_{n=1}^{\infty} nb_n^{(i)} P_n^1(\cos\theta) = 0, \quad \theta \in (\theta_0, \pi), \tag{3.153}$$

The finite energy condition (3.147) requires

$$\sum_{n=1}^{\infty} \frac{n}{2n+1} \left|a_n^{(i)}\right|^2 < \infty, \quad \sum_{n=1}^{\infty} \frac{n^2(n+1)}{2n+1} \left|b_n^{(i)}\right|^2 < \infty, \tag{3.154}$$

so that $\left\{a_n^{(i)}\right\}_{n=1}^{\infty} \in l_2(0)$ and $\left\{b_n^{(i)}\right\}_{n=0}^{\infty} \in l_2(2)$.

To solve Equations (3.152) and (3.153), set $x_n = nb_n^{(i)}$ and integrate (3.153) using Formula (B.49) (see Appendix) to obtain

$$\sum_{n=1}^{\infty} \frac{2n+1}{n(n+1)} x_n P_n^1(\cos\theta) = -\frac{3}{2}\sin\theta, \quad \theta \in (0, \theta_0) \tag{3.155}$$

$$\sum_{n=1}^{\infty} x_n P_n(\cos\theta) = c_1, \quad \theta \in (\theta_0, \pi) \tag{3.156}$$

where c_1 is the constant of integration. From the Dirichlet-Mehler representation for Legendre polynomials (1.149) we readily deduce representations of the same type for associated Legendre functions:

$$P_n^1(\cos\theta) = \frac{2\sqrt{2}}{\pi}\frac{1}{\sin\theta}\frac{n(n+1)}{2n+1}\int_0^\theta \frac{\sin(n+\frac{1}{2})\varphi\sin\varphi}{\sqrt{\cos\varphi-\cos\theta}}d\varphi. \tag{3.157}$$

Now, following the well-established procedure described in Section 2.1, transform (3.155) and (3.156) to the equations

$$\sum_{n=1}^{\infty} x_n \sin\left(n+\frac{1}{2}\right)\theta = \begin{cases} -\sin\frac{3}{2}\theta, & \theta\in(0,\theta_0) \\ c_1\sin\frac{1}{2}\theta, & \theta\in(\theta_0,\pi) \end{cases}. \tag{3.158}$$

Exploit orthogonality of the trigonometric functions on $(0,\pi)$ to obtain, for $m = 1, 2, ...,$

$$x_m = -R_{1m}(\theta_0) - c_1 R_{0m}(\theta_0) \tag{3.159}$$

and, corresponding to $m = 0$, an equation for c_1,

$$0 = -R_{10}(\theta_0) + [1 - R_{00}(\theta_0)]\, c_1, \tag{3.160}$$

where

$$R_{nm}(\theta_0) = 2\hat{Q}_{n-1,m-1}^{(\frac{1}{2},-\frac{1}{2})}(\cos\theta_0), \tag{3.161}$$

with $\hat{Q}_{nm}^{(-\frac{1}{2},\frac{1}{2})}$ denoting the usual normalised incomplete scalar product.

Thus, the final analytical form of the solution is

$$x_m = -\left\{R_{1m}(\theta_0) + \frac{R_{10}(\theta_0)}{1-R_{00}(\theta_0)}R_{0m}(\theta_0)\right\}. \tag{3.162}$$

From (3.161), it is evident that $x_m = O\left(m^{-1}\right)$ as $m \to \infty$; hence $b_m^{(i)} = O(m^{-2})$ as $m \to \infty$, and the obtained solution does in fact lie in $l_2(2)$.

The dual Equations (3.150) and (3.151) for the remaining coefficients $a_n^{(i)}$ may be solved in various ways. We start by integrating both equations:

$$\sum_{n=1}^{\infty} \frac{2n+1}{n+1} a_n^{(i)} P_n(\cos\theta) = -\frac{3}{2}\cos\theta + c_2, \quad \theta\in(0,\theta_0) \tag{3.163}$$

$$\sum_{n=1}^{\infty} \frac{a_n^{(i)}}{n+1} P_n^1(\cos\theta) = 0, \quad \theta\in(\theta_0,\pi) \tag{3.164}$$

where c_2 is an integration constant to be determined. In deducing (3.164) we used the well-known formula (see Appendix, (B.49) and (B.58))

$$P_{n+1}(x) - P_{n-1}(x) = -\frac{2n+1}{n(n+1)}\sqrt{1-x^2}P_n^1(x).$$

Integrate Equation (3.164) again to obtain

$$\sum_{n=1}^{\infty} \frac{a_n^{(i)}}{n+1} P_n(\cos\theta) = c_3, \quad \theta \in (\theta_0, \pi) \tag{3.165}$$

where c_3 is another constant of integration to be determined.

The dual series Equations (3.163) and (3.165) may be solved in various ways. We use a standard Abel integral transform to convert to equations with trigonometric kernels:

$$\sum_{n=1}^{\infty} \frac{2n+1}{n+1} a_n^{(i)} \cos\left(n+\frac{1}{2}\right)\theta = -\frac{3}{2}\cos\frac{3}{2}\theta + c_2\cos\frac{\theta}{2}, \theta \in (0, \theta_0) \tag{3.166}$$

$$\sum_{n=1}^{\infty} \frac{a_n^{(i)}}{n+1} \sin\left(n+\frac{1}{2}\right)\theta = c_3 \sin\frac{\theta}{2}, \quad \theta \in (\theta_0, \pi). \tag{3.167}$$

The dual Equations (3.166) and (3.167) are equivalent to two systems of functional equations,

$$\sum_{n=1}^{\infty} \frac{2n+1}{n+1} a_n^{(i)} \cos\left(n+\frac{1}{2}\right)\theta = \begin{cases} -\frac{3}{2}\cos\frac{3}{2}\theta + c_2\cos\frac{1}{2}\theta, & \theta \in (0, \theta_0) \\ c_3\cos\frac{1}{2}\theta, & \theta \in (\theta_0, \pi) \end{cases} \tag{3.168}$$

and

$$\sum_{n=1}^{\infty} \frac{a_n^{(i)}}{n+1} \sin\left(n+\frac{1}{2}\right)\theta = \begin{cases} -\frac{1}{2}\sin\frac{3}{2}\theta + c_2\sin\frac{1}{2}\theta, & \theta \in (0, \theta_0) \\ c_3\sin\frac{1}{2}\theta, & \theta \in (\theta_0, \pi) \end{cases}. \tag{3.169}$$

A retrospective justification for the differentiation process in obtaining (3.168) is needed, but none is needed for (3.169). It is obvious that the solution of the first equation lies in the required class (l_2),

$$\frac{2m+1}{m+1} a_m^{(i)} = -\frac{3}{2} Q_{1m}(\theta_0) + c_2 Q_{0m}(\theta_0) - c_3 Q_{0m}(\theta_0), \quad m \geq 1 \tag{3.170}$$

where $Q_{nm}(\theta_0) = \hat{Q}_{nm}^{(-\frac{1}{2},\frac{1}{2})}(\cos\theta_0)$ is the usual normalised incomplete scalar product.

By considering the product of (3.168) with $\cos\frac{1}{2}\theta$, the constants c_2 and c_3 are related by

$$-\frac{3}{2} Q_{10}(\theta_0) + c_2 Q_{00}(\theta_0) = c_3 \left[1 - Q_{00}(\theta_0)\right]. \tag{3.171}$$

If the constants c_2 and c_3 are arbitrarily chosen, the solution of Equation (3.169) does not lie in the required class. The correct solution is found by requiring the function to be continuous at the point $\theta = \theta_0$, leading to

$$-\frac{1}{2}\sin\frac{3}{2}\theta_0 + c_2 \sin\frac{\theta_0}{2} = c_3 \sin\frac{\theta_0}{2}. \tag{3.172}$$

From (3.170), (3.171), and (3.172) we finally deduce

$$a_m^{(i)} = -\frac{3}{2}\frac{m+1}{2m+1}\left\{Q_{1m}(\theta_0) - \frac{\sin\frac{3}{2}\theta_0}{3\sin\frac{1}{2}\theta_0}Q_{0m}(\theta_0)\right\}. \tag{3.173}$$

The closed form for the magnetostatic potential $\Psi^{(i)}(r,\theta,\varphi)$ is

$$\Psi^{(i)}(r,\theta,\varphi) = -\frac{3}{2}H_{\|}^{0}.a\sum_{n=1}^{\infty}\frac{n+1}{2n+1}Q_{1n}^{(1)}(\theta_0)\left(\frac{r}{a}\right)^n P_n(\cos\theta)$$
$$- H_{\perp}^{0}.a\cos\varphi\sum_{n=1}^{\infty}\frac{1}{n}R_{1n}^{(1)}(\theta_0)\left(\frac{r}{a}\right)^n P_n^1(\cos\theta) \tag{3.174}$$

where

$$R_{1n}^{(1)}(\theta_0) = R_{1n}(\theta_0) + \frac{R_{10}(\theta_0)}{1-R_{00}(\theta_0)}R_{0n}(\theta_0),$$

and

$$Q_{1n}^{(1)}(\theta_0) = Q_{1n}(\theta_0) - \frac{\sin\frac{3}{2}\theta_0}{3\sin\frac{1}{2}\theta_0}Q_{0n}(\theta_0).$$

A measure of screening effectiveness of the superconducting open spherical shell is the screening number (recall that α defines the direction of the external magnetic field),

$$K = H_0^{-1}H(0,\theta,\varphi) = \left(K_{\|}^2\cos^2\alpha + K_{\perp}^2\sin^2\alpha\right)^{\frac{1}{2}}, \tag{3.175}$$

where $H(0,\theta,\varphi)$ is the magnetic field at the centre of the shell, and $K_{\|}, K_{\perp}$ are screening numbers of the longitudinal and transverse magnetic field, respectively. It is evident that

$$K_{\|} = Q_{11}^{(1)}(\theta_0), \quad K_{\perp} = R_{11}^{(1)}(\theta_0). \tag{3.176}$$

Suppressing rather bulky details, the distribution of the magnetic field, which penetrates into the screen, when taken along the axis of the screen (with $q = r/a \le 1$) has components

$$\begin{aligned} H_r^{(i)}\left(q, {}^{0}_{\pi}, \varphi\right) &= \pm H_{\|}L(\pm q,\theta_0), \\ H_\theta^{(i)}\left(q, {}^{0}_{\pi}, \varphi\right) &= \mp H_{\perp}R(\pm q,\theta_0)\cos\varphi, \\ H_\varphi^{(i)}\left(q, {}^{0}_{\pi}, \varphi\right) &= H_{\perp}R(\pm q,\theta_0)\sin\varphi, \end{aligned} \tag{3.177}$$

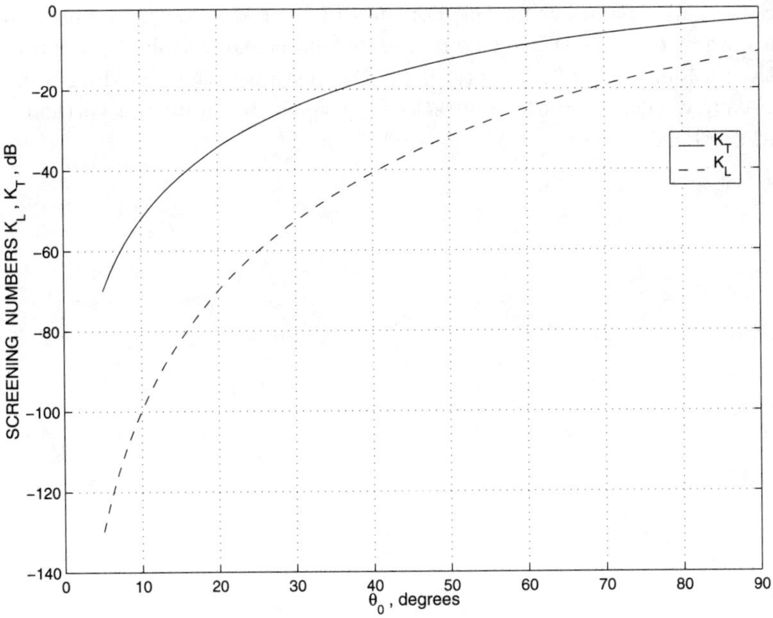

FIGURE 3.17. Longitudinal (K_L) and transversal (K_T) screening numbers for the spherically-shaped superconducting shell.

where

$$
\begin{aligned}
&2\pi t^2 R(t,\theta_0) \\
&\quad = -\frac{t}{2}\sin 2\theta_0 - t^{-1}\arctan\left[\frac{t\sin\theta_0}{1-t\cos\theta_0}\right] \\
&\quad + (1-t)^3\frac{\sin\theta_0}{1-2t\cos\theta_0+t^2} + 2t^2\arctan\left[\frac{\sin\theta_0}{t-\cos\theta_0}\right] \\
&+ \frac{R_{10}(\theta_0)}{1-R_{00}(\theta_0)}t\left\{\pi R_{00}(\theta_0) + \frac{(1-t)\sin\theta_0}{1-2t\cos\theta_0+t^2} + \arctan\left[\frac{\sin\theta_0}{t-\cos\theta_0}\right]\right\},
\end{aligned}
$$

and

$$
\begin{aligned}
&\frac{2\pi}{3}t^2 L(t,\theta_0) \\
&\quad = -\frac{2}{3t}\arctan\left[\frac{t\sin\theta_0}{1-t\cos\theta_0}\right] + \frac{(1+t^3)\sin\theta_0}{2(1-2t\cos\theta_0+t^2)} + \frac{1+t}{6}\sin\theta_0 \\
&\quad + \frac{2}{3}t^2\arctan\left[\frac{\sin\theta_0}{\cos\theta_0-t}\right] - \frac{\sin\frac{3}{2}\theta_0}{3\sin\frac{1}{2}\theta_0}\frac{t(1+t)}{2}\frac{\sin\theta_0}{1-2t\cos\theta_0+t^2}.
\end{aligned}
$$

It follows from the last formula that $L(-1,\theta_0) = 0$; this implies that $H_r(1,\pi,\varphi) = 0$, i.e., the boundary condition (3.144) holds at this point.

Some calculations using the Formula (3.176) are shown in Figure 3.17. These show that the transverse magnetic field is less well shielded compared with the longitudinal magnetic field. For instance, the shielding numbers of a cavity with $\theta_0 = 5°$ have ratio $K_{\perp}/K_{||} \backsim 10^3$ (note the vertical scale is in decibels).

4

Electrostatic Potential Theory for Open Spheroidal Shells

After spherical geometry, spheroidal geometry provides the simplest setting for three-dimensional potential theory. This chapter considers the potential surrounding various open spheroidal shell structures. It presents a significant extension and generalisation of the spherical shell studies because various combinations of cavity size and aspect ratio of the shell produce extremely interesting structures for physical and engineering applications; the hollow cylinder is one example.

As the ratio between the minor and major axes increases, a closed spheroidal surface takes widely differing shapes ranging from the disk through the oblate spheroid, to the sphere, through the prolate spheroid, to the limiting form of a thin cylinder of finite length or of a needle-shaped structure.

Whilst cutting slots in the spheroidal shell expands the possibilities of modelling of real physical objects, it increases the analytical complexity of the corresponding boundary-value problem. This accounts for the fact that, until now, only the simplest problems for conductors described in spheroidal coordinates have been analysed in detail, namely closed spheroids (see, for example, [26]) and spheroidal caps [12].

Nevertheless, significant progress can be made for axially symmetric structures in this setting. The Laplace operator separates in this coordinate system, so that dual or triple series equations can be constructed by enforcement of mixed boundary conditions on the conducting surface or the aperture as appropriate. As explained in Chapter 1, these equations are equivalent to (and can be reformulated as) a certain first-kind Fredholm integral equation. The original first-kind equations may be transformed to a Fredholm second-kind infinite matrix equation by the method of regularisation. As we have already seen, the regularised system of equations possesses many desirable features including rapid convergence of the solution, obtained by truncation methods, to the exact one, and guaranteed accuracy of computations.

As for the open spherical shell studies, we will consider spheroidal shells in which one or two apertures are introduced in an axisymmetric fashion. Prolate and oblate spheroids with such apertures will be discussed. After an introductory formulation (Section 4.1) of mixed boundary value problems in the spheroidal coordinate systems, we first examine the thin, perfectly conducting, prolate spheroidal shell with one circular hole (Section 4.2). The prolate spheroidal shell in which a longitudinal slot is introduced to produce a pair of equally sized spheroidal caps is then considered (Section 4.3). When the caps are oppositely charged, we may calculate the capacitance of the resulting condensor. The complementary structure, a prolate spheroidal shell with two symmetrically disposed circular holes, or *spheroidal barrel* is discussed in the following section (4.4); the hollow right circular cylinder may be viewed as a limiting case.

The next two sections examine the analogous structures for *oblate* spheroidal shells with two apertures: the oblate shell with a longitudinal slot, which produces a pair of equally sized spheroidal caps (Section 4.5), and the oblate spheroidal barrel (Section 4.6). In the final section, the capacitance of the various shells (when positively charged) and condensors (comprising oppositely charged components) are examined as a function of aspect ratio and aperture size.

In contrast to closed structures, there have been relatively few analytical studies of the electrostatic potential distribution surrounding three dimensional open structures with cavities and edges. Viewed as an example of a three-dimensional finite open conductor with a cavity, these canonical problems and their solutions can be used for the development and testing of approximate methods of general applicability in potential calculations.

4.1 Formulation of mixed boundary value problems in spheroidal geometry

As stated in the Introduction, we consider infinitely thin, perfectly conducting, open axisymmetric spheroidal shells (see Figure 4.1) charged to some electrostatic potential U. We shall use prolate and oblate spheroidal coordinates in the trigonometric coordinate form (α, β, φ) described in Sections 1.1.4 and 1.1.5. In both coordinate systems, the surface of each shell S_0 lies on a coordinate surface $\alpha = const = \alpha_0$ (which is a spheroid), whilst the interval of β defining S_0 depends on the particular structure. Thus, S_0 is defined by

$$\alpha = \alpha_0, \quad \varphi \in [0, 2\pi], \quad \text{and} \quad \beta \in I,$$

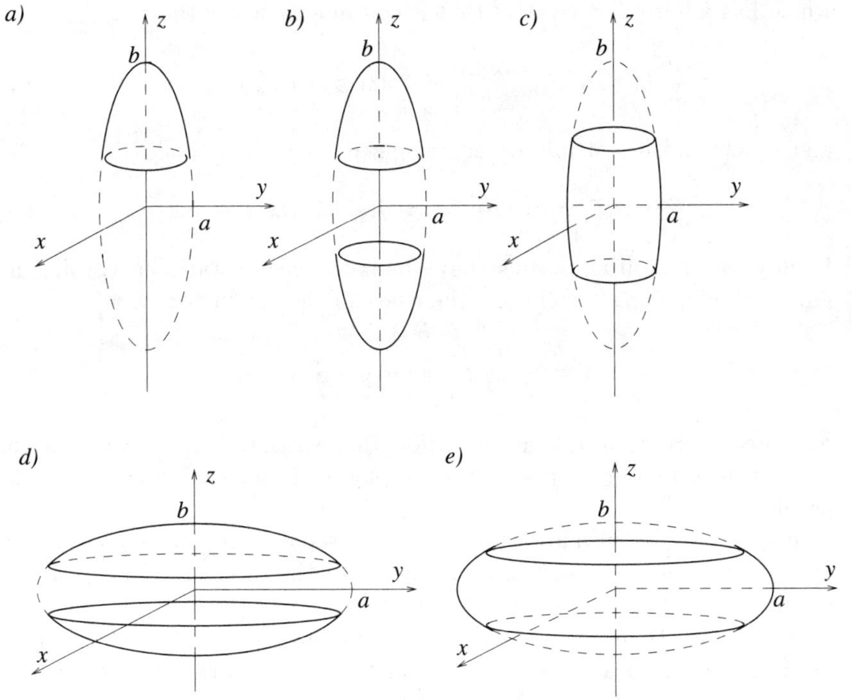

FIGURE 4.1. Spheroidal shell geometry: prolate and oblate

where I is a subinterval, or several disjoint subintervals of $[0, \pi]$; the complementary interval $I' = [0, \pi] \backslash I$ allows us to define the aperture or slot S_1 in the spheroidal surface by

$$\alpha = \alpha_0, \quad \varphi \in [0, 2\pi], \quad \text{and} \quad \beta \in I'.$$

Our aim is to construct the solution for electrostatic field potential distribution $U(\alpha, \beta, \varphi)$ near the charged open shell S_0 when the potential is specified in the form $U(\alpha_0, \beta, \varphi) = f(\beta, \varphi)$ (for $\beta \in I$) on the surface of the shell; we shall also calculate its associated capacitance and surface charge distribution.

This boundary value problem of potential theory for spheroidal conductors may be formulated as described in Section 1.3. Thus, we seek an electrostatic potential $U(\alpha, \beta, \varphi)$ that is harmonic in R^3,

$$\Delta U(\alpha, \beta, \varphi) = 0, \tag{4.1}$$

which satisfies the Dirichlet boundary condition on the surface of the conductor S_0,

$$U(\alpha_0 - 0, \beta, \varphi) = U(\alpha_0 + 0, \beta, \varphi) = f(\beta, \varphi) \quad \text{for} \quad \beta \in I, \varphi \in [0, 2\pi], \tag{4.2}$$

which has a normal derivative that is continuous across the slot S_1,

$$\frac{d}{d\alpha}U(\alpha,\beta,\varphi)|_{\alpha=\alpha_0-0}^{\alpha=\alpha_0+0}=0 \text{ for } \beta\in I',\varphi\in[0,2\pi], \tag{4.3}$$

and which vanishes at infinity according to

$$U(\alpha,\beta,\varphi)=O(r^{-1})=O(e^{-\alpha}) \quad \text{as } \alpha\to\infty. \tag{4.4}$$

Finally, the potential U must have bounded electrostatic energy in any finite volume of space including the edges of the conductor:

$$W=\frac{1}{2}\iiint_V |\text{grad}\, U|^2\, dV<\infty. \tag{4.5}$$

As noted in Section 1.3, any solution that satisfies all these conditions is necessarily unique and provides the physically relevant solution to this problem.

In spheroidal coordinates, the method of separation of variables for Laplace's equation leads to partial solutions of the form (1.31) or (1.35) in prolate or oblate coordinates, respectively.

We confine attention to axisymmetric potential distributions (so $\frac{\partial}{\partial\varphi}U=0$). Thus, the separation constant m of (1.72) or (1.74) is 0; furthermore the boundedness of the potential $U(\alpha,\beta)=U(\alpha,\beta,\phi)$ requires that the separation constant n be zero or a positive integer $n=0,1,2,\ldots$.

Thus, the solution that satisfies Laplace's equation, the continuity conditions on the boundary $\alpha=\alpha_0$ between the interior and exterior regions, and the decay condition at infinity, takes the following form in prolate spheroidal coordinates,

$$U(\alpha,\beta)=\sum_{n=0}^{\infty}C_n^{(p)}P_n(\cos\beta)\begin{cases}P_n(\cosh\alpha), & 0\le\alpha\le\alpha_0,\\ Q_n(\cosh\alpha)P_n(\cosh\alpha_0)/Q_n(\cosh\alpha_0), & \alpha>\alpha_0,\end{cases} \tag{4.6}$$

whilst in oblate spheroidal coordinates it takes the form

$$U(\alpha,\beta)=\sum_{n=0}^{\infty}C_n^{(o)}P_n(\cos\beta)\begin{cases}p_n(i\sinh\alpha), & 0\le\alpha\le\alpha_0,\\ q_n(i\sinh\alpha)p_n(i\sinh\alpha_0)/q_n(i\sinh\alpha_0), & \alpha>\alpha_0.\end{cases} \tag{4.7}$$

Here, $P_n(z)$, $Q_n(z)$ $(z\ge1)$ are the Legendre functions of the first and second kind, respectively, $P_n(\cos\beta)$ is a Legendre polynomial (with trigonometrical argument) and

$$p_n(z)=i^{-n}P_n(z), q_n(z)=i^{n+1}Q_n(z).$$

The unknown (Fourier) coefficients $\left\{C_n^{(p)}\right\}_{n=0}^{\infty}$ and $\left\{C_n^{(o)}\right\}_{n=0}^{\infty}$ are to be found.

Selecting the volume for integration V in (4.5) as the internal region of the spheroid ($\alpha \leq \alpha_0$), the prolate geometry coefficients must satisfy

$$W = \pi \frac{d}{2} \sinh \alpha_0 \sum_{n=0}^{\infty} \frac{1}{2n+1} \left| C_n^{(p)} \right|^2 \frac{d}{d\alpha} \left[P_n(\cosh \alpha) \right]^2 |_{\alpha=\alpha_0} < \infty, \quad (4.8)$$

whereas the oblate geometry coefficients must satisfy

$$W = \pi \frac{d}{2} \cosh \alpha_0 \sum_{n=0}^{\infty} \frac{1}{2n+1} \left| C_n^{(o)} \right|^2 \frac{d}{d\alpha} \left[p_n(i \sinh \alpha) \right]^2 |_{\alpha=\alpha_0} < \infty. \quad (4.9)$$

Taking into account the asymptotic behaviour of Legendre functions as $n \to \infty$ (see Appendix, (B.72) and (B.73)), it follows from (4.8) and (4.9) that the rescaled coefficients

$$A_n^{(p)} = C_n^{(p)} P_n(\cosh \alpha_0), A_n^{(o)} = C_n^{(o)} p_n(i \sinh \alpha_0) \quad (4.10)$$

belong to the functional space of square summable sequences l_2:

$$\left\{A_n^{(p)}\right\}_{n=0}^{\infty}, \left\{A_n^{(o)}\right\}_{n=0}^{\infty} \in l_2. \quad (4.11)$$

Thus, solutions to the potential problem will be sought in the following form for prolate spheroidal coordinates,

$$U(\alpha, \beta) = \sum_{n=0}^{\infty} A_n^{(p)} P_n(\cos \beta) \left\{ \begin{array}{l} P_n(\cosh \alpha)/P_n(\cosh \alpha_0), \alpha \leq \alpha_0 \\ Q_n(\cosh \alpha)/Q_n(\cosh \alpha_0), \alpha > \alpha_0 \end{array} \right\} \quad (4.12)$$

and for oblate spheroidal coordinates in the form

$$U(\alpha, \beta) = \sum_{n=0}^{\infty} A_n^{(o)} P_n(\cos \beta) \left\{ \begin{array}{l} p_n(i \sinh \alpha)/p_n(i \sinh \alpha_0), \alpha \leq \alpha_0 \\ q_n(i \sinh \alpha)/q_n(i \sinh \alpha_0), \alpha > \alpha_0 \end{array} \right\}. \quad (4.13)$$

Once the coefficients $A_n^{(p)}$ and $A_n^{(o)}$ are found, the electrostatic field potential $U(\alpha, \beta)$ is fully determined at any point of the space. Recall that *axisymmetric* problems are considered. The rigorous solution to be developed in the following sections makes it possible to analyse in detail the potential and electrostatic field near the conductor's edges.

The surface charge density σ accumulated on the conductor surface ($\alpha = \alpha_0, \beta \in I$) is defined by the jump in the normal component E_α of the electric field across the surface (cf. Equation (1.2)),

$$\sigma(\beta) = \frac{1}{4\pi} \left\{ E_\alpha(\alpha_0 + 0, \beta) - E_\alpha(\alpha_0 - 0, \beta) \right\}. \quad (4.14)$$

The normal component of the electric field $\overrightarrow{E} = -\operatorname{grad} U$ is

$$E_\alpha(\alpha, \beta)|_{\alpha=\alpha_0} = h_\alpha^{-1} \frac{d}{d\alpha} U(\alpha, \beta)|_{\alpha=\alpha_0}$$

(where h_α is the metric coefficient), so using (4.12), (4.13), and metric coefficients in spheroidal coordinates (see Section 1.1.4), the expression for σ in the prolate spheroidal system is

$$\sigma(\beta) = \frac{1}{4\pi} \frac{1}{\frac{d}{2}\sqrt{\sinh^2 \alpha_0 + \sin^2 \beta}} \sum_{n=0}^{\infty} \Lambda_n(\alpha_0) A_n^{(p)} P_n(\cos \beta), \tag{4.15}$$

where

$$\Lambda_n(\alpha_0) = \left[\sinh \alpha_0 \left(\frac{Q_n'(\cosh \alpha_0)}{Q_n(\cosh \alpha_0)} - \frac{P_n'(\cosh \alpha_0)}{P_n(\cosh \alpha_0)} \right) \right]^{-1}. \tag{4.16}$$

Employing the value of the Wronskian (B.69)

$$W(P_n, Q_n)(z) = P_n'(z) Q_n(z) - P_n(z) Q_n'(z) = (1 - z^2)^{-1},$$

we may simplify

$$\Lambda_n(\alpha_0) = [\sinh \alpha_0 P_n(\cosh \alpha_0) Q_n(\cosh \alpha_0)]^{-1}. \tag{4.17}$$

In the oblate spheroidal system, the charge density is

$$\sigma(\beta) = \frac{1}{4\pi} \frac{1}{\frac{d}{2}\sqrt{\cosh^2 \alpha_0 - \sin^2 \beta}} \sum_{n=0}^{\infty} \lambda_n(\alpha_0) A_n^{(o)} P_n(\cos \beta), \tag{4.18}$$

where the factor

$$\lambda_n(\alpha_0) = \{\cosh \alpha_0 q_n(i \sinh \alpha_0) p_n(i \sinh \alpha_0)\}^{-1} \tag{4.19}$$

arises from employing the value of the Wronskian of the pair p_n, q_n. It is worth noting that the surface charge density expressions (4.15) and (4.18) vanish for the range of β corresponding to the aperture surface.

The total charge Q on each isolated component of the conducting surface is obtained by integration of surface charge density σ over the component surface.

In considering particular problems, we will suppress the subscripts (p) and (o) on A_n when the context is unambiguous. In all calculations presented below, the semi-axial distance b is taken to be unity; thus, if the ratio a/b is specified, the interfocal distance d may be determined.

4.2 The prolate spheroidal conductor with one hole

Let us consider a prolate spheroidal shell S_0 with one circular hole determined by an angle β_0 so that S_0 is defined by

$$\alpha = \alpha_0, \quad 0 \leq \beta \leq \beta_0, \quad \text{and} \quad \phi \in [0, 2\pi].$$

When charged to unit potential, enforcement of the mixed boundary conditions upon Equations (4.12) determining the potential on the spheroidal shell produces the dual series equations

$$\sum_{n=0}^{\infty} A_n P_n(\cos\beta) = 1, \qquad \beta \in [0, \beta_0], \tag{4.20}$$

$$\sum_{n=0}^{\infty} A_n \Lambda_n(\alpha_0) P_n(\cos\beta) = 0, \qquad \beta \in (\beta_0, \pi]. \tag{4.21}$$

Equation (4.20) describes the potential on S_0, whereas (4.21) follows from the continuity of the normal derivative on the slot S_1 and

$$\Lambda_n(\alpha_0) = \left[\sinh\alpha_0 \left(\frac{Q_n'(\cosh\alpha_0)}{Q_n(\cosh\alpha_0)} - \frac{P_n'(\cosh\alpha_0)}{P_n(\cosh\alpha_0)}\right)\right]^{-1}. \tag{4.22}$$

As noted in Section 4.1, this simplifies to

$$\Lambda_n(\alpha_0) = \left[\sinh\alpha_0 P_n(\cosh\alpha_0) Q_n(\cosh\alpha_0)\right]^{-1}. \tag{4.23}$$

Let us introduce the parameter

$$\varepsilon_n = 1 - (2n+1)\sinh\alpha_0 P_n(\cosh\alpha_0) Q_n(\cosh\alpha_0). \tag{4.24}$$

The asymptotics of the Legendre functions (see (B.70) and (B.71)) show that ε_n is asymptotically small (as $n \to \infty$)

$$\varepsilon_n = O(n^{-2}) \quad \text{as } n \to \infty.$$

Define the new coefficients

$$x_n = \frac{\Lambda_n(\alpha_0) A_n}{(2n+1)} = \frac{A_n}{1-\varepsilon_n}, \tag{4.25}$$

so that $\{x_n\}_{n=0}^{\infty} \in l_2$. The system (4.20), (4.21) is thus converted to the standard form:

$$\sum_{n=0}^{\infty} x_n(1-\varepsilon_n) P_n(\cos\beta) = 1, \qquad \beta \in [0, \beta_0], \tag{4.26}$$

$$\sum_{n=0}^{\infty} (2n+1) x_n P_n(\cos\beta) = 0, \qquad \beta \in [\beta_0, \pi]. \tag{4.27}$$

This set of dual series equations has already been considered in Chapter 1; it is a special case of the general set considered in Section 2.1 with $\alpha = \beta = 0$, $m = 0$, $r_n = \varepsilon_n$, $q_n = 0$, $\eta = \frac{1}{2}$. For these specific parameters, the Abel integral transform method essentially employs the Mehler-Dirichlet integrals, and the following pair of equations is obtained:

$$\sum_{n=0}^{\infty} x_n(1-\varepsilon_n)\cos(n+\frac{1}{2})\beta = \cos\frac{\beta}{2}, \quad \beta \in [0,\beta_0], \tag{4.28}$$

$$\sum_{n=0}^{\infty} x_n \cos(n+\frac{1}{2})\beta = 0, \quad \beta \in [\beta_0,\pi]. \tag{4.29}$$

We may rewrite (4.28) and (4.29) as a Fourier series expression for a single function F that is piecewise defined on two subintervals of $[0,\pi]$,

$$F(\beta) = \sum_{n=0}^{\infty} x_n \cos(n+\frac{1}{2})\beta = \left\{ \begin{array}{ll} F_1(\beta) & \beta \in [0,\beta_0] \\ 0, & \beta \in [\beta_0,\pi] \end{array} \right\}, \tag{4.30}$$

where

$$F_1(\beta) = \cos\frac{1}{2}\beta + \sum_{n=0}^{\infty} x_n \varepsilon_n \cos(n+\frac{1}{2})\beta.$$

A standard argument utilising completeness and orthogonality properties of the trigonometric functions produces a second-kind system of linear algebraic equations for the coefficients $\{x_n\}_{n=0}^{\infty}$,

$$x_s - \sum_{n=0}^{\infty} x_n \varepsilon_n Q_{ns}(\beta_0) = Q_{0s}(\beta_0), \tag{4.31}$$

where $s = 0,1,2,...$, and $Q_{ns}(\beta_0) \equiv \hat{Q}_{ns}^{(-\frac{1}{2},\frac{1}{2})}(\cos\beta_0)$ is the usual normalised incomplete scalar product.

The system (4.31) has the form

$$(I-H)x = b$$

where H is a completely continuous operator on l_2; the norm of H may be bounded uniformly with respect to β_0 by

$$\|H\| \le \max_n |\varepsilon_n| = \varepsilon_0 = |1 - \sinh\alpha_0 Q_0(\cosh\alpha_0)|. \tag{4.32}$$

Considering that

$$Q_0(\cosh\alpha_0) = \frac{1}{2}\log\left[\frac{\cosh\alpha_0+1}{\cosh\alpha_0-1}\right] > \frac{1}{\cosh\alpha_0}, \tag{4.33}$$

the norm is bounded by

$$N \leq 1 - \tanh \alpha_0 < 1. \tag{4.34}$$

One or two iterations of the method of successive approximations provide an approximate analytical solution that is more accurate when α_0 is larger, i.e., the spheroid is closer in form to the sphere. When the eccentricity e is small ($e \ll 1$, $\alpha_0 \to \infty$) it is possible to show, using the hypergeometric representations of P_n, Q_n (see Appendix, (B.70) and (B.71)), that

$$\varepsilon_n = -\frac{e^2}{2(2n-1)(2n+3)} + O(e^4), \tag{4.35}$$

as $n \to \infty$. Accepting (4.35), the solution to (4.31) obtained by the method of successive approximations is

$$x_s = Q_{0s}(\beta_0) - \frac{e^2}{8}\frac{Q_{0s}(\beta_0)}{(s-\frac{1}{2})(s+\frac{3}{2})} - \frac{1}{2\pi}\frac{e^2}{8}\left(2\sin\frac{\beta_0}{2} + \frac{2}{3}\sin\frac{3}{2}\beta_0\right)\left[\frac{\cos(s-\frac{1}{2})\beta_0}{s-\frac{1}{2}} - \frac{\cos(s+\frac{3}{2})\beta_0}{s+\frac{3}{2}}\right] + O(e^4). \tag{4.36}$$

The corresponding approximation for the capacity $C = bx_0$ of the open charged spheroidal conductor is

$$C = \frac{b}{\pi}(\beta_0 + \sin\beta_0) + \frac{be^2}{24\pi}\left(4\beta_0 + \sin\beta_0 - 2\sin 2\beta_0 - \frac{1}{3}\sin 3\beta_0\right) + O(e^4). \tag{4.37}$$

The expression (4.37) coincides with the result [12] obtained by a different method. It agrees with the capacitance of a spherical shell when $e = 0$.

If the value of the eccentricity e is unrestricted, the solution to (4.31) is found by truncation to a finite system of linear algebraic equations that can be efficiently solved numerically. From a methodological point of view, it is worth demonstrating how to accelerate the convergence of the solution of the truncated system to the exact solution. The convergence rate depends upon the behaviour of the parameter ε_n. A more precise statement of its asymptotic behaviour is

$$\varepsilon_n = -\delta^2\left(n + \frac{1}{2}\right)^{-2} + O(n^{-4}), \text{ as } n \to \infty, \tag{4.38}$$

where $\delta^2 = (8\sinh\alpha_0)^{-1}$. With the aim of modifying the System (4.31), we introduce the new parameter

$$\varepsilon_n^* = \varepsilon_n + \delta^2(n + \frac{1}{2})^{-2}, \tag{4.39}$$

so that $\varepsilon_n^* = O(n^{-4})$ as $n \to \infty$. The transformation to be obtained is motivated by the observation that, if one neglects ε_n^*, the resulting dual series equations can be solved exactly. As explained in Section 2.1, the dual equations are then equivalent to a certain ordinary differential equation.

Let

$$g(\beta) = \sum_{n=0}^{\infty} x_n (n + \frac{1}{2})^{-2} \cos(n + \frac{1}{2})\beta, \tag{4.40}$$

and

$$f(\beta) = -\cos\frac{\beta}{2} - \sum_{n=0}^{\infty} x_n \varepsilon_n^* \cos(n + \frac{1}{2})\beta. \tag{4.41}$$

From Equation (4.28) we deduce the second order differential equation

$$g''(\beta) - \delta^2 g(\beta) = f(\beta), \qquad \beta \in [0, \beta_0]. \tag{4.42}$$

Solving this equation (with $g(0) = A, g'(0) = 0$) produces the following expression for g :

$$g(\beta) = A\cosh(\delta\beta) - \frac{\cosh(\delta\beta) - \cos\frac{\beta}{2}}{\delta^2 + \frac{1}{4}} - \sum_{n=0}^{\infty} x_n \varepsilon_n^* \frac{\cosh(\delta\beta) - \cos(n + \frac{1}{2})\beta}{\delta^2 + (n + \frac{1}{2})^2}, \tag{4.43}$$

where

$$A = \sum_{n=0}^{\infty} \frac{x_n}{(n + \frac{1}{2})^2}. \tag{4.44}$$

With the aid of these transformations, we may rewrite (4.28) and (4.29) in the final form

$$x_m - \sum_{n=0}^{\infty} x_n \varepsilon_n^* S_{nm}(\beta_0, \delta) = S_{0m}(\beta_0, \delta), \tag{4.45}$$

where $m = 0, 1, 2, \ldots$, and

$$S_{nm}(\beta_0, \delta) = \left\{ Q_{nm}(\beta_0) + \frac{\tau_n(\beta_0)}{\gamma(\beta_0, \delta)} \delta^2 V_m(\beta_0) \right\} \frac{(n + \frac{1}{2})^2}{(n + \frac{1}{2})^2 + \delta^2},$$

$$\tau_n(\beta_0) = \frac{1}{(n + \frac{1}{2})^2} \left\{ \cos(n + \frac{1}{2})\beta_0 - (\pi - \beta_0)(n + \frac{1}{2}) \sin(n + \frac{1}{2})\beta_0 \right\},$$

$$\gamma(\beta_0, \delta) = \cosh(\delta\beta_0) + (\pi - \beta_0)\delta \sinh(\delta\beta_0),$$

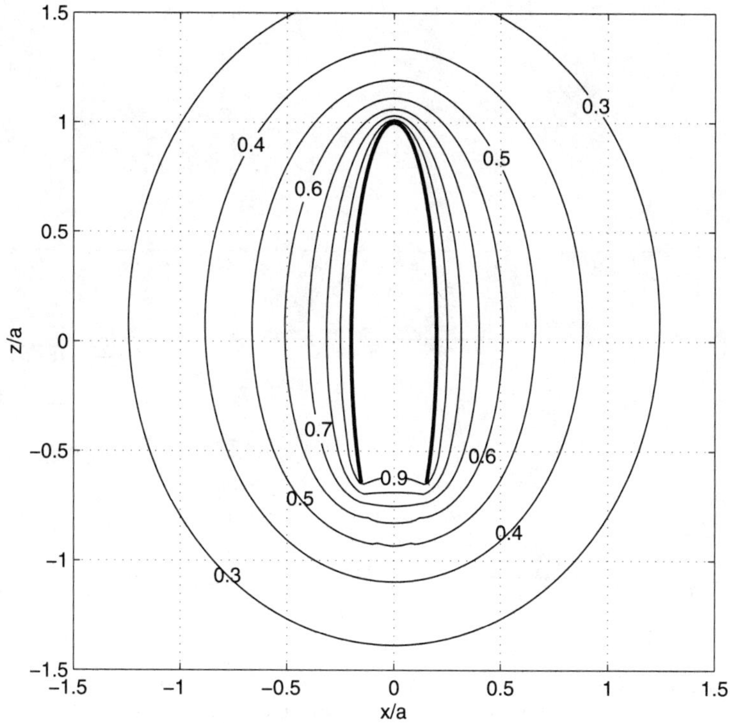

FIGURE 4.2. Electrostatic potential near a prolate spheroidal cap, charged to unit potential, with parameters $a/b = 0.2, \beta_0 = 130^0$. Truncation number $N_{tr} = 11$.

and

$$V_m(\beta_0) = \frac{2}{\pi} \frac{1}{(m+\frac{1}{2})^2 + \delta^2} \delta \cos(m + \frac{1}{2})\beta_0 \sinh(\delta\beta_0) + \\ \frac{2}{\pi} \frac{1}{(m+\frac{1}{2})^2 + \delta^2} (m + \frac{1}{2}) \sin(m + \frac{1}{2})\beta_0 \cosh(\delta\beta_0).$$

The truncation of the System (4.45) is much more rapidly convergent than the truncation of the System (4.31) because ε_n^* decays more rapidly to zero than does ε_n. By determining the asymptotic behaviour of ε_n^*, to $O(n^{-6})$ terms, this procedure may be repeated to obtain another system with a further accelerated convergence rate; however, the complicated form of the system coefficients hardly warrants the effort since satisfactory solutions can be derived from the systems already obtained.

We have computed the electrostatic field distribution surrounding infinitely thin prolate spheroidal conductors charged to unit potential by solving the system (4.31) numerically (taking into account (4.6) and (4.25)).

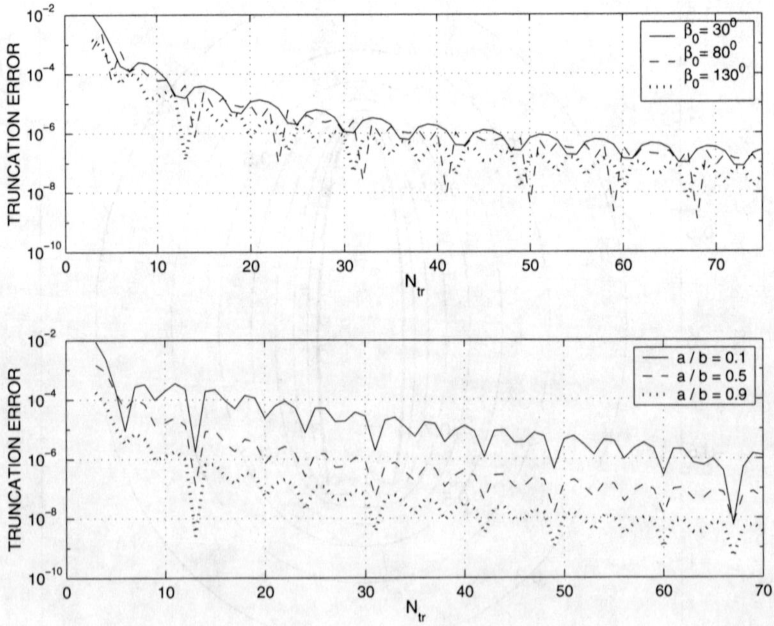

FIGURE 4.3. Normalised error $e(N_{tr})$ as a function of truncation number N_{tr} for the prolate spheroidal cap: (top) with aspect ratio $a/b = 0.5$ and varying β_0; and (bottom) with $\beta_0 = 130^o$ and varying aspect ratio a/b.

An example is shown in Figure 4.2; the ratio of minor to major axes, $a/b = \sinh\alpha_0/\cosh\alpha_0 = 0.2$, and the angular size β_0 of the aperture equals to 130°. The truncation number N_{tr} was chosen to be 11.

Computationally, the system (4.31) is very attractive. The solution of the truncated system converges to the exact solution (the solution of the infinite system) as $N_{tr} \to \infty$. The accuracy of calculations under truncation is illustrated in Figure 4.3, where normalised error is plotted as a function of truncation number. The error is estimated in the maximum norm sense as

$$e\left(N_{tr}\right) = \frac{\max_{n \le N_{tr}} \left| x_n^{N_{tr}+1} - x_n^{N_{tr}} \right|}{\max_{n \le N_{tr}} \left| x_n^{N_{tr}} \right|},$$

where $\left\{x_n^{N_{tr}}\right\}_{n=0}^{N_{tr}}$ denotes the solution to (4.31) truncated to N_{tr} equations. A study of truncated solution accuracy confirms that, in practice, for a wide range of geometrical parameters describing the conductor, the truncated coefficient set $\{x_n\}_{n=0}^{N_{tr}}$ may be obtained correctly to three digits, provided N_{tr} is approximately equal to 10. This accuracy is satisfactory for most calculations concerning the potential.

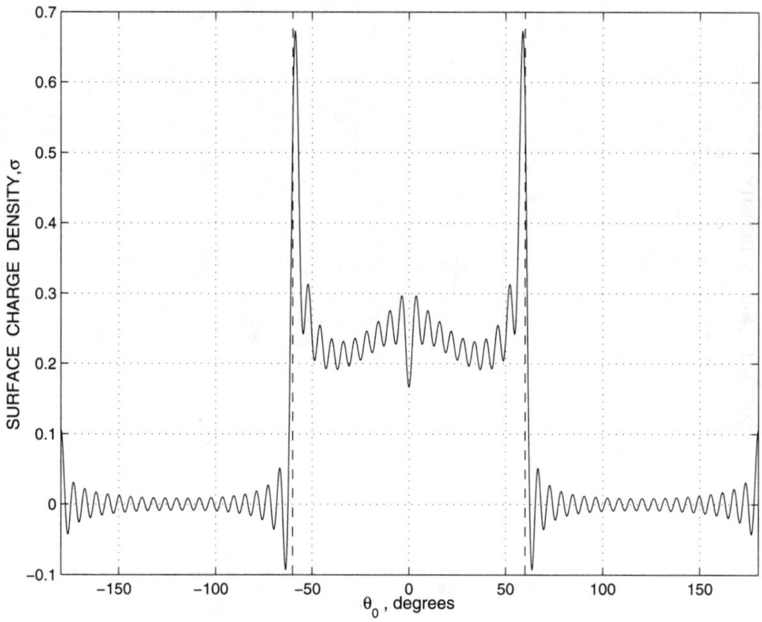

FIGURE 4.4. Surface charge density σ of a prolate spheroidal cap, charged to unit potential, with parameters $a/b = 0.5, \beta_0 = 60^o$. Truncation number $N_{tr} = 60$. The density was computed by simple summation of the Fourier series.

A correspondingly accurate calculation of the surface charge distribution requires more terms than for the potential, as is evident by comparing Equations (4.12) and (4.15), and taking into account the asymptotics (4.24) of the small parameter ε_n. Since the series is much less rapidly convergent than that for the potential, techniques to accelerate the convergence of the series are useful. An example of the surface charge distribution is shown in Figure 4.4 for the shell with ratio of minor to major axes, $a/b = \sinh \alpha_0 / \cosh \alpha_0 = 0.5$, and the angular size of the aperture $\beta_0 = 60^\circ$. The truncation number N_{tr} was chosen to be 60, and the values were computed by a simple summation of the truncated Fourier series, so that a continuous approximation to the surface charge is obtained. The oscillatory results are a manifestation of the familiar Gibbs' phenomenon; the surface charge should be zero outside the interval $[-\beta_0, \beta_0]$. If Cesàro summation is applied (see [9]), the oscillations are much suppressed, and one obtains the results of Figure 4.5. Except in the immediate vicinity of the edge a satisfactory representation of the surface charge is obtained.

It is possible to improve the situation by estimating the leading order of the coefficients in the infinite system and exploiting a known infinite sum which represents the discontinuity exactly. In terms of the coefficients x_n

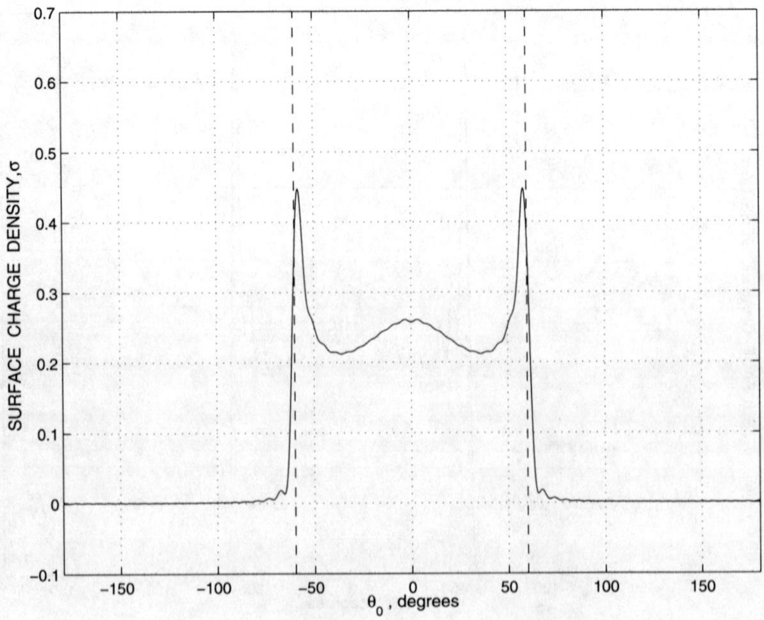

FIGURE 4.5. Surface charge density σ of a prolate spheroidal cap, charged to unit potential, with parameters $a/b = 0.5, \beta_0 = 60^o$. Truncation number $N_{tr} = 60$. The density was computed by Cesàro summation of the Fourier series.

defined in (4.25), the surface charge is

$$\sigma = \frac{1}{4\pi} \frac{1}{\frac{d}{2}\sqrt{\sinh^2 \alpha_0 + \sin^2 \beta}} \sum_{n=0}^{\infty} (2n+1)\, x_n P_n(\cos\beta), \tag{4.46}$$

where the coefficients x_n satisfy the System (4.31); in accordance with (4.27), σ vanishes when $\beta \in [\beta_0, \pi]$. Upon writing

$$Q_{sn}(\beta_0) = \frac{2}{\pi} \frac{\cos\left(s+\frac{1}{2}\right)\beta_0 \sin\left(n+\frac{1}{2}\right)\beta_0}{n+\frac{1}{2}} + \frac{2}{\pi} \frac{s+\frac{1}{2}}{n+\frac{1}{2}} R_{sn}(\beta_0), \tag{4.47}$$

where

$$R_{sn}(\beta_0) = \frac{1}{\pi}\left[\frac{\sin(s-n)\beta_0}{s-n} - \frac{\sin(s+n+1)\beta_0}{s+n+1}\right],$$

it is obvious that

$$Q_{sn}(\beta_0) = \frac{2}{\pi} \frac{\cos\left(s+\frac{1}{2}\right)\beta_0 \sin\left(n+\frac{1}{2}\right)\beta_0}{n+\frac{1}{2}} + O\left(n^{-2}\right) \tag{4.48}$$

as $n \to \infty$.

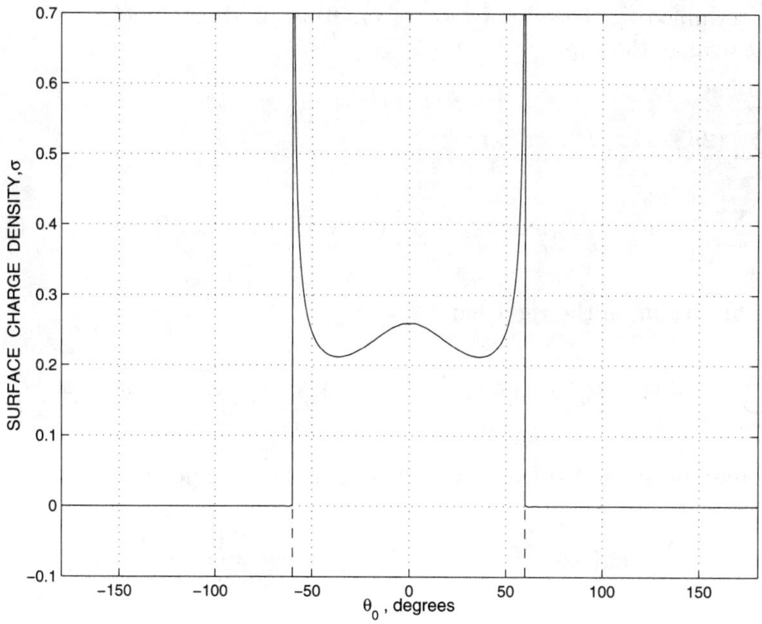

FIGURE 4.6. Surface charge density σ of a prolate spheroidal cap, charged to unit potential, with parameters $a/b = 0.5, \beta_0 = 60^o$. Truncation number $N_{tr} = 11$. The density was computed from Formula (4.53).

Consider the system derived from (4.31) by replacing $Q_{sn}(\beta_0)$ with the leading term in (4.48), i.e., neglecting the $O(n^{-2})$ term:

$$\widetilde{x}_n = \sum_{s=0}^{\infty} \widetilde{x}_s \varepsilon_s \frac{2}{\pi} \frac{\cos\left(s+\frac{1}{2}\right)\beta_0 \sin\left(n+\frac{1}{2}\right)\beta_0}{n+\frac{1}{2}} + \frac{2}{\pi} \frac{\cos\frac{1}{2}\beta_0 \sin\left(n+\frac{1}{2}\right)\beta_0}{n+\frac{1}{2}}, \tag{4.49}$$

where $n = 0, 1, 2, \ldots$. Its solution provides an asymptotic estimate for x_n as $n \to \infty$; it may be established that

$$x_n - \widetilde{x}_n = O\left(n^{-2}\right).$$

The special form of this asymptotic system allows us to determine its solution explicitly:

$$\widetilde{x}_n = \frac{2}{\pi} \frac{\sin\left(n+\frac{1}{2}\right)\beta_0}{n+\frac{1}{2}} D(\alpha_0, \beta_0) \tag{4.50}$$

where

$$D(\alpha_0, \beta_0) = \cos\frac{1}{2}\beta_0 + \sum_{s=0}^{\infty} \widetilde{x}_s \varepsilon_s \cos\left(s+\frac{1}{2}\right)\beta_0 \tag{4.51}$$

is determined by the substitution of (4.50) in (4.51).

Rearrange the summation in (4.46) as

$$\sum_{n=0}^{\infty}(2n+1)\,x_n P_n(\cos\beta_0) = \sum_{n=0}^{\infty}(2n+1)\,\widetilde{x}_n P_n(\cos\beta_0) + \sum_{n=0}^{\infty}(2n+1)\,(x_n - \widetilde{x}_n)\,P_n(\cos\beta_0). \quad (4.52)$$

The first term on the right-hand side is

$$\sum_{n=0}^{\infty}(2n+1)\,\widetilde{x}_n P_n(\cos\beta_0) = \frac{4}{\pi}D\,(\alpha_0,\beta_0)\sum_{n=0}^{\infty}P_n(\cos\beta_0)\sin\left(n+\frac{1}{2}\right)\beta_0$$

and may be evaluated from the well-known discontinuous series

$$\sum_{n=0}^{\infty}P_n(\cos\beta_0)\sin\left(n+\frac{1}{2}\right)\beta_0 = \frac{H\,(\beta_0-\beta)}{\sqrt{2\,(\cos\beta-\cos\beta_0)}}$$

derived from the Dirichlet-Mehler Formula (1.124). (H denotes the Heaviside function defined in Appendix A.) Thus the surface charge equals

$$\sigma = \frac{1}{4\pi}\frac{1}{\frac{d}{2}\sqrt{\sinh^2\alpha_0+\sin^2\beta}}\times \left\{\frac{2\sqrt{2}}{\pi}\frac{D\,(\alpha_0,\beta_0)}{\sqrt{\cos\beta-\cos\beta_0}}H\,(\beta_0-\beta) + \sum_{n=0}^{\infty}(2n+1)\,(x_n-\widetilde{x}_n)\,P_n(\cos\beta)\right\} \quad (4.53)$$

A calculation of the surface charge density using (4.53) is shown in Figure 4.6, using the coefficients $\{x_n\}_{n=0}^{N_{tr}}$ obtained by solving the system (4.31) by the truncation method with a truncation number N_{tr} equal to 11. Two features are apparent. The current singularity at the edges is accurately represented; and the summation in (4.53) has converged well. A sensitive test of the accuracy of this result with 11 terms is the magnitude of the calculated surface charge away from the conductor surface where the true surface charge vanishes. The maximum error (or deviation from zero) in this region is less than 0.5% of the value at the top of the cap. There is no visible improvement to the graphical results as N_{tr} is increased. Thus subtraction of an asymptotically correct estimate of the solution to the System (4.31) provides a much more rapidly convergent series than the first estimate obtained simply by truncation; this observation also remains true if the first estimate is replaced by an estimate obtained by Cesàro summation.

4.3 The prolate spheroidal conductor with a longitudinal slot

In this section we consider a prolate spheroidal surface in which a longitudinal slot has been cut, to produce two spheroidal caps of equal size; they are specified by

$$\alpha = \alpha_0, \beta \in (0, \beta_0) \cup (\pi - \beta_0, \pi), \phi \in (0, 2\pi).$$

The geometry is shown in Figure 4.1b. Assume that these two segments are charged to constant potentials U_1 and U_2, respectively.

Enforcement on (4.13) of the boundary conditions

$$U(\alpha_0 - 0, \beta) = U(\alpha_0 + 0, \beta) = U_1, \text{ for } \beta \in [0, \beta_0], \tag{4.54}$$

$$U(\alpha_0 - 0, \beta) = U(\alpha_0 + 0, \beta) = U_2, \text{ for } \beta \in [\pi - \beta_0, \pi], \tag{4.55}$$

and of the continuity of the normal derivative of the potential on the slot,

$$\frac{d}{d\alpha} U(\alpha, \beta)|_{\alpha=\alpha_0-0}^{\alpha=\alpha_0+0} = 0, \text{ for } \beta \in (\beta_0, \pi - \beta_0), \tag{4.56}$$

leads to the following symmetric triple series equations with Legendre polynomial kernels,

$$\begin{aligned} \sum_{n=0}^{\infty} A_n P_n(t) &= U_1, t \in (t_0, 1], \\ \sum_{n=0}^{\infty} \Lambda_n(\alpha_0) A_n P_n(t) &= 0, t \in (-t_0, t_0), \\ \sum_{n=0}^{\infty} A_n P_n(t) &= U_2, t \in [-1, -t_0), \end{aligned} \tag{4.57}$$

where $t = \cos\beta$, $t_0 = \cos\beta_0$. The system (4.57) is particular case of the equations of Type A described in Section 2.4.1 (Legendre polynomials are Jacobi polynomials $P_n^{(\alpha,\alpha)}$ with $\alpha = 0$), so the method described may be exploited to solve (4.57).

We now consider two particular cases, $U_1 = U_2 = 1$ and $U_1 = -U_2 = 1$. Obviously, cases with any other constant values of the potentials U_1 and U_2 can be deduced from these solutions. From a practical point of view, when $U_1 = U_2$, the two parts of the prolate spheroidal conductor with a longitudinal slot must to connected by a thin wire in order to allow charging to equal potential; however, we may assume that this wire is so thin that the influence of its electric field can be neglected. When $U_1 = -U_2$ this structure models a condensor or capacitor with plates in the form of spheroidal caps.

The symmetry property of Legendre polynomials,

$$P_n(-t) = (-1)^n P_n(t),$$

may be applied to establish two decoupled systems of dual series equations for the even ($l = 0$) and odd ($l = 1$) index coefficients, respectively, defined on $[-1, 0]$:

$$\begin{aligned} \sum_{n=0}^{\infty} A_{2n+l} P_{2n+l}(t) &= (-1)^l, \quad t \in [-1, -t_0), \\ \sum_{n=0}^{\infty} \Lambda_{2n+l}(\alpha_0) A_{2n+l} P_{2n+l}(t) &= 0, \quad t \in (-t_0, 0). \end{aligned} \tag{4.58}$$

The relation (2.131) connects Jacobi polynomials and Legendre polynomials,

$$P_{2n+l}(t) = t^l P_n^{(0,l-\frac{1}{2})}(2t^2 - 1),$$

so setting $u = 2t^2 - 1$, $u_0 = 2t_0^2 - 1$ we may transform (4.58) to dual series equations defined over the complete range $[-1, 1]$ of the new variable :

$$\sum_{n=0}^{\infty} \Lambda_{2n+l}(\alpha_0) A_{2n+l} P_n^{(0,l-\frac{1}{2})}(u) = 0, \quad u \in (-1, u_0), \tag{4.59}$$

$$\sum_{n=0}^{\infty} A_{2n+l} P_n^{(0,l-\frac{1}{2})}(u) = (-1)^l \left(\frac{1+u}{2}\right)^{-\frac{l}{2}}, \quad u \in (u_0, 1). \tag{4.60}$$

The dual series Equations (4.59) and (4.60) were considered in Section 2.1. Omitting some details let us illustrate the main stages of the argument in this particular case. The Abel integral representations for the Jacobi polynomials (1.171)–(1.174) are

$$\int_{-1}^{u} (1+t)^{l-\frac{1}{2}} P_n^{(0,l-\frac{1}{2})}(t)dt = \frac{\Gamma(n+l+\frac{1}{2})}{\sqrt{\pi}\Gamma(n+l+1)} \int_{-1}^{u} \frac{(1+x)^l P_n^{(-\frac{1}{2},l)}(x)dx}{(u-x)^{\frac{1}{2}}} \tag{4.61}$$

and

$$P_n^{(0,l-\frac{1}{2})}(u) = \frac{\Gamma(n+1)}{\sqrt{\pi}\Gamma(n+\frac{1}{2})} \int_{u}^{1} \frac{(1-x)^{-\frac{1}{2}} P_n^{(-\frac{1}{2},l)}(x)dx}{(x-u)^{\frac{1}{2}}}. \tag{4.62}$$

The functional equations are then converted to the following form:

$$\sum_{n=0}^{\infty} \Lambda_{2n+l}(\alpha_0) A_{2n+l} \frac{\Gamma(n+l+\frac{1}{2})}{\Gamma(n+l+1)} P_n^{(-\frac{1}{2},l)}(u) = 0, \; u \in (-1, u_0), \tag{4.63}$$

$$\sum_{n=0}^{\infty} A_{2n+l} \frac{\Gamma(n+1)}{\Gamma(n+\frac{1}{2})} P_n^{(-\frac{1}{2},l)}(u) = \frac{(-1)^l}{\sqrt{\pi}} \left(\frac{1+u}{2}\right)^{-l}, \quad u \in (u_0, 1). \tag{4.64}$$

A suitable small parameter may now be identified in the Equation (4.63) as

$$\varepsilon_{2n+l} = 1 - \frac{\Lambda_{2n+l}(\alpha_0)}{4} \frac{\Gamma(n+\frac{1}{2})\Gamma(n+l+\frac{1}{2})}{\Gamma(n+1)\Gamma(n+l+1)}. \tag{4.65}$$

It is asymptotically small: $\varepsilon_{2n+l} = O(n^{-2})$ as $n \to \infty$. The unknowns are rescaled according to

$$x_{2n+l} = A_{2n+l} \frac{\Gamma(n+1)}{\Gamma(n+\frac{1}{2})} \left\{ h_n^{(-\frac{1}{2},l)} \right\}^{\frac{1}{2}}, \tag{4.66}$$

where $\left\{ h_n^{(-\frac{1}{2},l)} \right\}^{\frac{1}{2}}$ is the norm of the Jacobi polynomials; thus $\{x_{2n+l}\}_{n=0}^{\infty} \in l_2$.

Equations (4.63) and (4.64) may now be written in the form

$$F(u) = \sum_{n=0}^{\infty} x_{2n+l} \hat{P}_n^{(-\frac{1}{2},l)}(u) = \left\{ \begin{array}{ll} F_1(u), & u \in (-1, u_0) \\ F_2(u), & u \in (u_0, 1) \end{array} \right\}, \tag{4.67}$$

where

$$\begin{aligned} F_1(u) &= \sum_{n=0}^{\infty} x_{2n+l} \varepsilon_{2n+l} \hat{P}_n^{(-\frac{1}{2},l)}(u), \\ F_2(u) &= (-1)^l \pi^{-\frac{1}{2}} 2^l (1+u)^{-l}. \end{aligned}$$

Exploiting orthogonality of the normalized Jacobi polynomials $\hat{P}_n^{(-\frac{1}{2},l)}$ leads, as usual, to the second-kind infinite system of linear algebraic equations for the unknowns $\{x_{2n+l}\}_{n=0}^{\infty}$,

$$\begin{aligned} &(1-\varepsilon_{2m+l}) x_{2m+l} + \sum_{n=0}^{\infty} x_{2n+l} \varepsilon_{2n+l} \hat{Q}_{nm}^{(-\frac{1}{2},l)}(u_0) \\ &= \begin{cases} 2^{\frac{3}{4}} \pi^{-\frac{1}{2}} \hat{Q}_{0m}^{(-\frac{1}{2},0)}(u_0), & \text{if } l = 0 \\ -2\pi^{-\frac{1}{2}} \left\{ (m+1)(m+\frac{1}{2}) \right\}^{-\frac{1}{2}} \sqrt{1-u_0} \hat{P}_m^{(\frac{1}{2},0)}(u_0), & \text{if } l = 1 \end{cases} \end{aligned} \tag{4.68}$$

where $m = 0, 1, 2, \ldots$, and $\hat{Q}_{nm}^{(-\frac{1}{2},l)}(u_0)$ is the incomplete scalar product of normalised Jacobi polynomials.

Because the matrix operator of the system (4.68) is a completely continuous perturbation of the identity, the sequence $\{x_{2n+l}\}_{n=0}^{\infty}$ is rapidly convergent and the truncation method is very efficient in solving this system numerically. The behaviour of the normalised error as a function of

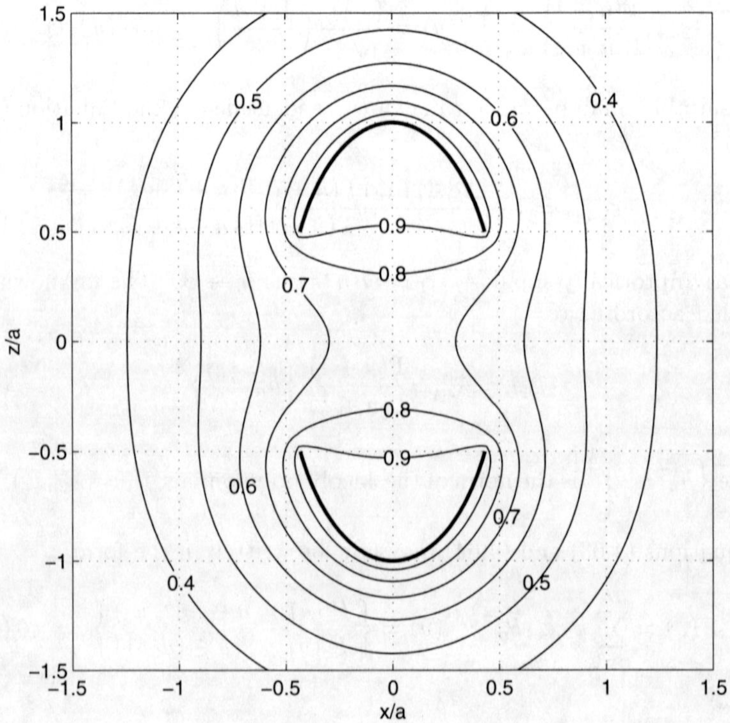

FIGURE 4.7. Electrostatic potential near a slotted prolate spheroidal shell, both components charged to unit potential. The geometrical parameters are $a/b = 0.5, \beta_0 = 60^o$. Truncation number $N_{tr} = 11$.

truncation number is very similar to that considered in the previous section (see Figures 4.3); typically, $N_{tr} = 10$ equations suffice to produce coefficient solutions with 3 correct digits for a wide range of aspect ratios (independent of aperture size). As an illustration of the numerical process, the distribution of electrostatic field potential near the spheroidal conductor with a longitudinal slot charged to unit potential ($U_1 = U_2 = 1$, $l = 0$ in (4.68)) is shown in Figure 4.7; the ratio of minor to major axes, $a/b = \sinh \alpha_0 / \cosh \alpha_0 = 0.5$ and the angular size of each cap is $\beta_0 = 60°$; and the system truncation number N_{tr} was taken to be 11.

The potential near the spheroidal condensor in which the upper and lower plates are charged to potentials $U_1 = 1$ and $U_2 = -1$ (so $l = 1$ in (4.68)) is displayed in Figure 4.8; the geometrical parameters are $a/b = 0.5, \beta_0 = 60°$, and a truncation number $N_{tr} = 11$ was used.

When $\beta_0 = \frac{\pi}{2}$ ($u_0 = -1$) the aperture in the conductor closes, becoming a closed spheroidal shell charged to unit potential ($l = 0$), the system (4.68)

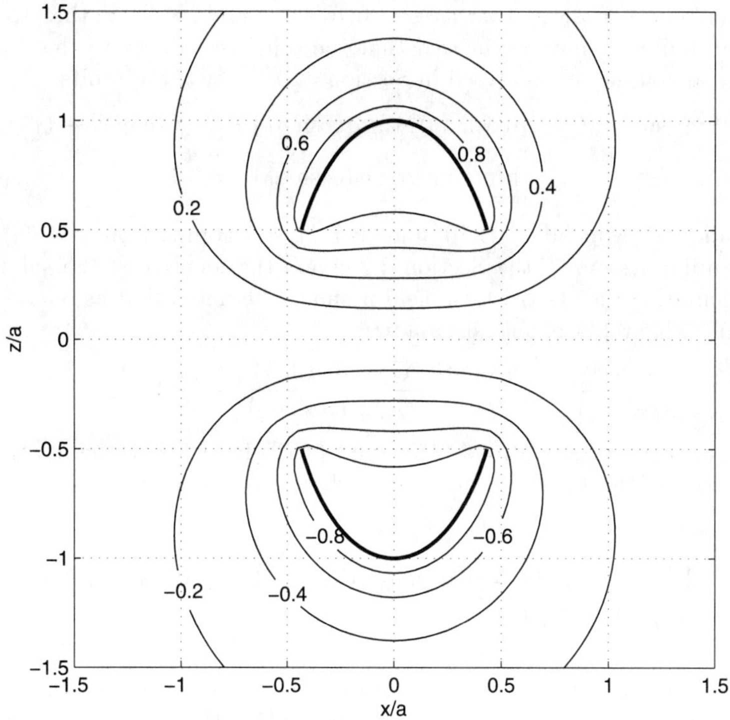

FIGURE 4.8. Electrostatic potential near the prolate spheroidal condenser, the plates charged to unit positive and negative potential. The geometrical parameters are $a/b = 0.5, \beta_0 = 60^o$. Truncation number $N_{tr} = 11$.

has the explicit solution

$$x_{2m} = (-1)^m \frac{2^{\frac{3}{4}}}{\sqrt{\pi}} \hat{Q}_{0m}^{(-\frac{1}{2},0)}(-1) = 0 \ (m > 0), \quad x_0 = \frac{2^{\frac{3}{4}}}{\sqrt{\pi}}, \tag{4.69}$$

from which follows the representation of the electrostatic potential in closed form:

$$U(\alpha, \beta) = \frac{Q_0(\cosh \alpha)}{Q_0(\cosh \alpha_0)} \quad \text{for } \alpha \geq \alpha_0, \beta \in [0, \pi]. \tag{4.70}$$

It is readily verified that this is indeed the correct potential.

Let us consider the transition from spheroid to sphere of radius a. Spheroidal coordinates (α, β, φ) degenerate to spherical coordinates $(r, \theta, \varphi_{sp})$ if the identifications

$$\theta = \beta, \varphi_{sp} \equiv \varphi, r = \frac{1}{2}\frac{d}{2}e^{\alpha}, a = \frac{1}{2}\frac{d}{2}e^{\alpha_0}$$

are made in such a way that as $\frac{d}{2} \to 0, \alpha \to \infty$, and $\alpha_0 \to \infty$, the products remain finite. It may be checked that the solution reduces to that for the spherical conductor (analysed in Section 3.2). In fact, the limits

$$\begin{aligned} \lim_{\alpha_0 \to \infty} \sinh \alpha_0 Q_{2n+l}(\cosh \alpha_0) P_{2n+l}(\cosh \alpha_0) &= (4n+2l+1)^{-1}, \\ \lim_{\alpha_0 \to \infty} \sinh \alpha_0 Q_0(\cosh \alpha_0) &= 1, \end{aligned} \tag{4.71}$$

are valid (see Appendix, (B.70) and (B.71)), so a comparison of (4.68) with the similar system in the Section 3.2 shows the identity of the solutions. In calculating the electrostatic field it should be noted that as $\alpha, \alpha_0 \to \infty$, the following replacement are made:

$$\frac{P_{2n+l}(\cosh \alpha)}{P_{2n+l}(\cosh \alpha_0)} \to \left(\frac{r}{a}\right)^{2n+l}, \frac{Q_{2n+l}(\cosh \alpha)}{Q_{2n+l}(\cosh \alpha_0)} \to \left(\frac{r}{a}\right)^{-2n-l-1}. \tag{4.72}$$

The limiting representations (4.71) and (4.72) follow from the asymptotic behaviour of the Legendre functions (when $\alpha, \alpha_0 \to \infty$, see [1]).

4.4 The prolate spheroidal conductor with two circular holes

In this section we consider the complementary structure to the slotted spheroid of the previous section, and suppose that the spheroidal conductor has two circular holes (see Figure 4.1(c)). The shell S_0 is defined by

$$\alpha = \alpha_0, \beta \in (\beta_0, \pi - \beta_0), \phi \in (0, 2\pi);$$

when $a/b \ll 1$, it may be visualised as a spheroidal cylinder. It is charged to unit potential, so

$$U(\alpha_0 - 0, \beta) = U(\alpha_0 + 0, \beta) = 1, \beta \in [\beta_0, \pi - \beta_0], \tag{4.73}$$

whereas the normal derivative of the potential is continuous on the apertures,

$$\frac{d}{d\alpha} U(\alpha, \beta)|_{\alpha=\alpha_0-0}^{\alpha=\alpha_0+0} = 0, \beta \in (0, \beta_0) \cup (\pi - \beta_0, \pi). \tag{4.74}$$

Enforcing the boundary conditions (4.73) and (4.74) on (4.12) produces a set of symmetric triple series equations of Type B (2.126)–(2.128) from which may be deduced (in the same way as for Equations (4.58)) the dual series equations defined over the half range $[-1, 0]$ (setting $t = \cos\beta, t_0 = \cos\beta_0$):

$$\begin{aligned} \sum_{n=0}^{\infty} \Lambda_{2n}(\alpha_0) A_{2n} P_{2n}(t) &= 0, \; t \in (-1, -t_0), \\ \sum_{n=0}^{\infty} A_{2n} P_{2n}(t) &= 1, \; t \in (-t_0, 0). \end{aligned} \tag{4.75}$$

Following the same argument as in Section 4.3, we may reduce the Equations (4.75) to dual series equations involving Jacobi polynomials defined over the interval $[-1,1]$. Setting $u = 2t^2 - 1$, $u_0 = 2t_0^2 - 1$, and $\beta_1 = \frac{\pi}{2} - \beta_0$, $u_1 = \cos 2\beta_1 = -u_0$, we obtain

$$\sum_{n=0}^{\infty}(-1)^n \Lambda_{2n}(\alpha_0) A_{2n} P_{2n}^{(-\frac{1}{2},0)}(u) = 0, \; u \in (-1, u_1), \tag{4.76}$$

$$\sum_{n=0}^{\infty}(-1)^n A_{2n} P_{2n}^{(-\frac{1}{2},0)}(u) = 1, \; u \in (u_1, 1). \tag{4.77}$$

The general treatment expounded in Section 2.4.2 of dual equations of this type, arising from Type B triple series, did not cover the pair (4.76), (4.77). Let us specifically demonstrate how to treat these equations. Before employing the integral representations of Abel's type for Jacobi polynomials, integrate the Equation (4.76) with the weight $(1-u)^{-\frac{1}{2}}$, using the variant (2.36) of Rodrigues' formula. (Although this integration complicates the solution process, it is absolutely necessary because a direct application of the integral representations of Abel type would result in the occurrence of the Jacobi polynomial kernels $P_n^{(-1,\frac{1}{2})}$ for which the theory developed in Section 2.1 is not valid.)

The transform method may now be applied in a standard manner, similar to that in the previous section, to obtain the expansion of some function F in a Fourier series over the complete orthogonal system of Jacobi polynomials $\left\{\hat{P}_n^{(0,\frac{3}{2})}\right\}_{n=1}^{\infty}$, piecewise defined over two subintervals of $[-1,1]$:

$$F(u) = \sum_{n=1}^{\infty} x_{2n} \hat{P}_{n-1}^{(0,\frac{3}{2})}(u) = \left\{ \begin{array}{ll} F_1(u), & u \in (u_1, 1) \\ F_2(u), & u \in (-1, u_1) \end{array} \right\}, \tag{4.78}$$

where

$$F_1(u) = 2\sqrt{2\pi}(1 - A_0)(1+u)^{-\frac{3}{2}} + \sum_{n=1}^{\infty} x_{2n}\varepsilon_{2n} \hat{P}_{n-1}^{(0,\frac{3}{2})}(u),$$

$$F_2(u) = -\frac{A_0\Lambda_0(\alpha_0)}{\sqrt{\pi}} \left\{ \frac{2}{(1+u)} + \frac{\sqrt{2}}{(1+u)^{\frac{3}{2}}} \ln\left[\frac{\sqrt{2}-\sqrt{1+u}}{\sqrt{2}+\sqrt{1+u}}\right] \right\}.$$

Here

$$x_{2n} = \frac{(-1)^n}{4} A_{2n}\Lambda_{2n}(\alpha_0) \frac{\Gamma(n+1)}{\Gamma(n+\frac{3}{2})} h_{n-1}^{(0,\frac{3}{2})} \left[h_{n-1}^{(1,\frac{1}{2})}\right]^{-\frac{1}{2}}, \tag{4.79}$$

and the asymptotically small parameter is

$$\varepsilon_{2n} = 1 - \frac{4}{\Lambda_{2n}(\alpha_0)} n(n+\frac{1}{2}) \left[\frac{\Gamma(n+\frac{1}{2})}{\Gamma(n+1)}\right]^2 = O(n^{-2}) \text{ as } n \to \infty. \tag{4.80}$$

The constant A_0 is determined by enforcing continuity on $F(u)$ at u_1,

$$A_0 = \frac{1}{g(u_1)}\left[1 + \sum_{n=1}^{\infty} x_{2n}\varepsilon_{2n}Q_n(u_1)\right], \tag{4.81}$$

where

$$g(u_1) = 1 - \frac{1}{\pi \sinh\alpha_0 Q_0(\cosh\alpha_0)}\left\{\sqrt{\frac{1+u_1}{2}} + \frac{1}{2}\log\left[\frac{\sqrt{2}-\sqrt{(1+u_1)}}{\sqrt{2}+\sqrt{(1+u_1)}}\right]\right\}$$

and

$$Q_n(u_1) = \frac{1}{\sqrt{\pi}}\left[\frac{1}{2}(1+u_1)\right]^{\frac{3}{2}}\left[n(n+\frac{1}{2})\right]^{-\frac{1}{2}}\hat{P}_{n-1}^{(0,\frac{3}{2})}(u_1).$$

The Equation (4.78) is now transformed in the same way as (4.67), taking into account (4.81). The final form of the i.s.l.a.e. is

$$x_{2m} + \sum_{n=1}^{\infty} x_{2n}\varepsilon_{2n}\left\{\hat{Q}_{n-1,m-1}^{(1,\frac{1}{2})}(u_1) - \frac{2\sqrt{2}Q_n(u_1)Q_m(u_1)}{g(u_1)\sinh\alpha_0 Q_0(\cosh\alpha_0)}\right\}$$
$$= \frac{2\sqrt{2}Q_m(u_1)}{g(u_1)\sinh\alpha_0 Q_0(\cosh\alpha_0)}, \tag{4.82}$$

where $m = 1, 2, \ldots$. The system (4.82) possesses the same features as the system (4.68). The norm of the completely continuous part H of the matrix operator in (4.82) is uniformly bounded (with respect to the parameters) by the estimate

$$\|H\| \le \max_n |\varepsilon_n| = \varepsilon_1.$$

The i.s.l.a.e. (4.82) is effectively solved numerically by the truncation method. The behaviour of solution accuracy as a function of truncation number is very similar to that described in the previous sections. Computed results of the potential distribution near the prolate spheroidal conductor with two holes when charged to the unit potential are shown in Figure 4.9. The geometrical parameters are $a/b = 0.5, \beta_0 = 30°$; a truncation number $N_{tr} = 11$ was used.

If $\beta_0 = 0$ (so that $\beta_1 = \frac{\pi}{2}, u_1 = -1$), the limiting case of a closed spheroidal shell is obtained; from (4.81) and (4.82) we see that

$$x_{2m} = 0\ (m = 1, 2, ...), \quad A_0 = 1.$$

Thus, the electrostatic potential near the closed spheroidal shell has the form

$$U(\alpha, \beta) = \frac{Q_0(\cosh\alpha)}{Q_0(\cosh\alpha_0)} \quad \text{for } \alpha \ge \alpha_0, \beta \in [0, \pi]. \tag{4.83}$$

This expression (4.83) agrees with the expression (4.70) that was obtained for the limiting case of the spheroidal shell with a closing narrow slot.

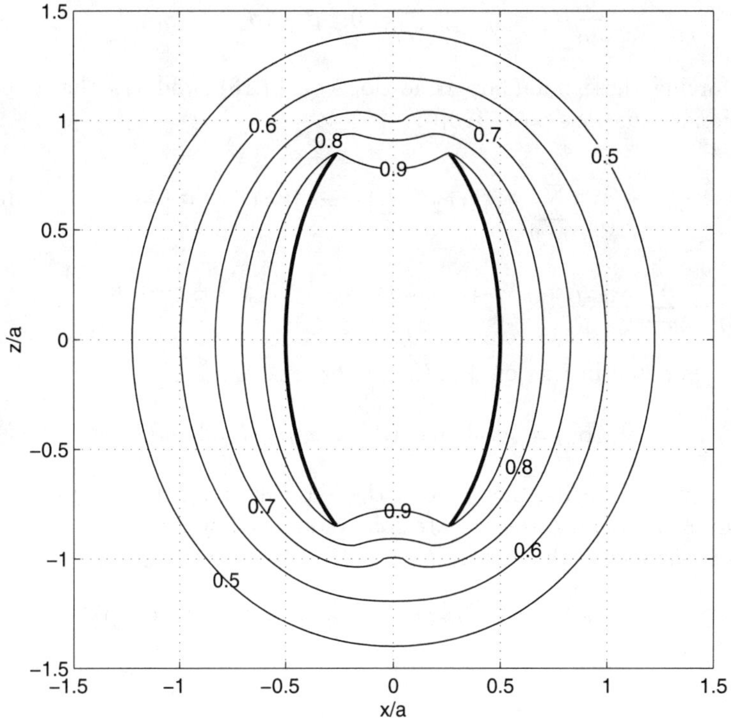

FIGURE 4.9. Electrostatic potential near a prolate spheroidal barrel charged to unit potential with geometrical parameters $a/b = 0.5, \beta_0 = 30^o$. Truncation number $N_{tr} = 11$.

4.5 The oblate spheroidal conductor with a longitudinal slot

In this section we consider an oblate spheroidal surface in which a longitudinal slot has been cut to produce two spheroidal caps of equal size; they are specified by

$$\alpha = \alpha_0, \beta \in (0, \beta_0) \cup (\pi - \beta_0, \pi), \phi \in (0, 2\pi)$$

The shell S_0 (see Figure 4.1(d)) is the oblate analogue of the structure considered in Section 4.3, and comprises two symmetrical oblate spheroidal segments that are assumed to be charged to the constant potential values $U_1 = 1$ and $U_2 = (-1)^l$ $(l = 0, 1)$. Then the mixed boundary conditions (similar to (4.54)–(4.56)) take the form

$$U(\alpha_0 - 0, \beta) = U(\alpha_0 + 0, \beta) = 1, \quad \beta \in [0, \beta_0], \tag{4.84}$$

$$U(\alpha_0 - 0, \beta) = U(\alpha_0 + 0, \beta) = (-1)^l, \quad \beta \in [\pi - \beta_0, \pi], \tag{4.85}$$

$$\frac{d}{d\alpha}U(\alpha,\beta)\Big|_{\alpha=\alpha_0-0}^{\alpha=\alpha_0+0}=0,\quad \beta\in(\beta_0,\pi-\beta_0)\,. \tag{4.86}$$

Enforcing these boundary conditions on (4.13) produces the following functional equations on $[-1,0]$:

$$\sum_{n=0}^{\infty} A_{2n+l}P_{2n+l}(t) = (-1)^l,\ t\in[-1,-t_0), \tag{4.87}$$

$$\sum_{n=0}^{\infty} \lambda_{2n+l}(\alpha_0)A_{2n+l}P_{2n+l}(t) = 0,\quad t\in(-t_0,0) \tag{4.88}$$

where, as noted in Section 4.1, the factor

$$\lambda_n(\alpha_0)=\{\cosh\alpha_0 q_n(i\sinh\alpha_0)p_n(i\sinh\alpha_0)\}^{-1} \tag{4.89}$$

arises from employing the value of the Wronskian of the pair p_n, q_n. This system is identical to the prolate spheroidal shell system (4.58) except for the replacement of the factor

$$\Lambda_{2n+l}(\alpha_0)=\{\sinh\alpha_0 Q_{2n+l}(\cosh\alpha_0)P_{2n+l}(\cosh\alpha_0)\}^{-1}$$

by

$$\lambda_{2n+l}(\alpha_0)=\{\cosh\alpha_0 q_{2n+l}(i\sinh\alpha_0)p_{2n+l}(i\sinh\alpha_0)\}^{-1} \tag{4.90}$$

in (4.88). With this replacement, the solution of the dual series Equations (4.87)–(4.88) is identical to that obtained in the prolate case yielding the i.s.l.a.e.

$$(1-\varepsilon_{2m+l})x_{2m+l}+\sum_{n=0}^{\infty} x_{2n+l}\varepsilon_{2n+l}\hat{Q}_{nm}^{(-\frac{1}{2},l)}(u_0)$$
$$=\begin{cases} 2^{\frac{3}{4}}\pi^{-\frac{1}{2}}\hat{Q}_{0m}^{(-\frac{1}{2},0)}(u_0), & \text{if } l=0,\\ -2\pi^{-\frac{1}{2}}\left\{(m+1)(m+\frac{1}{2})\right\}^{-\frac{1}{2}}\sqrt{1-u_0}\hat{P}_m^{(\frac{1}{2},0)}(u_0), & \text{if } l=1, \end{cases} \tag{4.91}$$

where $m=0,1,2,\ldots$,

$$\varepsilon_{2n+l}=1-\frac{\lambda_{2n+l}(\alpha_0)}{4}\frac{\Gamma(n+\frac{1}{2})\Gamma(n+l+\frac{1}{2})}{\Gamma(n+1)\Gamma(n+l+1)}=O(n^{-2})\ \text{as}\ n\to\infty, \tag{4.92}$$

and all the other definitions and relations are the same as in (4.68). The validity of the asymptotic estimate (4.92) is established by the behaviour of the functions $q_n(i\sinh\alpha_0), p_n(i\sinh\alpha_0)$ as $n\to\infty$ (see Appendix, (B.70) and (B.71)).

Switching to the complementary angle $\beta_1 = \frac{\pi}{2} - \beta_0$, with $u_1 = \cos 2\beta_1 = -u_0$, we set $y_{2n+l} = (-1)^n x_{2n+l}$, and use (B.170) to obtain another convenient form of the system,

$$y_{2m+l} - \sum_{n=0}^{\infty} y_{2n+l}\varepsilon_{2n+l}\hat{Q}_{nm}^{(l,-\frac{1}{2})}(u_1)$$
$$= \begin{cases} 2^{\frac{3}{4}}\pi^{-\frac{1}{2}}\left[\delta_{0m} - \hat{Q}_{0m}^{(0,-\frac{1}{2})}(u_1)\right], & \text{if } l = 0, \\ -2\pi^{-\frac{1}{2}}\sqrt{1+u_1}\left[(m+1)(m+\frac{1}{2})\right]^{-\frac{1}{2}}\hat{P}_m^{(0,\frac{1}{2})}(u_1), & \text{if } l = 1. \end{cases} \tag{4.93}$$

As an illustration of the numerical process, the spatial distribution of electrostatic field potential near the longitudinally slotted conductor, with both components charged to unit potential ($U_1 = U_2 = 1$, $l = 0$ in (4.93)), is shown in Figure 4.10; the ratio of major to minor axes is $a/b = \cosh\alpha_0/\sinh\alpha_0 = 2.0$, and the angular size of each component is $\beta_0 = 60°$; the system truncation number N_{tr} was chosen to be 11. As a function of truncation number, the accuracy of solutions to the system (4.93) after truncation has the same general behaviour as described for the prolate spheroidal shells considered in earlier sections.

When the components are oppositely charged, the structure acts as a condensor. The potential near the slotted oblate spheroidal shell, in which the upper and lower plates are charged to potentials $U_1 = 1$ and $U_2 = -1$ (so $l = 1$ in (4.93)), is displayed in Figure 4.11; the geometrical parameters are $a/b = 2$ and $\beta_0 = 60°$, and a truncation number $N_{tr} = 11$ was used. As expected, the electrostatic field is strongly confined to the interior.

The closed oblate spheroidal shell ($\beta_0 = \frac{\pi}{2}, l = 0$), charged to unit potential, has the explicit solution obtained from (4.93):

$$y_{2m} = \frac{1}{\sqrt{\pi}}2^{\frac{3}{4}}\delta_{0m}, \quad m \geq 0,$$

so the closed form of the potential distribution is

$$U(\alpha,\beta) = \frac{q_0(i\sinh\alpha)}{q_0(i\sinh\alpha_0)}, \quad \text{for } \alpha \geq \alpha_0, \beta \in [0,\pi].$$

This is in accord with the known solution [26].

Let us consider the transition from oblate spheroid to sphere of radius a. In a similar way to that discussed for the prolate case, oblate spheroidal coordinates (α,β,φ) degenerate to spherical coordinates (r,θ,φ_{sp}) if the identifications

$$\theta = \beta, \varphi_{sp} \equiv \varphi, r = \frac{1}{2}\frac{d}{2}e^{\alpha}, a = \frac{1}{2}\frac{d}{2}e^{\alpha_0}$$

are made in such a way that, as $\frac{d}{2} \to 0, \alpha \to \infty$ and $\alpha_0 \to \infty$, the products remain finite. It may be checked that the same solution as obtained for the

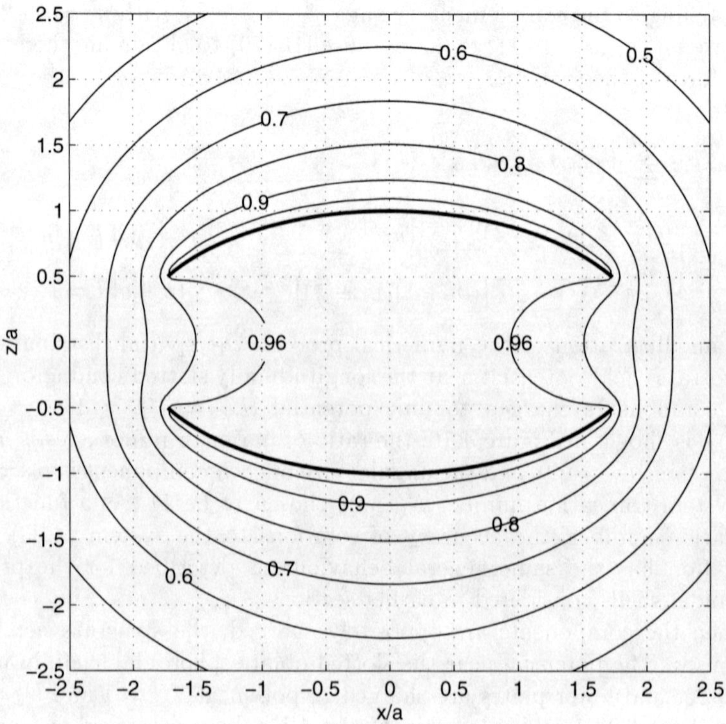

FIGURE 4.10. Electrostatic potential near a slotted oblate spheroidal shell, both components charged to unit potential. The geometrical parameters are $a/b = 2, \beta_0 = 60^o$. Truncation number $N_{tr} = 11$.

spherical conductor (Section 3.2) is found. In fact, the limits

$$\begin{aligned} \lim_{\alpha_0 \to \infty} \cosh \alpha_0 q_{2n+l}(i \sinh \alpha_0) p_{2n+l}(i \sinh \alpha_0) &= (4n + 2l + 1)^{-1}, \\ \lim_{\alpha_0 \to \infty} \cosh \alpha_0 q_0(i \sinh \alpha_0) &= 1, \end{aligned} \tag{4.94}$$

are valid (see Appendix, (B.70) and (B.71)), so a comparison of (4.93) with the similar system in Section 3.2 shows the identity of the solutions. In calculating the electrostatic field it should be noted that as $\alpha, \alpha_0 \to \infty$, the following replacements are made:

$$\frac{p_{2n+l}(i \sinh \alpha)}{p_{2n+l}(i \sinh \alpha_0)} \to \left(\frac{r}{a}\right)^{2n+l}, \frac{q_{2n+l}(i \sinh \alpha)}{q_{2n+l}(i \sinh \alpha_0)} \to \left(\frac{r}{a}\right)^{-2n-l-1}. \tag{4.95}$$

The limiting representations (4.94) and (4.95) follow from the asymptotic behaviour of the Legendre functions (when $\alpha, \alpha_0 \to \infty$, see [1]).

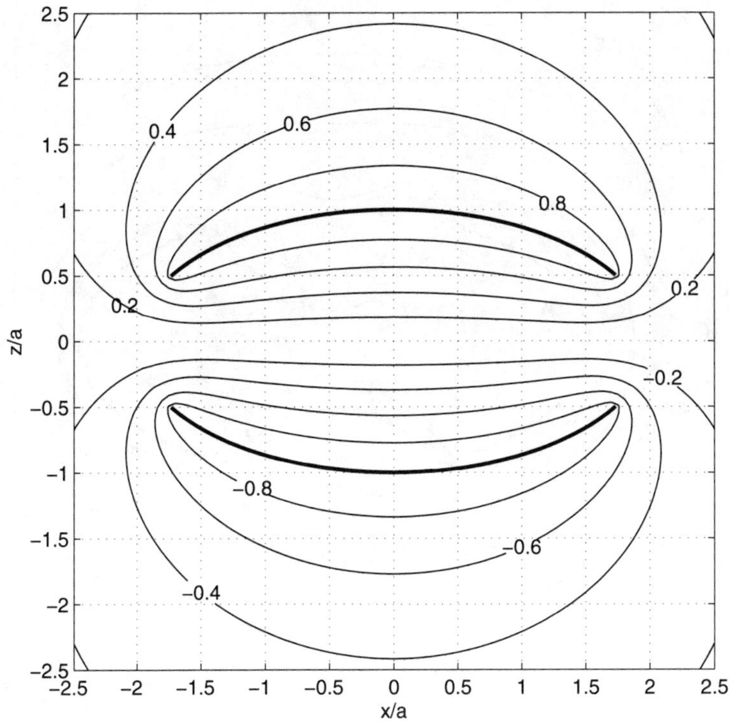

FIGURE 4.11. Electrostatic potential near the oblate spheroidal condenser, the plates charged to unit positive and negative potential. The geometrical parameters $a/b = 2, \beta_0 = 60^o$. Truncation number $N_{tr} = 11$.

4.6 The oblate spheroidal conductor with two circular holes

In this section we consider the structure complementary to the slotted oblate spheroid of the previous section. The geometry of an oblate spheroidal shell with two equal circular holes is shown in Figure 4.1(e). The shell S_0 is defined by

$$\alpha = \alpha_0, \beta \in (\beta_0, \pi - \beta_0), \phi \in (0, 2\pi);$$

and is assumed to be charged to unit potential. The mixed boundary conditions are

$$U(\alpha_0 - 0, \beta) = U(\alpha_0 + 0, \beta) = 1, \beta \in (\beta_0, \pi - \beta_0) \tag{4.96}$$

and

$$\frac{d}{d\alpha}U(\alpha, \beta)|_{\alpha=\alpha_0-0}^{\alpha=\alpha_0+0} = 0, \beta \in (0, \beta_0) \cup (\pi - \beta_0, \pi). \tag{4.97}$$

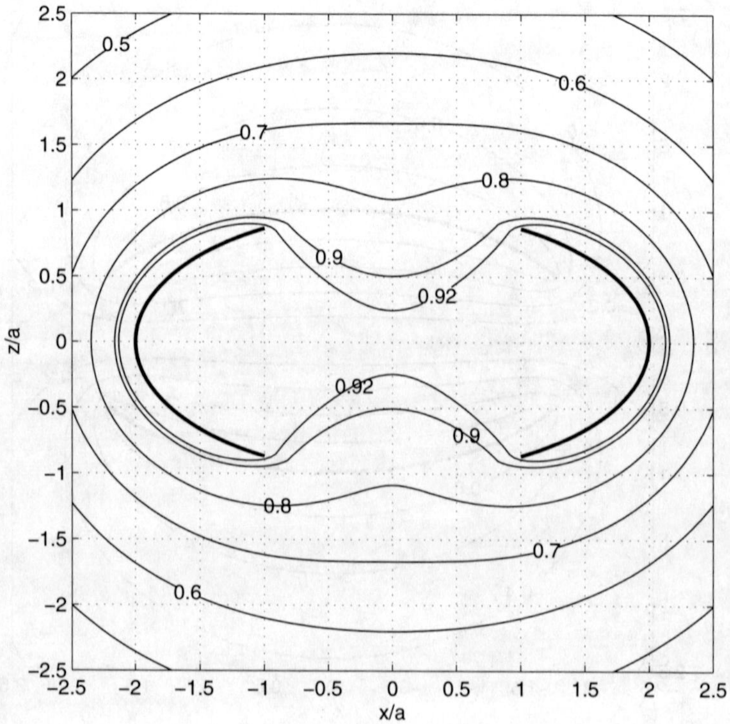

FIGURE 4.12. Electrostatic potential near an oblate spheroidal barrel charged to unit potential with geometrical parameters $a/b = 2, \beta_0 = 30^o$. Truncation number $N_{tr} = 11$.

Enforcement of the boundary conditions (4.96) and (4.97) on (4.13) produces symmetric triple series equations; a standard argument reduces these to the following dual series equations defined over $[-1, 0]$, where $t = \cos\beta$, $t_0 = \cos\beta_0$:

$$\sum_{n=0}^{\infty} \lambda_{2n}(\alpha_0) A_{2n} P_{2n}(t) = 0, t \in (-1, -t_0), \tag{4.98}$$

$$\sum_{n=0}^{\infty} A_{2n} P_{2n}(t) = 1, t \in (-t_0, 0) \tag{4.99}$$

where the factor

$$\lambda_{2n}(\alpha_0) = \{\cosh\alpha_0 q_{2n}(i\sinh\alpha_0) p_{2n}(i\sinh\alpha_0)\}^{-1} \tag{4.100}$$

arises from employing the value of the Wronskian of the pair p_n, q_n.

This system is identical to the prolate spheroidal shell system (4.75), except for the replacement of the factor $\Lambda_{2n}(\alpha_0)$ by $\lambda_{2n}(\alpha_0)$ in (4.98). With this replacement, the solution of the dual series Equations (4.98) and (4.99) is identical to that obtained in the prolate case. Thus, *mutatis mutandis,* we obtain the i.s.l.a.e.

$$x_{2m} + \sum_{n=1}^{\infty} x_{2n}\varepsilon_{2n} \left\{ \hat{Q}^{(1,\frac{1}{2})}_{n-1,m-1}(u_1) - \frac{2\sqrt{2}Q_n(u_1)Q_m(u_1)}{g(u_1)\cosh\alpha_0 q_0(i\sinh\alpha_0)} \right\} = \frac{2\sqrt{2}Q_m(u_1)}{g(u_1)\cosh\alpha_0 q_0(i\sinh\alpha_0)}, \tag{4.101}$$

where $m = 1, 2, \ldots$, $u_1 = \cos 2\beta_1$, $\beta_1 = \frac{\pi}{2} - \beta_0$, and

$$x_{2n} = \frac{(-1)^n}{4} A_{2n}\lambda_{2n}(\alpha_0)\frac{\Gamma(n+1)}{\Gamma(n+\frac{3}{2})} h^{(0,\frac{3}{2})}_{n-1}(u)\left\{h^{(1,\frac{1}{2})}_{n-1}(u)\right\}^{-\frac{1}{2}}, \tag{4.102}$$

$$\varepsilon_{2n} = 1 - \frac{4}{\lambda_{2n}(\alpha_0)} n(n+\frac{1}{2})\left[\frac{\Gamma(n+\frac{1}{2})}{\Gamma(n+1)}\right]^2 = O(n^{-2}) \text{ as } n \to \infty,$$

$$g(u_1) = 1 - \frac{1}{\pi\cosh\alpha_0 q_0(i\sinh\alpha_0)}\left\{\sqrt{\frac{1+u_1}{2}} + \frac{1}{2}\ln\left[\frac{\sqrt{2}-\sqrt{1+u_1}}{\sqrt{2}+\sqrt{1+u_1}}\right]\right\},$$

$$A_0 = \frac{1}{g(u_1)}\left[1 + \sum_{n=1}^{\infty} x_{2n}\varepsilon_{2n}Q_n(u_1)\right], \tag{4.103}$$

and

$$Q_n(u_1) = \frac{1}{\sqrt{\pi}}\left[\frac{1}{2}(1+u_1)\right]^{\frac{3}{2}}\left[n(n+\frac{1}{2})\right]^{-\frac{1}{2}} \hat{P}^{(0,\frac{3}{2})}_{n-1}(u_1).$$

Solving the system (4.101) numerically by the truncation method, and employing the rescaling (4.102), we may find the distribution of the electrostatic potential near the conductor by the formula (4.13). An example of the computed potential near an oblate spheroidal conductor with two apertures and charged to unit potential is shown in Figure 4.12. The ratio of major to minor axes is $a/b = \cosh\alpha_0/\sinh\alpha_0 = 2$ and the angular size of the aperture is $\beta_0 = 30^o$; the system truncation number N_{tr} was chosen to be 19. The potential decreases rather uniformly with distance from the structure.

4.7 Capacitance of spheroidal conductors

The surface charge density σ accumulated on the conductor surface ($\alpha = \alpha_0$) is defined by the jump (4.14) in the normal component E_α of the electric

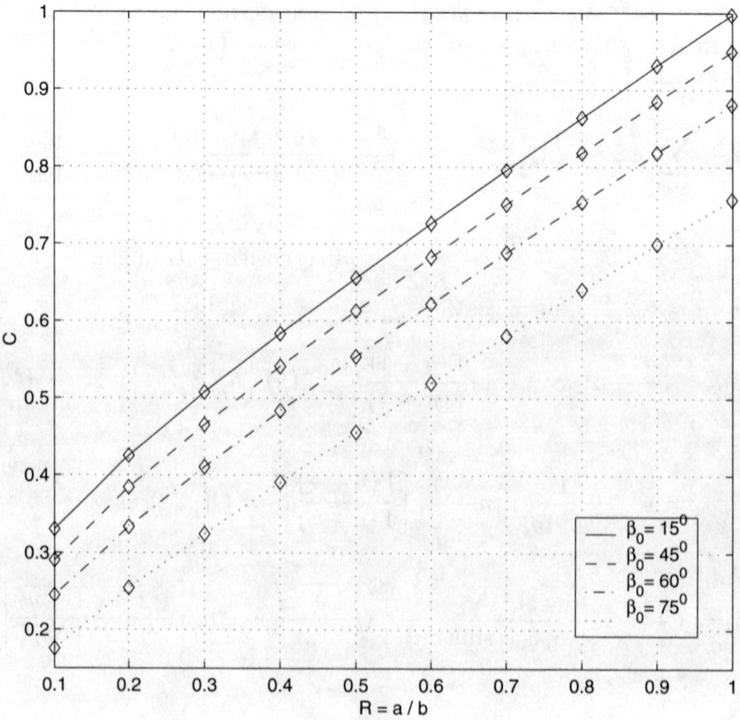

FIGURE 4.13. Capacitance of the prolate spheroidal barrel, as a function of aspect ratio a/b, for varying aperture sizes β_0.

field across the surface, and is given by the expressions (4.15) and (4.18) for the prolate and oblate systems, respectively. As noted in Section 4.1, the total charge Q on each isolated component of the conducting surface is obtained by integration of surface charge density σ over the component surface.

4.7.1 Open spheroidal shells

The total charge on an open spheroidal shell comprising a single component S is, in prolate coordinates, equal to

$$Q = \iint_S \sigma dS = \int_{\beta=0}^{\pi} \int_{\phi=0}^{2\pi} \sigma h_\phi h_\beta d\phi d\beta = \frac{d}{2} \frac{A_0^{(p)}}{Q_0(\cosh \alpha_0)}, \tag{4.104}$$

or, in oblate coordinates, equal to

$$Q = \iint_S \sigma dS = \int_{\beta=0}^{\pi} \int_{\phi=0}^{2\pi} \sigma h_\phi h_\beta d\phi d\beta = \frac{d}{2} \frac{A_0^{(o)}}{q_0(i \sinh \alpha_0)}. \tag{4.105}$$

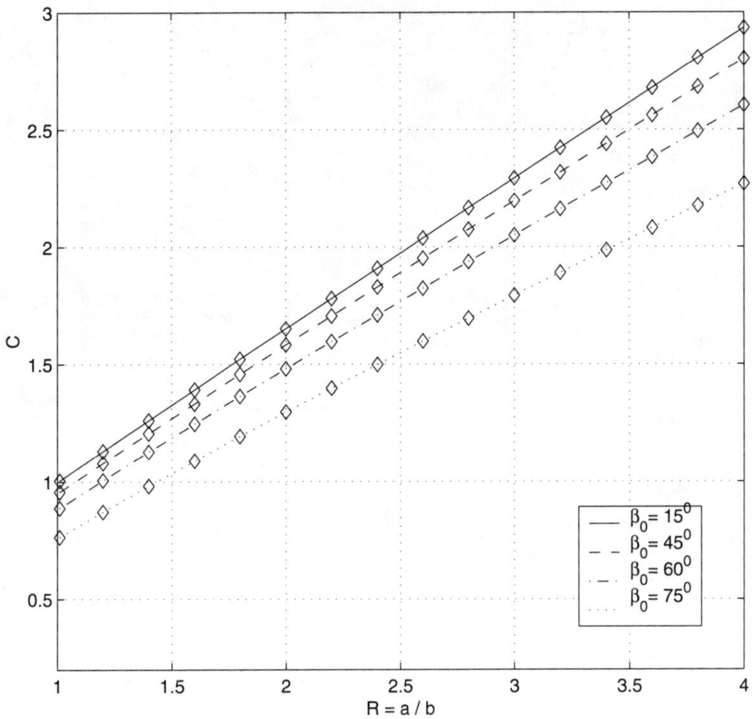

FIGURE 4.14. Capacitance of the oblate spheroidal barrel, as a function of aspect ratio a/b, for varying aperture sizes β_0.

In calculating these integrals we may take the range of β to be $[0, \pi]$ without affecting the result of integration because, as noted above, the expression for surface charge density vanishes over the aperture region. We recall that if the potential of an isolated conductor is equal to unity ($U(\alpha_0, \beta)|_{\beta \in S_0} = 1$), its capacitance C and the charge Q are numerically equal.

The capacitance C (4.104) of the prolate spheroidal shell with two symmetrical circular holes (the barrel) was computed over a wide range of the geometrical parameters a/b and β_0 (the coefficient $A_0^{(p)}$ was found from Formula (4.81)); representative results are presented in Figure 4.13 (the geometrical scale is set by $b = 1$, and so $\frac{d}{2} = \operatorname{sech} \alpha_0$).

In the oblate case, the capacitance C (4.105) was computed from (4.103); representative results are presented in Figure 4.14 (where the geometrical scale is set by $b = 1$, and so $\frac{d}{2} = \operatorname{cosech} \alpha_0$). We recall that a/b is the ratio of minor to major semi-axes of the prolate spheroid, or the ratio of major to minor semi-axes of the oblate spheroid; in both prolate and oblate systems, β_0 defines the angular size of each aperture surface $S_1(\alpha_0, \beta) : \beta \in$

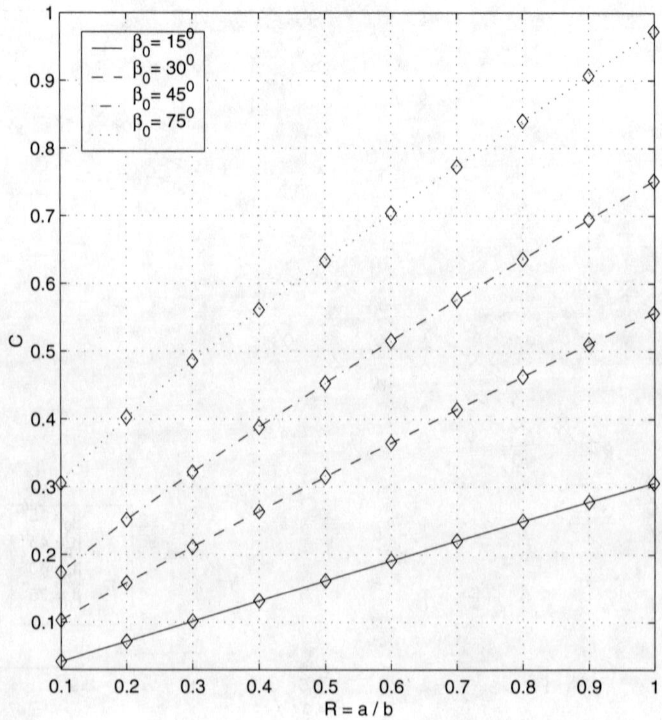

FIGURE 4.15. Capacitance of the slotted prolate spheroidal shell, as a function of aspect ratio a/b, for varying angular size β_0 of components, each charged to unit potential.

$[0, \beta_0] \cup [\pi - \beta_0, \pi]$. The capacitance is an increasing function of aspect ratio and an increasing function of component size.

The total charge on a pair of open spheroidal caps composed of two components S_0 both charged to unit potential ($U_1 = U_2 = 1$) may also be calculated from (4.104) and (4.105) for the prolate and oblate cases, respectively. Numerical results for the capacitance C are presented in Figure 4.15 for the prolate case (the coefficient $A_0^{(p)}$ is found by solving (4.68)) and in Figure 4.16 for the oblate case (the coefficient $A_0^{(o)}$ is found from (4.93)). β_0 defines the angular size of the each component of S_0 ($\beta \in [0, \beta_0]$), and $b = 1$. The capacitance is an increasing function of aspect ratio and of cap size.

When the spheroidal shell with a longitudinal slot degenerates to a closed spheroidal shell ($\beta_0 = \frac{\pi}{2}$), the capacitances of the prolate and oblate closed shells obtained from (4.104) and (4.105) are explicitly calculated to be,

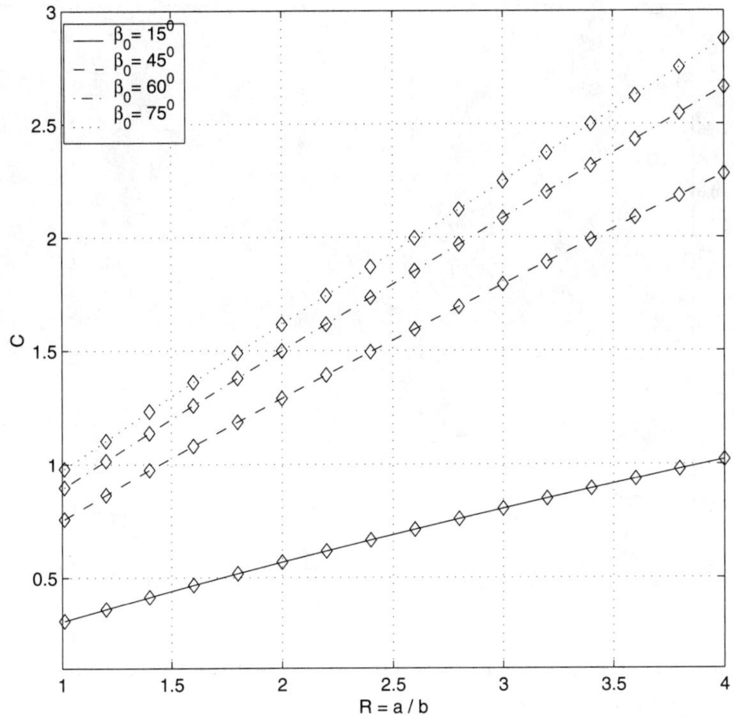

FIGURE 4.16. Capacitance of the slotted oblate spheroidal shell, as a function of aspect ratio a/b, for varying angular size β_0 of components, each charged to unit potential.

respectively,

$$C^{(p)} = \frac{d}{2}\frac{1}{Q_0(\cosh\alpha_0)}, \quad C^{(o)} = \frac{d}{2}\frac{1}{q_0(i\sinh\alpha_0)}. \tag{4.106}$$

It is easy to show that this is identical to that obtained in [26] by another method.

4.7.2 Spheroidal condensors

Consider the condensor formed from oppositely charged plates in the form of spheroidal segments (Figure 4.1(b), 4.1(d)); the upper and lower surfaces are charged to potentials $U_1 = 1$ and $U_2 = -1$, respectively. The charge Q^+ of the positively charged plate is found by the integration of surface charge density σ, given in (4.104) and (4.105) for the prolate and oblate shells, respectively, over the plate surface $S_0 = S_0(\alpha_0, \beta, \varphi)$, where the intervals for integration over β, ϕ are, respectively, $[0, \beta_0]$ and $[0, 2\pi]$. However, we

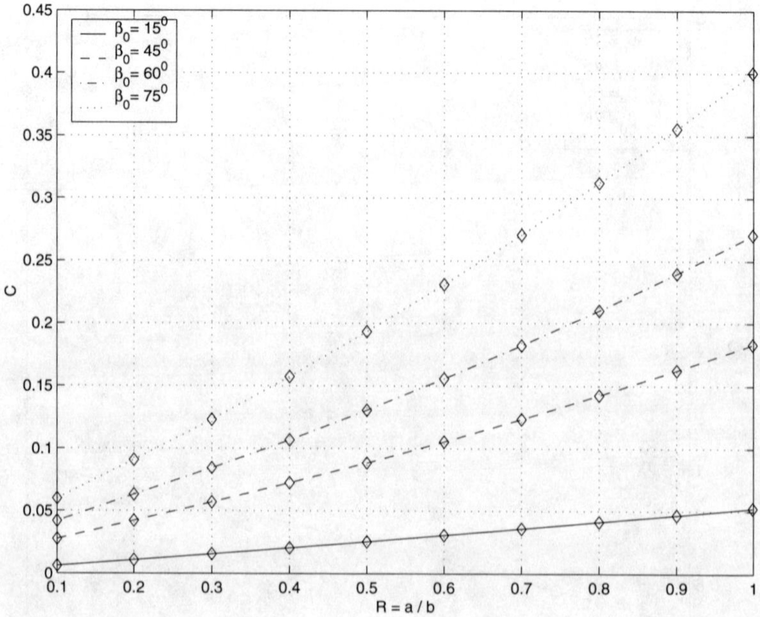

FIGURE 4.17. Capacitance of the prolate spheroidal condenser, as a function of aspect ratio a/b, for varying *plate* size β_0.

may take the interval for integration over β to be $[0, \frac{\pi}{2}]$ because over the slot (defined by $\beta \in [\beta_0, \pi - \beta_0]$), the charge equals zero.

As a result in the prolate case we obtain,

$$Q^+ = \frac{d}{2}\frac{1}{\sqrt{\pi}}\sum_{n=0}^{\infty}\frac{(-1)^n A_{2n+1}}{4Q_{2n+1}(\cosh\alpha_0)P_{2n+1}(\cosh\alpha_0)}\frac{\Gamma(n+\frac{1}{2})}{\Gamma(n+2)} \tag{4.107}$$

where $\{A_{2n+1}\}_{n=0}^{\infty}$ is the solution of the system (4.68) with $l = 1$ and employing the rescaling (4.66); in the oblate case, we obtain

$$Q^+ = \frac{d}{2}\frac{1}{\sqrt{\pi}}\sum_{n=0}^{\infty}\frac{(-1)^n A_{2n+1}}{4q_{2n+1}(i\sinh\alpha_0)p_{2n+1}(i\sinh\alpha_0)}\frac{\Gamma(n+\frac{1}{2})}{\Gamma(n+2)} \tag{4.108}$$

where $\{A_{2n+1}\}_{n=0}^{\infty}$ is the solution of the system (4.91) with $l = 1$ and employing the same rescaling (4.66). The capacitance of the condensor C is then given by the expression

$$C = \left|\frac{Q^+}{U_1 - U_2}\right|.$$

The computed capacitance C of various prolate spheroidal condensors is presented in Figure 4.17, whilst that of the oblate spheroidal condensors is

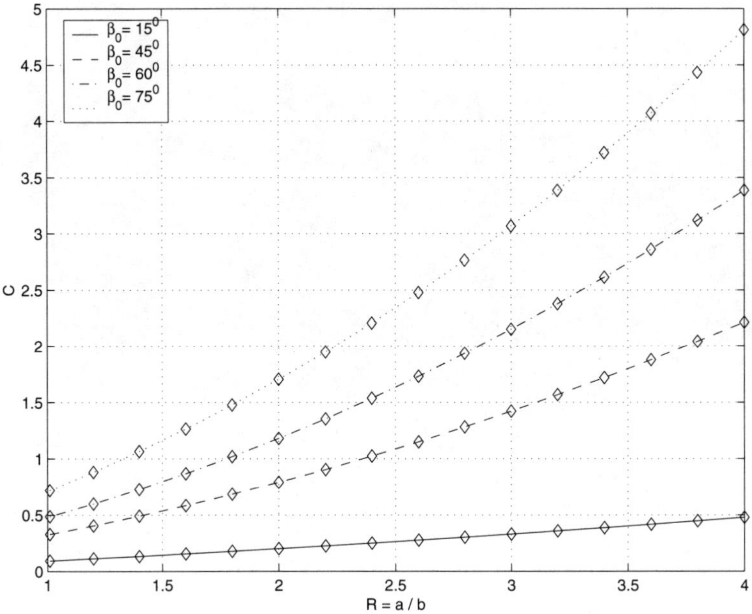

FIGURE 4.18. Capacitance of the oblate spheroidal condenser, as a function of aspect ratio a/b, for varying *plate* size β_0.

presented in Figure 4.18; β_0 is the angular size of each capacitor plate. In both cases, the capacitance is an increasing function of aspect ratio and of angular size of the capacitor plates.

5

Charged Toroidal Shells

Toroidal surfaces provide an interesting canonical class of conductors that illustrate methods for determining potential distributions and the surrounding electrostatic fields when they are charged. The field surrounding a closed torus and its associated capacitance has previously been calculated [36, 26]. However, our interest is in the effect of slots or apertures that might be opened in the surface. If some degree of symmetry is retained, substantive progress with analytic and semi-analytic methods can be made.

Thus, we first consider charged toroidal conductors with slots introduced so that axial symmetry is preserved, as shown in Figure 5.2. The potential is then determined by solving dual or triple series equations with trigonometric kernels. The standard tools provided by the Abel integral transform approach allow us to regularise the series equations and calculate the electrostatic potential by solving an infinite system of linear algebraic equations of the second kind. Surface charge density and capacitance of these conductors are then readily computed. The matrix operator of this system is a completely continuous perturbation of the identity (in the sequence space l_2); this guarantees fast convergence of the truncated system solution to that of infinite system, as has already been demonstrated by similar systems arising from spherical and spheroidal shells (Chapters 3 and 4).

The toroidal coordinate system (α, β, φ) introduced in Section 1.1.7 provides a convenient system for formulating the potential distribution generated by open charged toroidal surfaces as a mixed boundary value problem. If c is the scale factor, the coordinate surface $\alpha = \alpha_0$ defines a torus with

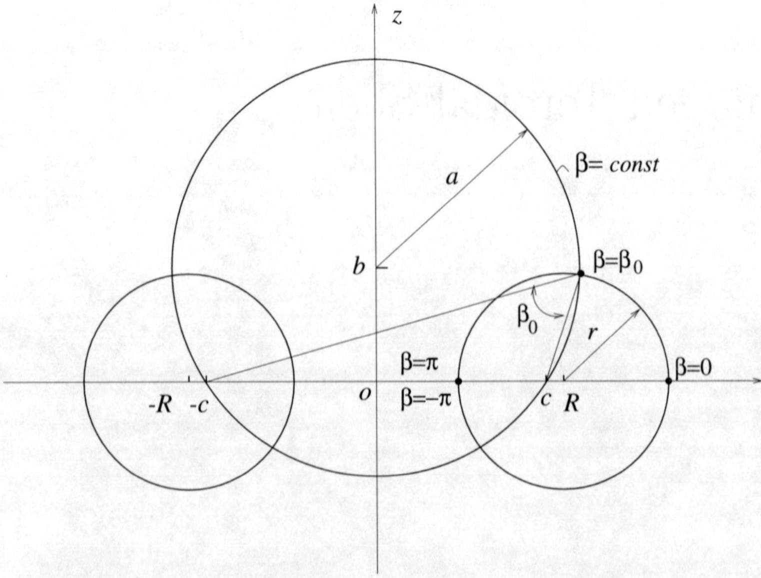

FIGURE 5.1. The toroidal coordinate system in cross-section.

minor radius $r = c\,\text{cosech}\,\alpha_0$ and major radius $R = c\coth\alpha_0$,

$$\left(\sqrt{x^2+y^2} - c\coth\alpha\right)^2 + z^2 = c^2\,\text{cosec}^2\,\alpha;$$

its interior and exterior are respectively specified by the intervals (α_0, ∞) and $[0, \alpha_0)$ for α, whilst β and φ range over their full intervals of definition $[-\pi, \pi]$. (See Figure 5.1.) In all our numerical calculations the scale factor c is chosen so that $r = 1$.

We consider the potential distribution surrounding toroidal surfaces with various types of slots or apertures. Fix α_0 and consider the toroidal surface $\alpha = \alpha_0$ (see Figure 5.2(a)). First, we calculate the potential surrounding various axially symmetric structures obtained by cutting axisymmetric slots in this surface. In Section 5.2, the single slot (see Figure 5.2(b)) is examined. The slot may be described by a fixed parameter β_0; the connected portion of the toroidal surface given by $\beta \in [-\pi, -\beta_0] \cup [\beta_0, \pi]$ is removed from the complete torus. The introduction of two types of (axisymmetric) slots is considered in the following two sections: *transversal* slots (Section 5.3, see Figure 5.2(c)) that remove part of the conductor surface so that the remaining segments are specified by

$$\alpha = \alpha_0, \quad \beta \in [-\beta_0, \beta_0] \cup [\pi - \beta_0, \pi] \cup [-\pi, -(\pi - \beta_0)], \tag{5.1}$$

and *longitudinal* slots (Section 5.4, see Figure 5.2(d)), in which the segment (5.1) is removed from the full torus $\alpha = \alpha_0$. Capacitances are briefly examined in Section 5.5.

The calculation becomes more complicated when apertures are introduced so that axial symmetry is broken. The final Section (5.6) describes one such structure that can be solved semi-analytically – the degenerate toroidal shell, with equal major and minor radii, from which an azimuthal sector is removed (see Figure 5.6). In cylindrical polars (ρ, θ, ϕ), this toroid has equation

$$(\rho - a)^2 + z^2 = a^2, \quad \varphi \in [-\pi, \pi],$$

and the azimuthal sector of angular semi-width ϕ_0,

$$(\rho - a)^2 + z^2 = a^2, \quad \varphi \in [-\phi_0, \phi_0],$$

is removed. The potential distribution is determined for this structure, as well as for the degenerate toroidal surface from which multiple azimuthal sectors are removed (see Figure 5.7). The approach invokes the principle of Kelvin inversion (in a sphere) to transform the problem to a set of dual series equations dependent upon a continuous spectral parameter.

This final calculation is a very significant extension of analytic and semi-analytic techniques to the determination of the three-dimensional potential distribution surrounding *nonsymmetric* open conducting surfaces.

5.1 Formulation of mixed boundary value problems in toroidal geometry

We consider the potential distribution surrounding the toroidal surface $\alpha = \alpha_0$ into which one or more axisymmetric slots are introduced; such a surface may be specified by

$$\alpha = \alpha_0, \quad \beta \in I_0, \quad \varphi \in [0, 2\pi],$$

where I_0 is a subinterval, or disjoint union of several subintervals of $[0, 2\pi]$. The mixed boundary value problem for the potential theory surrounding such a slotted toroidal conductor is formulated as follows. Find the function U that is harmonic in R^3,

$$\Delta U(\alpha, \beta, \varphi) = 0, \tag{5.2}$$

that satisfies the Dirichlet boundary conditions on that part of toroidal surface S_0 occupied by the conductor, specifying the potential f on S_0,

$$U(\alpha_0 - 0, \beta, \varphi) = U(\alpha_0 + 0, \beta, \varphi) = f(\beta, \varphi), \text{ for } \beta \in I_0, \varphi \in [-\pi, \pi], \tag{5.3}$$

that has continuous normal derivative on the aperture surface S_1,

$$\frac{d}{d\alpha} U(\alpha, \beta, \varphi)|_{\alpha=\alpha_0-0}^{\alpha=\alpha_0+0} = 0, \text{ for } \beta \in [-\pi, \pi] \backslash I_0, \varphi \in [-\pi, \pi], \tag{5.4}$$

and that vanishes at infinity,

$$U(\alpha,\beta,\varphi) = O\left(|\overrightarrow{r}|^{-1}\right) \quad \text{as } |\overrightarrow{r}| = \left(x^2+y^2+z^2\right)^{\frac{1}{2}} \to \infty,$$

i.e., U vanishes as $\alpha \to 0$ and $\beta \to 0$. Also, the electrostatic energy in any volume of space including edges of the conductor must be bounded:

$$W = \iiint_V |\operatorname{grad} U|^2 \, dV < \infty. \tag{5.5}$$

In toroidal coordinates, the Laplace equation (see Section 1.2.7) admits separation of variables and has solution in the form:

$$\frac{U(\alpha,\beta,\varphi)}{\sqrt{2\cosh\alpha - 2\cos\beta}} = \sum_{m=0}^{\infty}\sum_{n=m}^{\infty}\left[A_{nm}P^m_{n-\frac{1}{2}}(\cosh\alpha) + B_{nm}Q^m_{n-\frac{1}{2}}(\cosh\alpha)\right]\begin{Bmatrix}\cos n\beta\cos m\varphi \\ \sin n\beta \sin m\varphi\end{Bmatrix} \tag{5.6}$$

where $P^m_{n-\frac{1}{2}}(\cosh\alpha), Q^m_{n-\frac{1}{2}}(\cosh\alpha)$ are toroidal functions, and A_{nm}, B_{nm} are constants to be determined by the mixed boundary conditions. The separation constants n, m are integers because U is periodic in the coordinates β and φ. We restrict attention to axisymmetric problems so that only those terms with $m = 0$ are retained in (5.6); moreover, the open shell structure will be assumed to be symmetric about the xy plane, so that any dependence upon terms involving $\sin n\beta$ in (5.6) is avoided (the interval I_0 is therefore symmetric about the origin). Considering the asymptotic behaviour of the functions $P_{n-\frac{1}{2}}(\cosh\alpha)$ and $Q_{n-\frac{1}{2}}(\cosh\alpha)$ at the singular points ($\alpha = 0, \alpha \to \infty$), solutions of the type (5.6), which decay appropriately at infinity and are continuous across the toroidal surface $\alpha = \alpha_0$, have the following form in the interior ($\alpha \geq \alpha_0$) and exterior ($0 \leq \alpha < \alpha_0$) regions

$$\frac{U(\alpha,\beta)}{\sqrt{2\cosh\alpha_0 - 2\cos\beta}} = \sum_{n=0}^{\infty} C_n \cos n\beta \begin{cases} Q_{n-\frac{1}{2}}(\cosh\alpha), & \alpha \geq \alpha_0, \\ Q_{n-\frac{1}{2}}(\cosh\alpha_0)P_{n-\frac{1}{2}}(\cosh\alpha)/P_{n-\frac{1}{2}}(\cosh\alpha_0), & \alpha < \alpha_0. \end{cases} \tag{5.7}$$

The constants C_n are to be determined by enforcement of the mixed boundary conditions (5.3) and (5.4).

5.2 The open charged toroidal segment

The toroidal shell with one slot or toroidal segment is shown in Figure 5.2(b); it occupies the region $\alpha = \alpha_0, \beta \in [-\beta_0, \beta_0]$ whilst the slot is defined

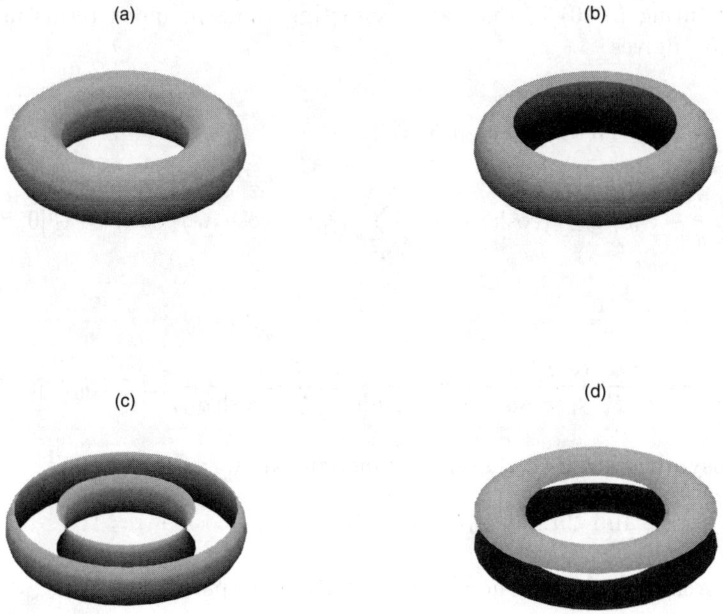

FIGURE 5.2. The torus (a), and various toroidal shells: (b) single slot, (c) two transversal slots and (d) two longitudinal slots.

by $\alpha = \alpha_0, \beta \in [-\pi, -\beta_0) \cup (\beta_0, \pi]$. If the segment is charged to unit potential, enforcement of the boundary conditions (5.3) and (5.4) produces the following ,

$$\sum_{n=0}^{\infty} C_n Q_{n-\frac{1}{2}}(\cosh \alpha_0) \cos n\beta = (2\cosh \alpha_0 - 2\cos \beta)^{-\frac{1}{2}}, \beta \in [0, \beta_0], \quad (5.8)$$

$$\sum_{n=0}^{\infty} C_n \frac{1}{\sinh \alpha_0 P_{n-\frac{1}{2}}(\cosh \alpha_0)} \cos n\beta = 0, \beta \in (\beta_0, \pi], \quad (5.9)$$

where the value of the Wronskian of $P_{n-\frac{1}{2}}$ and $Q_{n-\frac{1}{2}}$ (see Appendix, (B.69)) has been employed.

The toroidal asymmetry factor appearing on the right-hand side of (5.8) has an expansion in a Fourier series

$$(2\cosh \alpha_0 - 2\cos \beta)^{-\frac{1}{2}} = \frac{1}{\pi} \sum_{n=0}^{\infty} (2 - \delta_{n0}) Q_{n-\frac{1}{2}}(\cosh \alpha_0) \cos n\beta. \quad (5.10)$$

Substituting (5.10) in (5.8) and extracting the zero index terms in (5.8) and (5.9) gives

$$\sum_{n=1}^{\infty} C_n Q_{n-\frac{1}{2}}(\cosh\alpha_0)\cos n\beta =$$
$$\left(\frac{1}{\pi} - C_0\right) Q_{-\frac{1}{2}}(\cosh\alpha_0) + \frac{2}{\pi}\sum_{n=1}^{\infty} Q_{n-\frac{1}{2}}(\cosh\alpha_0)\cos n\beta, \beta \in [0,\beta_0], \tag{5.11}$$

$$\sum_{n=1}^{\infty} \frac{C_n \cos n\beta}{\sinh\alpha_0 P_{n-\frac{1}{2}}(\cosh\alpha_0)} = -\frac{C_0}{\sinh\alpha_0 P_{-\frac{1}{2}}(\cosh\alpha_0)}, \beta \in (\beta_0,\pi]. \tag{5.12}$$

The asymptotics of the Legendre functions allows us to estimate

$$\lim_{n\to\infty} 2n\sinh\alpha_0 P_{n-\frac{1}{2}}(\cosh\alpha_0) Q_{n-\frac{1}{2}}(\cosh\alpha_0) = 1; \tag{5.13}$$

we therefore introduce the asymptotically small parameter

$$\varepsilon_n = 1 - 2n\sinh\alpha_0 P_{n-\frac{1}{2}}(\cosh\alpha_0) Q_{n-\frac{1}{2}}(\cosh\alpha_0) = O(n^{-2}) \text{ as } n \to \infty. \tag{5.14}$$

Rescaling the unknowns

$$x_n = \frac{C_n}{2nP_{n-\frac{1}{2}}(\cosh\alpha_0)} = \frac{\sinh\alpha_0}{1-\varepsilon_n} C_n Q_{n-\frac{1}{2}}(\cosh\alpha_0),$$

we convert Equations (5.8) and (5.9) to the form

$$\sum_{n=1}^{\infty}\left\{x_n(1-\varepsilon_n) - \frac{2}{\pi}\sinh\alpha_0 Q_{n-\frac{1}{2}}(\cosh\alpha_0)\right\}\cos n\beta$$
$$= \sinh\alpha_0\left(\frac{1}{\pi} - C_0\right) Q_{-\frac{1}{2}}(\cosh\alpha_0), \ \beta \in [0,\beta_0], \tag{5.15}$$

$$\sum_{n=1}^{\infty} n x_n \cos n\beta = -\frac{C_0}{2P_{-\frac{1}{2}}(\cosh\alpha_0)}, \ \beta \in (\beta_0,\pi]. \tag{5.16}$$

The standard procedure for solving such series equations involving cosine kernels has been described in Section 2.2 in some detail (see Equations (2.39) and (2.40)). Making the necessary identification of terms, the solution may directly be deduced from (2.62) to be as stated below in (5.21). Let us sketch briefly some of the main steps in its deduction. It employs the replacement of cosine functions by Jacobi polynomials given by (1.151) and

(1.152). A necessary preliminary step is the integration of both equations to increase the indices of the Jacobi polynomials so that the methods of Chapter 2 are applicable. The variant (2.36) of Rodrigues' formula may be applied after the insertion of (1.151) in (5.15) and (5.16); equivalently, we may directly integrate these equations to obtain

$$\sum_{n=1}^{\infty}\left\{x_n\left(1-\varepsilon_n\right)-\frac{2}{\pi}\sinh\alpha_0 Q_{n-\frac{1}{2}}(\cosh\alpha_0)\right\}\frac{\sin n\beta}{n}$$
$$=\beta\sinh\alpha_0\left(\frac{1}{\pi}-C_0\right)Q_{-\frac{1}{2}}(\cosh\alpha_0),\qquad \beta\in[0,\beta_0],\quad (5.17)$$

$$\sum_{n=1}^{\infty}x_n\sin n\beta=(\pi-\beta)\frac{C_0}{2P_{-\frac{1}{2}}(\cosh\alpha_0)},\qquad \beta\in(\beta_0,\pi].\qquad(5.18)$$

Setting $z=\cos\beta$, $z_0=\cos\beta_0$, and employing the formula (1.153) produces

$$\sum_{n=1}^{\infty}x_n\frac{\Gamma(n+1)}{\Gamma(n+\frac{1}{2})}P_{n-1}^{(\frac{1}{2},\frac{1}{2})}(z)$$
$$=\frac{C_0}{P_{-\frac{1}{2}}(\cosh\alpha_0)(1-z^2)^{\frac{1}{2}}\sqrt{\pi}}\left[\arcsin z+\frac{\pi}{2}\right],\quad z\in(-1,z_0),\quad(5.19)$$

$$\sum_{n=1}^{\infty}\left[x_n\left(1-\varepsilon_n\right)-\frac{2}{\pi}\sinh\alpha_0 Q_{n-\frac{1}{2}}(\cosh\alpha_0)\right]\frac{\Gamma(n)}{\Gamma(n+\frac{1}{2})}P_{n-1}^{(\frac{1}{2},\frac{1}{2})}(z)$$
$$=\frac{2\sinh\alpha_0}{(1-z^2)^{\frac{1}{2}}\sqrt{\pi}}\left(\frac{1}{\pi}-C_0\right)Q_{-\frac{1}{2}}(\cosh\alpha_0)\left[\frac{\pi}{2}-\arcsin z\right],\quad z\in(z_0,1).$$
$$(5.20)$$

From the Abel integral representation (1.171) expressing $P_{n-1}^{(\frac{1}{2},\frac{1}{2})}$ in terms of $P_{n-1}^{(0,1)}$, and its companion (1.172) expressing $P_{n-1}^{(0,1)}$ in terms of $P_{n-1}^{(\frac{1}{2},\frac{1}{2})}$, we derive the infinite system of linear algebraic equations of the second kind for the rescaled unknowns $y_n=\sqrt{2n}x_n$ in the standard way described previously:

$$y_m-\sum_{n=1}^{\infty}y_n\varepsilon_n Q_{nm}(z_0)=$$
$$\frac{2\sinh\alpha_0}{\pi}\sum_{n=1}^{\infty}\sqrt{2n}Q_{n-\frac{1}{2}}(\cosh\alpha_0)Q_{nm}(z_0)+$$
$$2\sinh\alpha_0 Q_{-\frac{1}{2}}(\cosh\alpha_0)\frac{(1+z_0)}{\pi t}\frac{\hat{P}_{m-1}^{(0,1)}(z_0)}{m}.\quad(5.21)$$

Here

$$t = t_1 - \ln\left(\frac{1-z_0}{2}\right) \tag{5.22}$$

where

$$t_1 = 2\sinh\alpha_0 P_{-\frac{1}{2}}(\cosh\alpha_0) Q_{-\frac{1}{2}}(\cosh\alpha_0), \tag{5.23}$$

$$Q_{nm}(z_0) = \left[\hat{Q}^{(1,0)}_{n-1,m-1}(z_0) + \frac{(1+z_0)^2}{t}\frac{\hat{P}^{(0,1)}_{n-1}(z_0)\hat{P}^{(0,1)}_{m-1}(z_0)}{nm}\right] \tag{5.24}$$

and

$$C_0 = \frac{t_1}{t\pi} + \frac{1}{t}\sum_{n=1}^{\infty}\left[\frac{1}{n}y_n\varepsilon_n + \frac{2}{\pi}\sqrt{\frac{2}{n}}\sinh\alpha_0 Q_{n-\frac{1}{2}}(\cosh\alpha_0)\right] \times$$
$$P_{-\frac{1}{2}}(\cosh\alpha_0)\,(1+z_0)\,\hat{P}^{(0,1)}_{n-1}(z_0). \tag{5.25}$$

Note that C_0 is found by enforcement of a continuity condition of the function at the point $z = z_0$.

From the solution of the system of Equations (5.21), we may find the Fourier coefficients of the series (5.7) and thus calculate the potential U and the associated electrostatic field near the charged toroidal segment. An example is shown in Figure 5.3. Recall that the scale factor c is chosen so that the minor radius $r = c\,\mathrm{cosech}\,\alpha_0$ equals 1.

5.3 The toroidal shell with two transversal slots

This section begins the examination of toroidal surfaces with two axially symmetric slots. The geometry of a toroidal shell with two transversal slots is shown in Figure 5.2(c). The conducting surface is specified by

$$\alpha = \alpha_0, \beta \in [-\pi, -(\pi - \beta_0)] \cup [-\beta_0, \beta_0] \cup [\pi - \beta_0, \pi].$$

If the toroidal segments are charged to unit potential, enforcement on (5.7) of the boundary conditions (5.3) (unit potential on the surface), and (5.4) (continuity of the normal derivative on the slots) leads to the following triple series equations with the trigonometric kernels to be solved for the unknown coefficients C_n,

$$\sum_{n=0}^{\infty} C_n Q_{n-\frac{1}{2}}(\cosh\alpha_0)\cos n\beta = (2\cosh\alpha_0 - 2\cos\beta)^{-\frac{1}{2}},$$
$$\beta \in [0, \beta_0] \cup [\pi - \beta_0, \pi], \tag{5.26}$$

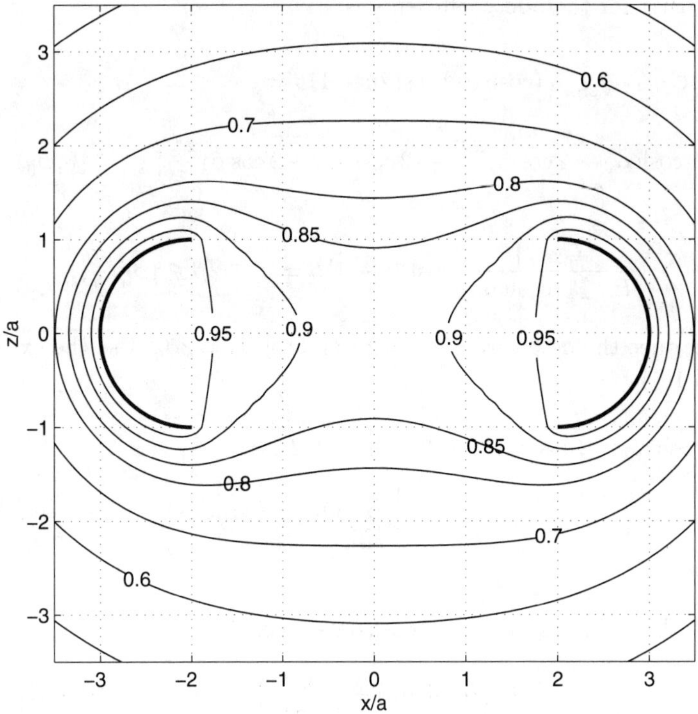

FIGURE 5.3. Electrostatic potential surrounding the charged toroidal segment with radii $r = 1, R = 2$, and $\beta_0 = 60^o$.

$$\sum_{n=0}^{\infty} C_n \frac{1}{\sinh \alpha_0 P_{n-\frac{1}{2}}(\cosh \alpha_0)} \cos n\beta = 0, \quad \beta \in (\beta_0, \pi - \beta_0). \tag{5.27}$$

The property, $\cos n(\pi - \beta) = (-1)^n \cos n\beta$, allows us to decouple even and odd index coefficients and obtain the following pair of dual series equations defined on the half interval $\left[0, \frac{\pi}{2}\right]$. The system for the even coefficients is

$$\sum_{n=0}^{\infty} C_{2n} Q_{2n-\frac{1}{2}}(\cosh \alpha_0) \cos 2n\beta =$$
$$\frac{1}{2}\left\{(2\cosh\alpha_0 - 2\cos\beta)^{-\frac{1}{2}} + (2\cosh\alpha_0 + 2\cos\beta)^{-\frac{1}{2}}\right\}, \ \beta \in (0, \beta_0) \tag{5.28}$$

$$\sum_{n=0}^{\infty} \frac{C_{2n}}{P_{2n-\frac{1}{2}}(\cosh \alpha_0)} \cos 2n\beta = 0, \qquad \beta \in (\beta_0, \frac{\pi}{2}), \tag{5.29}$$

whilst that for the odd coefficients is

$$\sum_{n=0}^{\infty} C_{2n+1} Q_{2n+\frac{1}{2}}(\cosh\alpha_0)\cos(2n+1)\beta =$$
$$\frac{1}{2}\left\{(2\cosh\alpha_0 - 2\cos\beta)^{-\frac{1}{2}} - (2\cosh\alpha_0 + 2\cos\beta)^{-\frac{1}{2}}\right\}, \beta \in (0,\beta_0) \quad (5.30)$$

$$\sum_{n=0}^{\infty} \frac{C_{2n+1}}{P_{2n+\frac{1}{2}}(\cosh\alpha_0)}\cos(2n+1)\beta = 0, \qquad \beta \in (\beta_0, \frac{\pi}{2}). \quad (5.31)$$

Introduce the new variable $\theta = 2\beta$ and set $\theta_0 = 2\beta_0$. Use the expansions (cf. (5.10))

$$(2\cosh\alpha_0 - 2\cos\beta)^{-\frac{1}{2}} - (2\cosh\alpha_0 + 2\cos\beta)^{-\frac{1}{2}}$$
$$= \frac{4}{\pi}\sum_{n=0}^{\infty} Q_{2n+\frac{1}{2}}(\cosh\alpha_0)\cos(n+\frac{1}{2})\theta, \quad (5.32)$$

$$(2\cosh\alpha_0 - 2\cos\beta)^{-\frac{1}{2}} + (2\cosh\alpha_0 + 2\cos\beta)^{-\frac{1}{2}}$$
$$= \frac{2}{\pi}Q_{-\frac{1}{2}}(\cosh\alpha_0) + \frac{4}{\pi}\sum_{n=1}^{\infty} Q_{2n-\frac{1}{2}}(\cosh\alpha_0)\cos n\theta, \quad (5.33)$$

to obtain the following pair of dual equations defined on the full interval of the variable $[0,\pi]$. The system for the even coefficients is

$$\sum_{n=1}^{\infty} C_{2n} Q_{2n-\frac{1}{2}}(\cosh\alpha_0)\cos n\theta$$
$$= \left(\frac{1}{\pi} - C_0\right) Q_{-\frac{1}{2}}(\cosh\alpha_0) + \frac{2}{\pi}\sum_{n=1}^{\infty} Q_{2n-\frac{1}{2}}(\cosh\alpha_0)\cos n\theta,\ \theta \in (0,\theta_0), \quad (5.34)$$

$$\sum_{n=1}^{\infty} \frac{C_{2n}}{P_{2n-\frac{1}{2}}(\cosh\alpha_0)}\cos n\theta = -\frac{C_0}{P_{-\frac{1}{2}}(\cosh\alpha_0)}, \qquad \theta \in (\theta_0, \pi), \quad (5.35)$$

whilst that for the odd coefficients is

$$\sum_{n=0}^{\infty} C_{2n+1} Q_{2n+\frac{1}{2}}(\cosh\alpha_0)\cos(n+\frac{1}{2})\theta$$
$$= \frac{2}{\pi}\sum_{n=0}^{\infty} Q_{2n+\frac{1}{2}}(\cosh\alpha_0)\cos(n+\frac{1}{2})\theta, \quad \theta \in (0,\theta_0), \quad (5.36)$$

$$\sum_{n=0}^{\infty} \frac{C_{2n+1}}{P_{2n+\frac{1}{2}}(\cosh \alpha_0)} \cos(n+\frac{1}{2})\theta = 0, \qquad \theta \in (\theta_0, \pi). \tag{5.37}$$

The Equations (5.34) and (5.35) are very similar to the equations (5.8) and (5.9) considered in the previous section. Setting $z = \cos\theta$, $z_0 = \cos\theta_0$, we may immediately deduce that the regularised system for the even coefficients is

$$y_{2m} - \sum_{n=1}^{\infty} y_{2n}\varepsilon_{2n} Q_{nm}(z_0) = \frac{4\sinh\alpha_0}{\pi} \sum_{n=1}^{\infty} \sqrt{2n} Q_{2n-\frac{1}{2}}(\cosh\alpha_0) Q_{nm}(z_0) + 4\sinh\alpha_0 Q_{-\frac{1}{2}}(\cosh\alpha_0) \frac{(1+z_0)}{\pi t} \frac{\hat{P}_{m-1}^{(0,1)}(z_0)}{m} \tag{5.38}$$

where t, t_1 and $Q_{nm}(z_0)$ are defined by (5.22)–(5.24),

$$y_{2n} = \frac{C_{2n}}{\sqrt{2n} P_{2n-\frac{1}{2}}(\cosh\alpha_0)},$$

$$C_0 = \frac{t_1}{t\pi} + P_{-\frac{1}{2}}(\cosh\alpha_0) \frac{(1+z_0)}{t} \times \sum_{n=1}^{\infty} \left[\frac{1}{n} y_{2n}\varepsilon_{2n} + \frac{4}{\pi}\sqrt{\frac{2}{n}} \sinh\alpha_0 Q_{2n-\frac{1}{2}}(\cosh\alpha_0) \right] \hat{P}_{n-1}^{(0,1)}(z_0),$$

and

$$\varepsilon_{2n} = 1 - 4n\sinh\alpha_0 P_{2n-\frac{1}{2}}(\cosh\alpha_0) Q_{2n-\frac{1}{2}}(\cosh\alpha_0) = O(n^{-2}) \quad \text{as } n \to \infty.$$

Let us now turn to the equations (5.36) and (5.37). The latter series (5.37) is nonuniformly convergent and we integrate it to obtain the uniformly convergent series equations

$$\sum_{n=0}^{\infty} \frac{C_{2n+1}}{P_{2n+\frac{1}{2}}(\cosh\alpha_0)} \frac{\Gamma(n+1)}{\Gamma(n+\frac{3}{2})} P_n^{(\frac{1}{2},-\frac{1}{2})}(z) = \frac{a}{\sqrt{1-z}}, \quad z \in (-1, z_0) \tag{5.39}$$

$$\sum_{n=0}^{\infty} C_{2n+1} Q_{2n+\frac{1}{2}}(\cosh\alpha_0) \frac{\Gamma(n+1)}{\Gamma(n+\frac{1}{2})} P_n^{(-\frac{1}{2},\frac{1}{2})}(z) = \frac{2}{\pi} \sum_{n=0}^{\infty} Q_{2n+\frac{1}{2}}(\cosh\alpha_0) \frac{\Gamma(n+1)}{\Gamma(n+\frac{1}{2})} P_n^{(-\frac{1}{2},\frac{1}{2})}(z), z \in (z_0, 1) \tag{5.40}$$

where we have replaced $\sin\left(n+\frac{1}{2}\right)\theta$ by its representation (1.154) in terms of the Jacobi polynomial $P_n^{\left(-\frac{1}{2},\frac{1}{2}\right)}$; a is a constant that will be determined later. Now apply the Abel integral transform technique, employing the integral representation (1.172) for $P_n^{(0,0)} \equiv P_n$ in terms of $P_n^{\left(\frac{1}{2},-\frac{1}{2}\right)}$, and the companion representation (1.171) for P_n in terms of $P_n^{\left(-\frac{1}{2},\frac{1}{2}\right)}$; from Equations (5.39) and (5.40) we may deduce

$$\sum_{n=0}^{\infty} \frac{C_{2n+1}}{\left(n+\frac{1}{2}\right) P_{2n+\frac{1}{2}}(\cosh\alpha_0)} P_n(z) = a\sqrt{\frac{2}{\pi}} Q_{-\frac{1}{2}}(z), \qquad z \in (-1, z_0), \tag{5.41}$$

$$\sum_{n=0}^{\infty} C_{2n+1} Q_{2n+\frac{1}{2}}(\cosh\alpha_0) P_n(z) = \frac{2}{\pi} \sum_{n=0}^{\infty} Q_{2n+\frac{1}{2}}(\cosh\alpha_0) P_n(z), \quad z \in (z_0, 1). \tag{5.42}$$

Let

$$x_{2n+1} = \frac{C_{2n+1}}{2\left(n+\frac{1}{2}\right) P_{2n+\frac{1}{2}}(\cosh\alpha_0)}.$$

As shown previously, the parameter

$$\varepsilon_{2n+1} = 1 - 2(2n+1)\sinh\alpha_0 P_{2n+\frac{1}{2}}(\cosh\alpha_0) Q_{2n+\frac{1}{2}}(\cosh\alpha_0) \tag{5.43}$$

is asymptotically small as $n \to \infty$: $\varepsilon_{2n+1} = O(n^{-2})$. The rescaled unknowns satisfy

$$\sum_{n=0}^{\infty} x_{2n+1} P_n(z) = \frac{1}{2} a \sqrt{\frac{2}{\pi}} Q_{-\frac{1}{2}}(z), \qquad z \in (-1, z_0), \tag{5.44}$$

$$\sum_{n=0}^{\infty} \left(1 - \varepsilon_{2n+1}\right) x_{2n+1} P_n(z) = \frac{\sinh\alpha_0}{\pi} \sum_{n=0}^{\infty} Q_{2n+\frac{1}{2}}(\cosh\alpha_0) P_n(z), \qquad z \in (z_0, 1). \tag{5.45}$$

Rearranging (5.41) and (5.42) gives

$$\sum_{n=0}^{\infty} x_{2n+1} P_n(z) = \begin{Bmatrix} F_1(z), & z \in (-1, z_0) \\ F_2(z), & z \in (z_0, 1) \end{Bmatrix}, \tag{5.46}$$

where

$$\begin{aligned} F_1(z) &= a\,(2\pi)^{-\frac{1}{2}} \sinh\alpha_0 Q_{-\frac{1}{2}}(z), \\ F_2(z) &= \sum_{n=0}^{\infty} x_{2n+1}\varepsilon_{2n+1} P_n(z) + 4\pi^{-1} \sinh\alpha_0 \sum_{n=0}^{\infty} Q_{2n+\frac{1}{2}}(\cosh\alpha_0) P_n(z). \end{aligned}$$

The constant a is determined by the continuity requirement on the function on the left-hand side of (5.46) at the point $z = z_0$:

$$a = \frac{\sqrt{2\pi}}{\sinh \alpha_0 Q_{-\frac{1}{2}}(z_0)} \sum_{n=0}^{\infty} \left[x_{2n+1}\varepsilon_{2n+1} + \frac{4}{\pi} \sinh \alpha_0 Q_{2n+\frac{1}{2}}(\cosh \alpha_0) \right] P_n(z_0). \tag{5.47}$$

After rescaling the unknowns via $y_{2n+1} = \left(n + \frac{1}{2}\right)^{-\frac{1}{2}} x_{2n+1}$, we obtain the following infinite system of linear algebraic equations

$$y_{2m+1} - \sum_{n=0}^{\infty} y_{2n+1}\varepsilon_{2n+1} \left[\hat{Q}_{nm}^{(0,0)}(z_0) + R_{nm}(z_0) \right] =$$

$$\frac{4}{\pi} \sinh \alpha_0 \sum_{n=0}^{\infty} Q_{2n+\frac{1}{2}}(\cosh \alpha_0)(n + \frac{1}{2})^{-\frac{1}{2}} \left[\hat{Q}_{nm}^{(0,0)}(z_0) + R_{nm}(z_0) \right], \tag{5.48}$$

where $m = 0, 1, 2, \ldots,$

$$R_{nm}(z_0) = \frac{\hat{P}_n(z_0)}{Q_{-\frac{1}{2}}(z_0)} I_m, \; I_m = \int_{-1}^{z_0} Q_{-\frac{1}{2}}(z) \hat{P}_m(z) dz, \tag{5.49}$$

and $\hat{P}_n$ is the normalised Legendre polynomial. The integrals I_m are readily computed (see Appendix, (B.97)):

$$\begin{aligned} I_0 &= -2\left(Q_{\frac{1}{2}}(z_0) - z_0 Q_{-\frac{1}{2}}(z_0) \right), \\ I_m &= \frac{z_0^2 - 1}{m(m+1) + \frac{1}{4}} \left(Q_{-\frac{1}{2}}(z_0) P'_m(z_0) - Q'_{-\frac{1}{2}}(z_0) P_m(z_0) \right), \; m > 0, \end{aligned} \tag{5.50}$$

where we note that the Legendre functions $Q_{\pm\frac{1}{2}}$ are simply expressed in term of complete elliptic integrals (see Appendix, (B.80) and (B.82)).

The solution of the systems (5.38) and (5.48) yields the Fourier coefficients of the series (5.7), and thus the potential and the associated electrostatic field may be calculated.

Computationally, the systems (5.38) and (5.48) enjoy the same advantages as the regularised systems considered in Chapter 4. As noted above, the Legendre functions of half-integer index $P_{\pm\frac{1}{2}}, Q_{\pm\frac{1}{2}}$ are simply expressed in terms of complete elliptic integrals (see Appendix, (B.77)–(B.82)). Recurrence relations for the matrix elements of these systems are readily developed, so numerical values of the unknown Fourier coefficients and the electrostatic field may be computed very efficiently. Four correct digits in the values of the coefficients $\{x_n\}_{n=0}^{\infty}$ are guaranteed by a choice of truncation number N_{tr} not exceeding 20. Some computed examples of the electrostatic potential are given in Figures 5.4 and 5.5 for the toroidal conductor with two transversal slots having radii $r = 1, R = 2$, and angular parameter β_0 equal to 60° and 30°, respectively.

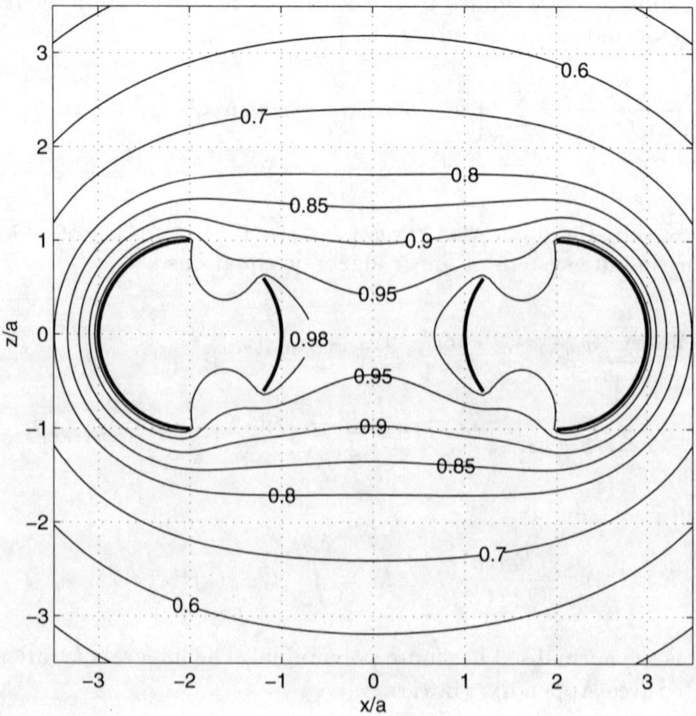

FIGURE 5.4. The charged toroidal shell with two transversal slots; the radii are $r = 1, R = 2$, and $\beta_0 = 60°$.

5.4 The toroidal shell with two longitudinal slots

This section continues the examination of toroidal surfaces with two axially symmetric slots. In particular, we consider the surface complementary to that of the previous section, where the locations of conducting surface and slots are interchanged and consider a toroidal surface with longitudinal slots (see Figure 5.2(d)) defined by

$$\alpha = \alpha_0, \beta \in [-(\pi - \beta_0), -\beta_0] \cup [\beta_0, \pi - \beta_0],$$

so that the slots occupy the region

$$\alpha = \alpha_0, \beta \in [-\pi, -(\pi - \beta_0)] \cup [-\beta_0, \beta_0] \cup [\pi - \beta_0, \pi].$$

Assume that the segments are charged to unit potential. Then enforcement of the boundary conditions on (5.7) produces the symmetric triple series

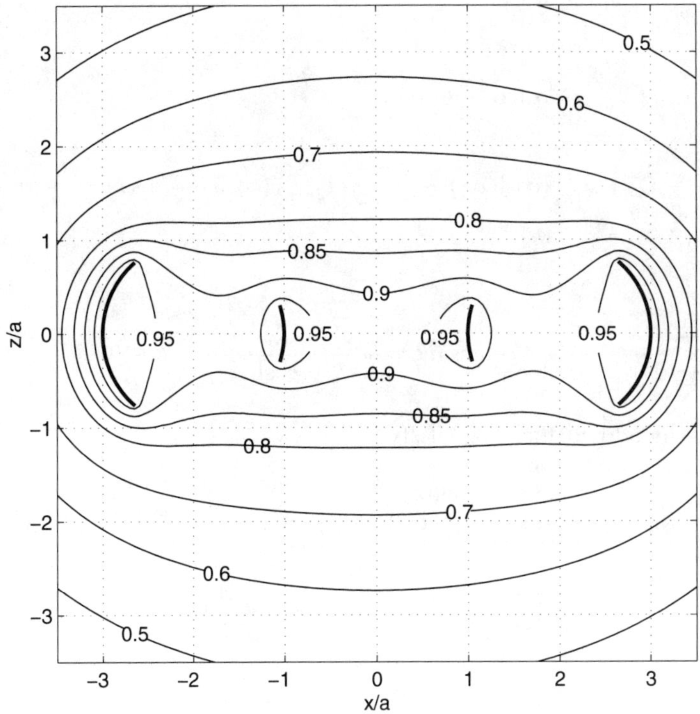

FIGURE 5.5. The charged toroidal shell with two transversal slots; the radii are $r = 1, R = 2$, and $\beta_0 = 30^o$.

equations

$$\sum_{n=0}^{\infty} C_n Q_{n-\frac{1}{2}}(\cosh\alpha_0)\cos n\beta = (2\cosh\alpha_0 - 2\cos\beta)^{-\frac{1}{2}}, \ \beta \in [\beta_0, \pi - \beta_0], \tag{5.51}$$

$$\sum_{n=0}^{\infty} \frac{C_n}{\sinh\alpha_0 P_{n-\frac{1}{2}}(\cosh\alpha_0)} \cos n\beta = 0, \ \beta \in (0, \beta_0) \cup (\pi - \beta_0, \pi). \tag{5.52}$$

As in the previous section, these triple series equations may be converted to a decoupled pair of dual series equations for even and odd coefficients. Introducing the new variable $\theta = 2\beta$ and setting $\theta_0 = 2\beta_0$, the even coefficients satisfy

$$\sum_{n=1}^{\infty} C_{2n} Q_{2n-\frac{1}{2}}(\cosh\alpha_0)\cos n\theta =$$
$$\left(\frac{1}{\pi} - C_0\right) Q_{-\frac{1}{2}}(\cosh\alpha_0) + \frac{2}{\pi}\sum_{n=1}^{\infty} Q_{2n-\frac{1}{2}}(\cosh\alpha_0)\cos n\theta,\ \theta \in (\theta_0, \pi), \tag{5.53}$$

$$\sum_{n=1}^{\infty} \frac{C_{2n}}{P_{2n-\frac{1}{2}}(\cosh\alpha_0)}\cos n\theta = -\frac{C_0}{P_{-\frac{1}{2}}(\cosh\alpha_0)}, \qquad \theta \in (0, \theta_0), \tag{5.54}$$

whilst the odd coefficients satisfy

$$\sum_{n=0}^{\infty} C_{2n+1} Q_{2n+\frac{1}{2}}(\cosh\alpha_0)\cos(n+\frac{1}{2})\theta =$$
$$\frac{2}{\pi}\sum_{n=0}^{\infty} Q_{2n+\frac{1}{2}}(\cosh\alpha_0)\cos(n+\frac{1}{2})\theta, \qquad \theta \in (\theta_0, \pi), \tag{5.55}$$

$$\sum_{n=0}^{\infty} \frac{C_{2n+1}}{P_{2n+\frac{1}{2}}(\cosh\alpha_0)}\cos(n+\frac{1}{2})\theta = 0, \qquad \theta \in (0, \theta_0). \tag{5.56}$$

We first consider the system (5.55)–(5.56) for the odd coefficients and convert it to dual series equations involving the Jacobi polynomials $P_n^{(-\frac{1}{2}\frac{1}{2})}$ with $z = \cos\theta$, $(z_0 = \cos\theta_0)$,

$$\sum_{n=0}^{\infty} \frac{C_{2n+1}}{P_{2n+\frac{1}{2}}(\cosh\alpha_0)} \frac{\Gamma(n+1)}{\Gamma(n+\frac{1}{2})} P_n^{(-\frac{1}{2},\frac{1}{2})}(z) = 0, \quad z \in (z_0, 1), \tag{5.57}$$

$$\sum_{n=0}^{\infty} C_{2n+1} Q_{2n+\frac{1}{2}}(\cosh\alpha_0)\frac{\Gamma(n+1)}{\Gamma(n+\frac{1}{2})} P_n^{(-\frac{1}{2},\frac{1}{2})}(z) =$$
$$\frac{2}{\pi}\sum_{n=0}^{\infty} Q_{2n+\frac{1}{2}}(\cosh\alpha_0)\frac{\Gamma(n+1)}{\Gamma(n+\frac{1}{2})} P_n^{(-\frac{1}{2},\frac{1}{2})}(z),\ z \in (-1, z_0). \tag{5.58}$$

The Abel transform technique may be employed with the integral representation (1.172) for $P_n^{(0,0)} \equiv P_n$ in terms of $P_n^{(\frac{1}{2},-\frac{1}{2})}$, its companion representation (1.171) for P_n in terms of $P_n^{(-\frac{1}{2},\frac{1}{2})}$, and the representation (1.172) for $P_n^{(-\frac{1}{2},\frac{1}{2})}$ in terms of P_n. The asymptotically small parameter

ε_{2n+1} defined by (5.43), appears and, arguing as in the last section, we obtain

$$\sum_{n=0}^{\infty} x_{2n+1}\hat{P}_n(z) = \left\{ \begin{array}{ll} 0, & z \in (z_0, 1) \\ F_2(z), & z \in (-1, z_0) \end{array} \right\}, \tag{5.59}$$

where

$$F_2(z) = \sum_{n=0}^{\infty} x_{2n+1}\varepsilon_{2n+1}\hat{P}_n(z) + \frac{4}{\pi}\sinh\alpha_0 \sum_{n=0}^{\infty} Q_{2n+\frac{1}{2}}(\cosh\alpha_0)\hat{P}_n(z),$$

and the rescaled Fourier coefficients

$$x_{2n+1} = C_{2n+1}\left\{ 2\sqrt{n+\frac{1}{2}}P_{2n+\frac{1}{2}}(\cosh\alpha_0) \right\}^{-1} \tag{5.60}$$

belong to l_2. Invoking completeness and orthogonality of the normalised Legendre polynomials, we deduce from (5.59) the following infinite system of linear algebraic equations of the second kind (its matrix operator is a completely continuous perturbation of the identity in l_2):

$$(1-\varepsilon_{2m+1})x_{2m+1} + \sum_{n=0}^{\infty} x_{2n+1}\varepsilon_{2n+1}\hat{Q}_{nm}^{(0,0)}(z_0) = d_m - \sum_{n=0}^{\infty} d_n \hat{Q}_{nm}^{(0,0)}(z_0), \tag{5.61}$$

where $m = 0, 1, 2, \ldots$, and

$$d_n = \frac{4}{\pi}\sinh\alpha_0 \sqrt{n+\frac{1}{2}}Q_{2n+\frac{1}{2}}(\cosh\alpha_0).$$

The system (5.53)–(5.54) for even coefficients is solved in a similar way, and the rescaled coefficients

$$x_{2n} = (-1)^n C_{2n}\left\{ \sqrt{2n}P_{2n-\frac{1}{2}}(\cosh\alpha_0) \right\}^{-1}$$

satisfy the i.s.l.a.e.

$$x_{2m} - \sum_{n=1}^{\infty} x_{2n}\varepsilon_{2n}Q_{nm}(z_0) =$$
$$\frac{4\sinh\alpha_0}{\pi}\sum_{n=1}^{\infty}(-1)^n\sqrt{2n}Q_{2n-\frac{1}{2}}(\cosh\alpha_0)Q_{nm}(z_0) +$$
$$\frac{4\sinh\alpha_0}{\pi t}Q_{-\frac{1}{2}}(\cosh\alpha_0)(1+z_0)\frac{\hat{P}_{m-1}^{(0,1)}(z_0)}{m} \tag{5.62}$$

where $z_0 = \cos\theta_0$, and t, t_1 and $Q_{nm}(z_0)$ are defined by (5.22) and (5.24), and

$$C_0 = \frac{t_1}{t\pi} + P_{-\frac{1}{2}}(\cosh\alpha_0)\frac{(1+z_0)}{t} \times \sum_{n=1}^{\infty}\left[\frac{1}{n}x_{2n}\varepsilon_{2n} + \frac{4}{\pi}\sqrt{\frac{2}{n}}\sinh\alpha_0 Q_{2n-\frac{1}{2}}(\cosh\alpha_0)\right]\hat{P}_{n-1}^{(0,1)}(z_0),$$

and

$$\varepsilon_{2n} = 1 - 4n\sinh\alpha_0 P_{2n-\frac{1}{2}}(\cosh\alpha_0)Q_{2n-\frac{1}{2}}(\cosh\alpha_0) = O(n^{-2}) \text{ as } n \to \infty.$$

The closed toroidal shell is a special limiting case. It corresponds to setting $\beta_0 = \frac{\pi}{2}$ in (5.26) and (5.27) for the conductor with transversal slots, or to setting $\beta_0 = 0$ in (5.51) and (5.52) for the conductor with longitudinal slots. In these cases, the corresponding regularised systems (5.38), (5.48), or (5.61), (5.62) have solutions in explicit form.

Noting that $\hat{Q}_{nm}^{(1,0)}(-1) = \delta_{nm}$, the solution to (5.48) with $\beta_0 = \frac{\pi}{2}$ ($z_0 = -1$) is

$$y_{2m+1} = \left\{\pi(m+\frac{1}{2})^{\frac{3}{2}}P_{2m+\frac{1}{2}}(\cosh\alpha_0)\right\}^{-1}, \quad m \geq 0,$$

and the solution to (5.38) is

$$y_{2m} = \sqrt{2}\left\{\pi\sqrt{m}P_{2m-\frac{1}{2}}(\cosh\alpha_0)\right\}^{-1}, \quad m \geq 0.$$

Thus

$$C_0 = \pi^{-1}, C_n = 2\pi^{-1}, \quad \text{for } n = 1, 2, ... \tag{5.63}$$

Substituting this solution in (5.7) produces a potential that coincides with the earlier published solution of [36]. Identical results are obtained by solving the systems (5.61) and (5.62).

The computational properties of the systems (5.38) and (5.48) and (5.61)–(5.62) are rather similar, and as for the transversal slots, numerical values of the unknown Fourier coefficients and the electrostatic field may be computed very efficiently, correct to four digits with a choice of truncation number N_{tr} not exceeding 20.

5.5 Capacitance of toroidal conductors

Following the same argument as in Section 4.7, the capacitance of the open toroidal conductor in terms of the Fourier coefficients C_n in (5.7) is

$$C = 2c\left\{C_0\frac{Q_{-\frac{1}{2}}(\cosh\alpha_0)}{P_{-\frac{1}{2}}(\cosh\alpha_0)} + \sum_{n=1}^{\infty}C_n\frac{Q_{n-\frac{1}{2}}(\cosh\alpha_0)}{P_{n-\frac{1}{2}}(\cosh\alpha_0)}\right\}. \tag{5.64}$$

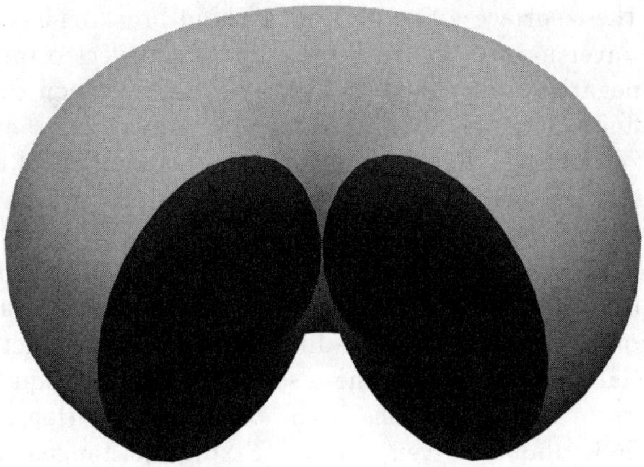

FIGURE 5.6. A degenerate toroidal shell with one azimuthal cut.

Substitution of the explicit solution (5.63) for the closed toroidal conductor in (5.64) produces an expression for capacitance that coincides with the published result of [26].

5.6 An open toroidal shell with azimuthal cuts

The determination of the potential distribution surrounding the slotted toroidal conductors considered in previous sections was significantly facilitated by their axial symmetry. The symmetry permitted the problem to be formulated in terms of an appropriate set of dual or triple series equations. The situation becomes more complicated when slots are cut in the shell so that axial symmetry is broken. In this section we derive some new results for a class of conductors without axial symmetry, in particular for the perfectly conducting shell that is part of a degenerate torus (in which the major and minor radii are equal) that may be viewed as an incomplete body of revolution (see Figure 5.6).

An essential preliminary step is provided by the method of inversion in a sphere, so that Bouwkamp's theorem (see Chapter 3) may be exploited. Some axially symmetric situations are relatively easily analysed by this approach, such as the spherical cap (Section 3.4). Also, potential problems for asymmetric spherical conductors (such as the asymmetric *barrel* or the pair of asymmetric caps) may be symmetrised by an inversion process prior to solution of the electrostatic problem (Section 3.3). Moreover, the connection formally described in [77] and [3] between some classes of dual integral equations and dual series equations has the inversion method at its root.

Inversion has previously been used for studying charged closed conductors of rather *exotic* form, such as degenerate tori [7] or spindles [51]. Cutting holes in these surfaces of revolution, without breaking axial symmetry, leads, under inversion, to the determination of the electrostatic field produced by a negative unit charge, located on the inversion centre, in the presence of finite or semi-infinite grounded cylinders (in the case of the torus), or of open semi-infinite grounded cones (in the case of the spindle). These problems are thereby reduced to the solution of certain well-studied dual series or integral equations.

In this section, we focus on conductors with azimuthal openings that break the axial symmetry, and thereby demonstrate an essential and significant extension to the class of three-dimensional open conducting surfaces whose potential is obtainable by these semi-analytic techniques.

The degenerate toroidal surface is the body of revolution generated by revolving a circle about a given tangent. Fixing this tangent to be the z-axis in the cylindrical coordinate system (ρ, φ, z), and taking the circle radius to be a units, the closed surface has the equation $(\rho - a)^2 + z^2 = a^2$, $\varphi \in [-\pi, \pi]$. We first consider open toroidal shells having one azimuthal cut, or hole, of semi-width φ_1, specified by

$$(\rho - a)^2 + z^2 = a^2; \varphi \in [-\pi, -\varphi_1] \cup [\varphi_1, \pi].$$

(See Figure 5.6.) Subsequently, open toroidal shells with multiple azimuthal cuts symmetrically disposed as shown in Figure 5.7 will be examined.

We wish to determine the electrostatic potential when such open shells are charged to unit constant potential. Let M denote the origin of the coordinate system and consider inversion of the toroidal shell in a sphere of radius $2a$ centred at M. From Bouwkamp's theorem (Section 3.4), the problem is equivalent to the determination of the electrostatic field produced by a negative unit charge, located at M, in the presence of a semi-infinite grounded cylinder having one or more longitudinal slots. (See figure 5.8.)

The equivalent problem may be formulated as a set of dual series equations involving trigonometric functions with unknown Fourier coefficients. However, in contrast to the axially symmetric problems previously investigated, the coefficients depend on some spectral parameter ν. For apertures of arbitrary angle size, regularisation of the dual series equations transforms them to an infinite system of linear algebraic equations of the second kind for the modified Fourier coefficients. The Fredholm nature of the matrix operator, at each fixed value of the spectral parameter ν, makes it possible to use a truncation method effectively to obtain a finite number of Fourier coefficients numerically.

An approximate formula for capacitance can be obtained for three limiting cases: the narrow cut ($\varphi_1 \ll 1$), a narrow skew ring ($\varphi_0 = \pi - \varphi_1 \ll 1$), and a large number of cuts ($N \gg 1$). Some representative numerical results

FIGURE 5.7. A degenerate toroidal shell with four azimuthal cuts.

are presented to demonstrate the efficacy of the analysis, and to check the accuracy of the approximate formulae derived in the limiting cases.

5.6.1 The toroidal shell with one azimuthal cut.

Consider first the toroidal shell with a single opening arising from an azimuthal cut. Let U be the potential associated with the field induced by a unit negative charge, located at M, on the infinite circular cylinder of radius $2a$ with a longitudinal slot of angular semi-width φ_1. The potential must satisfy Laplace's equation, together with the boundary conditions, edge conditions, and a decay condition at infinity. In cylindrical coordinates, the potential therefore has the form

$$U(\rho, \varphi, z) = U^0 + \int_0^\infty d\nu \cos \nu z \sum_{n=0}^{\infty} A_n(\nu) \cos(n\varphi) \begin{cases} I_n(\nu\rho), & 0 \le \rho < 2a \\ I_n(2\nu a) K_n(\nu\rho)/K_n(2\nu a), & \rho > 2a \end{cases} \tag{5.65}$$

where I_n, K_n are the modified Bessel functions, $U^0 = -(\rho^2 + z^2)^{-\frac{1}{2}}$ is the electrostatic potential of the free space negative unit charge located at M, and $\{A_n(\nu)\}_{n=0}^{\infty}$ is the sequence of unknown Fourier coefficients, which are functions of the spectral parameter ν.

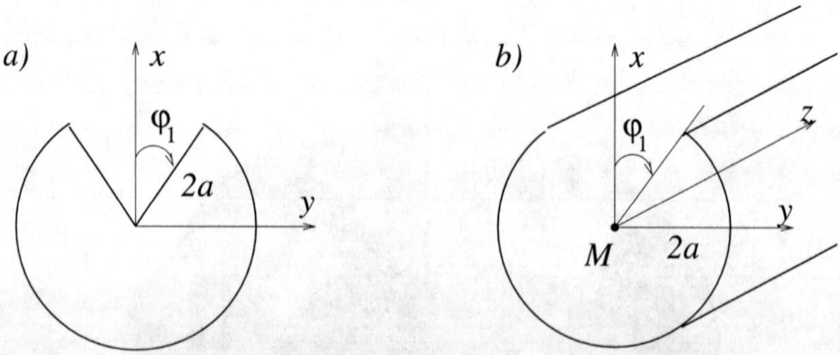

FIGURE 5.8. (a) The degenerate toroid with one azimuthal cut (top view) and (b) the slotted infinite cylinder, its image under inversion.

Using the mixed boundary conditions for the surface

$$U(2a-0,\varphi,z)=U(2a+0,\varphi,z)=0,\quad \varphi\in(\varphi_1,\pi), \tag{5.66}$$

and on the aperture

$$\frac{\partial U}{\partial\rho}(2a-0,\varphi,z)=\frac{\partial U}{\partial\rho}(2a-0,\varphi,z),\quad \varphi\in(0,\varphi_1), \tag{5.67}$$

and applying the Fourier cosine transform, the following dual series equations result:

$$\sum_{n=0}^{\infty}A_n(\nu)\frac{\cos(n\varphi)}{K_n(2\nu a)} = 0, \qquad \varphi\in(0,\varphi_1), \tag{5.68}$$

$$\sum_{n=0}^{\infty}A_n(\nu)I_n(2\nu a)\cos(n\varphi) = \frac{2}{\pi}K_0(2\nu a),\quad \varphi\in(\varphi_1,\pi). \tag{5.69}$$

We now proceed, as usual, to transform this basic set of equations, which are of first kind, to a Fredholm matrix equation of the second kind. The main difference to previous analysis is that the coefficients A_n are functions of the spectral parameter ν.

The standard approach is to replace the cosine kernels $\cos(n\varphi)$ by Jacobi polynomials $P_n^{(-\frac{1}{2},-\frac{1}{2})}$. It is then necessary to integrate these equations, using the variant (2.36) Rodrigues' formula, so that the techniques described in Chapter 2 are applicable. Equivalently, we may first integrate the pair (5.68) and (5.69), and then replace the sine kernels by the Jacobi polynomials $P_n^{(\frac{1}{2},\frac{1}{2})}$ to obtain

$$\sum_{n=1}^{\infty} A_n(\nu) I_n(2\nu a) \frac{\Gamma(n)}{\Gamma(n+\frac{1}{2})} P_{n-1}^{(\frac{1}{2},\frac{1}{2})}(t) = \frac{2\left[2I_0(2\nu a)A_0 - \frac{4}{\pi}K_0(2\nu a)\right]}{\sqrt{\pi}\,(1-t^2)^{\frac{1}{2}}} \left(\arcsin t + \frac{\pi}{2}\right), \; t \in (-1, t_1), \quad (5.70)$$

$$\sum_{n=1}^{\infty} \frac{A_n(\nu)}{nK_n(2\nu a)} \frac{\Gamma(n+1)}{\Gamma(n+\frac{1}{2})} P_{n-1}^{(\frac{1}{2},\frac{1}{2})}(t) = - \frac{A_0\,(\pi - 2\arcsin t)}{K_0(2\nu a)\sqrt{\pi}\,(1-t^2)^{\frac{1}{2}}} \quad t \in (t_1, 1), \quad (5.71)$$

where $t = \cos(\varphi)$ and $t_1 = \cos(\varphi_1)$. Define the new unknown quantities

$$M_n = A_n / \{nK_n(2\nu a)\}$$

that are to be determined. The asymptotics of the modified Bessel functions (see Appendix, (B.159) and (B.160)) show that the parameter

$$\varepsilon_n = 1 - 2nI_n(2\nu a)K_n(2\nu a) \quad (5.72)$$

is asymptotically small: $\varepsilon_n = O(n^{-2})$ as $n \to \infty$. The Equations (5.70) and (5.71) become

$$(1-t^2)^{\frac{1}{2}} \sqrt{\pi} \sum_{n=1}^{\infty} M_n \frac{\Gamma(n+1)}{\Gamma(n+\frac{1}{2})} P_{n-1}^{(\frac{1}{2},\frac{1}{2})}(t) = - \frac{A_0}{K_0(2\nu a)} (\pi - 2\arcsin t), \quad t \in (t_1, 1), \quad (5.73)$$

$$(1-t^2)^{\frac{1}{2}} \sqrt{\pi} \sum_{n=1}^{\infty} M_n (1-\varepsilon_n) \frac{\Gamma(n)}{\Gamma(n+\frac{1}{2})} P_{n-1}^{(\frac{1}{2},\frac{1}{2})}(t) = 2\left[2I_0(2\nu a)A_0 - \frac{4}{\pi}K_0(2\nu a)\right] \left(\arcsin t + \frac{\pi}{2}\right), \; t \in (-1, t_1). \quad (5.74)$$

The Abel transform method may now be applied. To make the rate of convergence of the terms in series (5.73) and (5.74) equal, Equation (5.73) is integrated using the particular case of (1.171),

$$(1-t)^{\frac{3}{2}} P_{n-1}^{(\frac{3}{2},-\frac{1}{2})}(t) = \left(n + \frac{1}{2}\right) \int_t^1 (1-x)^{\frac{1}{2}} P_{n-1}^{(\frac{1}{2},\frac{1}{2})}(x)dx, \quad (5.75)$$

to obtain

$$(1-t)^{\frac{3}{2}}\sqrt{\pi}\sum_{n=1}^{\infty} M_n \frac{\Gamma(n+1)}{\Gamma(n+\frac{3}{2})} P_{n-1}^{(\frac{3}{2},-\frac{1}{2})}(t) = \frac{A_0}{K_0(2\nu a)}\left[2\left(\pi - 2\arcsin t\right)(1+t)^{\frac{1}{2}} - 8(1-t)^{\frac{1}{2}}\right], \; t \in (t_1, 1). \quad (5.76)$$

Using the Abel-type integral representation (1.172) for Jacobi polynomials $P_{n-1}^{(\frac{1}{2},\frac{1}{2})}$ in terms of $P_{n-1}^{(1,0)}$ and the particular case of (1.171),

$$P_{n-1}^{(\frac{3}{2},-\frac{1}{2})}(t) = (1-t)^{-\frac{3}{2}} \frac{\Gamma(n+\frac{3}{2})}{\sqrt{\pi}\Gamma(n+1)} \int_t^1 \frac{(1-x)\, P_{n-1}^{(1,0)}(x)}{(x-t)^{\frac{1}{2}}} dx,$$

a standard argument shows that

$$\sum_{n=1}^{\infty} M_n (1-t) P_{n-1}^{(1,0)}(t) = \begin{cases} F_1(t), & t \in (-1, t_1) \\ F_2(t), & t \in (t_1, 1) \end{cases} \quad (5.77)$$

where

$$\begin{aligned} F_1(t) &= \sum_{n=1}^{\infty} M_n \varepsilon_n (1-t) P_{n-1}^{(1,0)}(t) + 2\left[2I_0(2\nu a)A_0 - \frac{4}{\pi}K_0(2\nu a)\right], \\ F_2(t) &= 2\ln\left(\frac{1}{2}(1+t)\right) A_0/K_0(2\nu a). \end{aligned}$$

A familiar orthogonality argument produces the matrix equation

$$\begin{aligned} M_m(1-\varepsilon_m) + m\sum_{n=1}^{\infty} M_n \varepsilon_n Q_{n-1,m-1}^{(1,0)}(t_1) = \\ 2\left[2I_0(2\nu a)A_0 - \frac{4}{\pi}K_0(2\nu a)\right](1+t_1)\, P_{m-1}^{(0,1)}(t_1) - \\ \frac{2A_0}{K_0(2\nu a)}\left[(1+t_1)\ln\left(\frac{1+t_1}{2}\right) P_{m-1}^{(0,1)}(t_1) + \frac{1-t_1}{m} P_{m-1}^{(1,0)}(t_1)\right] \end{aligned} \quad (5.78)$$

holding for all indices $m = 1, 2, \ldots$.

The system (5.78) has an infinite number of solutions if the constant A_0 has an arbitrary value. A unique solution is obtained by requiring that the function on the left-hand side of (5.77) is continuous at the point $t = t_1$. Hence

$$A_0 = \frac{4\pi^{-1}K_0^2(2\nu a) - \frac{1}{2}(1-t_1)K_0(2\nu a)\sum_{n=1}^{\infty} M_n \varepsilon_n P_{n-1}^{(1,0)}(t_1)}{2I_0(2\nu a)K_0(2\nu a) - \ln\left(\frac{1}{2}(1+t_1)\right)}. \quad (5.79)$$

A combination of (5.78) and (5.79) yields the final form of the Fredholm matrix equation of second kind for the unknown Fourier coefficients M_n:

$$M_m(1-\varepsilon_m) + \sum_{n=1}^{\infty} M_n \varepsilon_n B_{nm}(t_1) = -\frac{4}{\pi} \frac{(1-t_1)K_0(2\nu a)}{2I_0(2\nu a)K_0(2\nu a) - \ln\left(\frac{1}{2}(1+t_1)\right)} \frac{P_{m-1}^{(1,0)}(t_1)}{m} \tag{5.80}$$

where $m = 1, 2, ...,$ and

$$B_{nm}(t_1) = \frac{n}{2}\left\{Q_{n-1,m-1}^{(0,1)}(t_1) - \frac{(1-t_1)^2 P_{n-1}^{(1,0)}(t_1) P_{m-1}^{(1,0)}(t_1)}{nm\left(2I_0(2\nu a)K_0(2\nu a) - \ln\left(\frac{1}{2}(1+t_1)\right)\right)}\right\}.$$

Here it should be noted that we used the relationship (cf. (B.172))

$$Q_{n-1,m-1}^{(1,0)}(t_1) = -\frac{(1-t_1)(1+t_1)}{m} P_{n-1}^{(1,0)}(t_1) P_{m-1}^{(0,1)}(t_1) + \frac{n}{m} Q_{n-1,m-1}^{(0,1)}(t_1).$$

This completes the regularisation of the original pair of dual series equations (5.68) and (5.69). Computationally, system (5.80) is very attractive; it may be rapidly solved by a truncation method with predetermined accuracy for every value of ν, whatever the angular measure of the hole may be. The electrostatic field is then found from (5.65) as a Fourier cosine transform of the coefficients $A_n(\nu)$.

Finally, the capacitance of the conductor, as a function of the angular semi-width φ_1, is

$$C = C(\varphi_1) = 4a^2 \int_0^{\infty} A_0(\nu) d\nu. \tag{5.81}$$

The logarithmic singularity of K_0 affects the numerical calculations, and the expression should be transformed to

$$C = \frac{4a}{\pi} \int_0^{\infty} \{I_0(x)\}^{-2} dx + \frac{2a}{\pi} \ln\left(\frac{1+t_1}{2}\right) \int_0^{\infty} \{I_0(x) L_0(x)\}^{-1} dx$$
$$- a(1-t_1) \int_0^{\infty} \left\{\sum_{n=1}^{\infty} M_n(x)\varepsilon_n(x) P_{n-1}^{(1,0)}(t_1)\right\} \{L_0(x)\}^{-1} dx, \tag{5.82}$$

where

$$L_0(x) = I_0(x) - \frac{1}{2K_0(x)} \ln\left(\frac{1+t_1}{2}\right).$$

This depends upon the identity (derived by an integration by parts)

$$\int_0^{\infty} \frac{K_0(x)}{I_0(x)} dx = \int_0^{\infty} \frac{dx}{I_0^2(x)}.$$

5.6.2 The toroidal shell with multiple cuts

The potential surrounding a toroidal conductor having 2^N $(N = 1, 2, ...)$ equal azimuthal cuts may be analysed in the same way. The structure is displayed in Figure 5.7. Let φ_1 be the semi-width of each cut: thus $\varphi_1 + \varphi_0 = 2^{-N}\pi$, where φ_0 is the angular semi-width of each of the 2^N conducting sectors. Taking into account the symmetrical location of the cuts and the identity $\cos n\varphi = (-1)^n \cos(n(\pi - \varphi))$, it is easy to show that the pair of equations corresponding to (5.68) and (5.69) take the special form

$$\sum_{n=1}^{\infty} A_{2^N n}(\nu) \frac{\cos(n\theta)}{K_{2^N n}(2\nu a)} = -\frac{A_0(\nu)}{K_0(2\nu a)}, \quad \theta \in (0, \theta_1) \tag{5.83}$$

$$\sum_{n=1}^{\infty} A_{2^N n}(\nu) I_{2^N n}(2\nu a) \cos(n\theta) = \frac{2}{\pi} K_0(2\nu a) - I_0(2\nu a) A_0(\nu), \ \theta \in (\theta_1, \pi) \tag{5.84}$$

where $\theta = 2^N \varphi$ and $\theta_1 = 2^N \varphi_1$. Note that $A_k = 0$ unless k is an integral multiple of 2^N.

Using the same solution scheme considered above, and introducing the rescaled unknowns

$$M_{2^N n} = A_{2^N n} / \left\{ n K_{2^N n}(2\nu a) \right\},$$

we obtain the following matrix equation of second kind,

$$M_{2^N m}\left(1 - \varepsilon_{2^N m}\right) + \sum_{n=1}^{\infty} M_{2^N n} \varepsilon_{2^N n} B_{n,m}(u_1) = \frac{-4(1-u_1) K_0(2\nu a) P_{m-1}^{(1,0)}(u_1)}{2^N m\pi \left(2 I_0(2\nu a) K_0(2\nu a) - \ln\left(\frac{1}{2}(1+u_1)\right)\right)}, \tag{5.85}$$

where $m = 1, 2, ...$, $u = \cos\theta$, $u_1 = \cos\theta_1$, and

$$B_{n,m}(u_1) = \frac{n}{2} Q_{n-1,m-1}^{(0,1)}(u_1) - \frac{n(1-u_1)^2 P_{n-1}^{(1,0)}(u_1) P_{m-1}^{(1,0)}(u_1)}{2^{N+1} m \left(2 I_0(2\nu a) K_0(2\nu a) - \ln\left(\frac{1}{2}(1+u_1)\right)\right)},$$

$$A_0 = \frac{4\pi^{-1} K_0^2(2\nu a) - \frac{1}{2}(1-u_1) K_0(2\nu a) \sum_{n=1}^{\infty} M_{2^N n} \varepsilon_{2^N n} P_{n-1}^{(1,0)}(u_1)}{2 I_0(2\nu a) K_0(2\nu a) - 2^{-N} \ln\left(\frac{1}{2}(1+u_1)\right)}, \tag{5.86}$$

and

$$\varepsilon_{2^N n} = 1 - 2n.2^N I_{2^N n}(2\nu a) K_{2^N n}(2\nu a) = O(\left(2^N n\right)^{-2}), \text{ as } n \to \infty. \tag{5.87}$$

When $N = 0$ (a single cut), the pairs of Equations (5.79) and (5.80) and (5.85) and (5.86) are equivalent. It is clear that (5.85) enjoys the same mathematical and computational properties as obtained for (5.80), arising from its form as a Fredholm matrix equation of second kind.

5.6.3 Limiting cases

The pairs of Equations (5.79) and (5.80) and (5.85) and (5.86) have approximate analytical solutions in three limiting cases: the toroidal surface with a narrow single cut ($\varphi_1 \ll 1$), the toroidal surface with a large number of cuts ($N \gg 1$), and the narrow skew ring (in which the angle $\varphi_0 = \pi - \varphi_1$ satisfies $\varphi_0 \ll 1$).

When the cut in the torus is narrow ($\varphi_1 \ll 1, t_1 = \cos\varphi_1 \to 1$), the Fourier coefficients of the system (5.80) have the behaviour $M_n \sim O(\varphi_1^2)$, and it follows that

$$C = C_0\left(1 - \frac{1}{8}\varphi_1^2\right) + O\left(\varphi_1^4\right), \tag{5.88}$$

where

$$C_0 = \frac{4a}{\pi}\int_0^\infty \frac{K_0(x)}{I_0(x)}dx = \frac{4a}{\pi}\int_0^\infty \frac{dx}{I_0^2(x)} = 1.74138027a \tag{5.89}$$

is the capacity of the closed toroidal conductor [51]. Capacitance values obtained from this formula agree well with results of computations on the system (5.78), at least for cuts of angle φ_1 not exceeding 30°.

When the toroidal shell has a large number of symmetrically placed cuts ($N \gg 1$) it is easy to show that its capacitance is

$$C = C_0\left[1 + \frac{1}{2^N}\ln\left(\cos\frac{\theta_1}{2}\right)\right]\left[1 + O\left(2^{-2N}\right)\right]. \tag{5.90}$$

When $N \to \infty$, expression (5.90) reduces to the expression for the capacity C_0 of the fully closed conductor. When the cuts are narrow, formula (5.90) is computationally very accurate because both approximations for multiple holes and for narrow cuts work together.

When the angular semi-width $\varphi_0 = \pi - \varphi_1$ of the ring is small ($\varphi_0 \ll 1$), the approximate expression for capacity of this *skew* ring is

$$C_{ring}(\varphi_0) = \frac{4a}{\pi}\int_0^\infty \frac{K_0^2(x)}{I_0(x)K_0(2\nu a) - \ln\left(\frac{1}{2}\varphi_0\right)}dx. \tag{5.91}$$

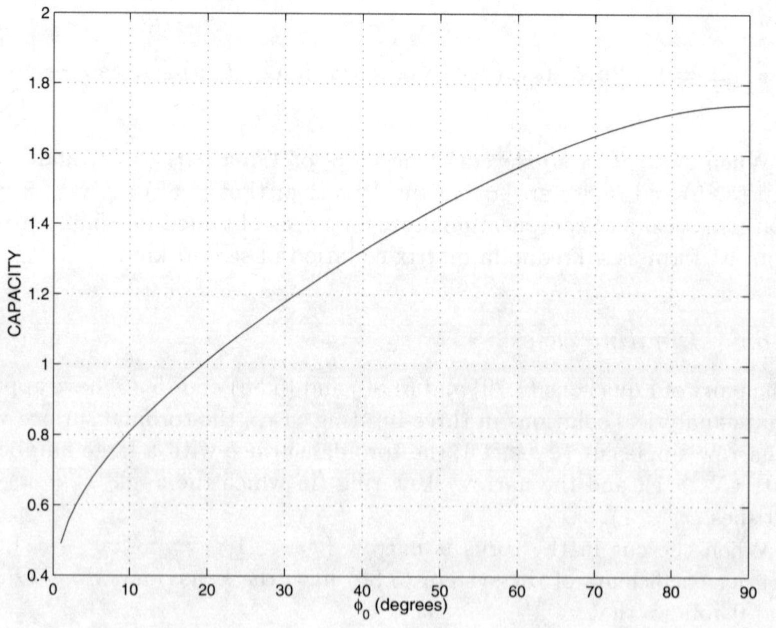

FIGURE 5.9. Capacitance of an open toroidal shell with azimuthal cuts.

We make two remarks about the expression (5.91). First, it has a logarithmic singularity near $x = 0$ which should be addressed in any numerical integration. Second, the infinite range of integration may be truncated to $(0, 4e^{-\gamma-1}\varphi_0^{-1})$ with an error $O(\exp(-2\varphi_0^{-1}))$. Values of capacity computed according to (5.91) agree well, in the range $0 < \varphi_0 \leq 10°$, with the numerical results obtained from (5.82) (employing the solution of the system (5.80)).

Numerical values for the capacity of a toroidal conductor having radius $a = 1$ and $k = 2^N$ cuts may be obtained by solution of (5.78) or (5.85) as appropriate. These systems are truncated to a finite number of equations and, after numerical solution, the value of A_0 may be determined from (5.79) or (5.86) as appropriate. The capacity C is then calculated according to (5.82) by repeating the calculation for $A_0(\nu)$ for a suitable range of ν. Selected results are shown in Table 5.1 (φ_1 is the angular semi-width of each cut in degrees); for single cut ($k = 1$), a graph of capacity C as a function of $\varphi_0 = \pi - \varphi_1$ is shown in Figure 5.9. It was found that the maximum size of a system to be solved did not exceed 10 equations. In the case of a multiply-cut conductor, it was enough to solve only one equation, provided $k = 2^N \geq 4$.

By examining systems of respective orders one and ten, the seven decimal place results displayed in Table 5.2 exemplify, when $k = 4$, how the accuracy of the computed capacity C_t depends upon the number t of equations solved

φ_1(deg.)	$k = 1$	$k = 2$	$k = 4$	$k = 16$	$k = 64$
0.1	1.741380	1.741379	1.741378	1.741370	1.741338
1.0	1.74131	1.74125	1.74112	1.74032	1.73692
10	1.7349	1.7285	1.7154	1.5804	
30	1.6893	1.6358	1.5095		
90	1.3912				
150	0.9173				
170	0.6749				
175	0.5800				
179	0.4397				
179.9	0.3282				
179.99	0.26194				

TABLE 5.1. Computed capacity of a toroid with k azimuthal cuts of angular semi-width φ_1.

φ_1	C_1	C_{10}	$\lvert C_{10} - C_1 \rvert$
30^0	1.5094431	1.5095232	$\approx 8 \cdot 10^{-5}$
1^0	1.7411151	1.7411151	$< 10^{-7}$

TABLE 5.2. Computed capacitance values for a toroidal shell with $k = 4$ cuts.

after truncation of system (5.85) to a finite system. As a consequence, an iteration method may be successfully used to refine accuracy.

6

Potential Theory for Conical Structures with Edges

Conical structures are distinctively different from the spheroidal and toroidal structures considered in previous chapters. Electrostatic fields induced by a point source in the vicinity of the conical tip possess singularities unique to this class. On the other hand, the open or hollow conical frustrum produced by removal of the tip region exhibits an interesting range of geometries, from the flat, annular disc to the hollow, circular cylinder. In this chapter, we treat a selection of potential problems that are most distinctive of conically-shaped thin conductors with edges. The selection is not exhaustive, but is intended to indicate the class of conical structures that might be successfully analysed by this approach.

By way of introduction, we first consider the related two-dimensional calculation of the electrostatic field generated by a pair of oppositely charged strips that are not coplanar or parallel; the structure is a two-dimensional analogue of the conical frustrum. After considering the infinitely long cone, the electrostatic field of the open conical frustrum is investigated in Section 6.2. The potential is determined by a set of dual integral equations: a notable aspect of their solution is the use of the Mehler-Fock transform in the regularisation process. The resultant second-kind Fredholm integral equations are well conditioned and possess the familiar properties conducive to the straightforward application of standard numerical methods.

The next section (6.3) examines the spindle, which is the image of the cone under inversion in a centre located on the conical axis (but not on the vertex). The potential of both spindle and cone are intimately related by Bouwkamp's theorem. Cutting a sectoral slot in the cone corresponds to opening an azimuthal or longitudinal slot on the spindle surface. Both structures are interesting because of the departure from the axial symmetry evident in previously considered conductors. The dual series equations describing the potential of the slotted cone are regularised; the capacitance of the associated slotted spindle is obtained. Whilst these potential problems have been studied previously, their solutions are rather less well known, especially when the slots break the axial symmetry of the conductor. As for the toroid with azimuthal slots considered in Chapter 5, this represents a significant extension of analytic and semi-analytic techniques to determining the potential distribution surrounding nonsymmetric open conducting surfaces. In this context, the hollow spindle with a slot is particularly instructive, because it uses most of the mathematical tools set forth in this book.

The final section (6.4) considers the confluent case of the slotted spindle in which the open conducting surface becomes a spherical shell with a longitudinal slot. This class of nonsymmetric apertures on the sphere complements the earlier studies on axially symmetric open spherical conductors.

6.1 Non-coplanar oppositely charged infinite strips

Let us consider the electrostatic field due to a pair of oppositely charged infinite strips that are not coplanar or parallel. This may be viewed as a conducting wedge with sections removed symmetrically from each arm, as shown in Figure 6.1. The strips lie on half-planes emanating from the origin and are symmetric with respect to the x-axis. In cylindrical polar coordinates (ρ, φ, z), the positively charged strip is described by $\rho \in (a, b)$, $\varphi = \varphi_0$, and the negatively charged strip by $\rho \in (a, b)$; $\varphi = 2\pi - \varphi_0$. The electrostatic potential $\psi(\rho, \varphi, z)$ is independent of z; the problem to be solved is two-dimensional, $\psi \equiv \psi(\rho, \varphi)$.

It is convenient to introduce the dimensionless radial coordinate $r = \rho/(ab)^{\frac{1}{2}}$; Laplace's equation becomes

$$\frac{1}{r}\frac{\partial}{\partial r}\left(r\frac{\partial \psi(r,\varphi)}{\partial r}\right) + \frac{1}{r^2}\frac{\partial^2 \psi(r,\varphi)}{\partial \varphi^2} = 0. \tag{6.1}$$

The geometry of the problem forces us to seek discontinuous solutions in the variable φ, and imposes conditions on the separation constants when the method of separation of variables is used to construct total solutions of the Laplace equation. In particular, the boundedness of the potential at the origin and at infinity imply that in each of the regions $\varphi < \varphi_0$ and $\varphi > \varphi_0$ it has the form

$$\psi(r,\varphi) = \int_0^\infty \{C(\tau)\cos\tau\sigma + D(\tau)\sin\tau\sigma\}\{A(\tau)e^{-\tau\varphi} + B(\tau)e^{\tau\varphi}\}\,d\tau, \tag{6.2}$$

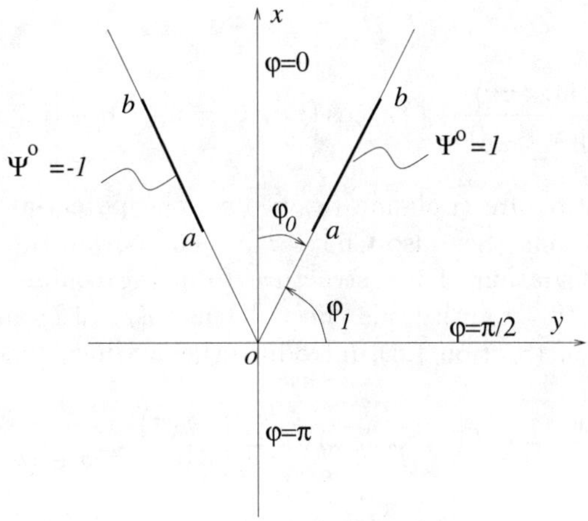

FIGURE 6.1. Oppositely charged infinite strips, not coplanar.

where

$$\sigma = \log r, \tag{6.3}$$

and A, B, C, and D are unknown functions to be determined.

Due to the symmetry it is clear that

$$\psi(r, 0) = \psi(r, \pi) = 0, \qquad r \in (0, \infty).$$

Enforcing a continuity condition at $\varphi = \varphi_0$, the desired form of solution is

$$\psi(\sigma, \varphi) = \int_0^\infty d\tau \{f(\tau) \cos \tau\sigma + g(\tau) \sin \tau\sigma\} F(\tau, \varphi), \tag{6.4}$$

where

$$F(\tau, \varphi) = \begin{cases} \sinh(\tau\varphi), & \varphi < \varphi_0 \\ \sinh(\tau\varphi_0) \sinh[\tau(\pi - \varphi)] / \sinh[\tau(\pi - \varphi_0)], & \varphi > \varphi_0 \end{cases}$$

and f, g are unknown functions to be determined.

The mixed boundary conditions to be enforced on the representation are

$$\frac{\partial \psi}{\partial \varphi}(\sigma, \varphi_0 - 0) = \frac{\partial \psi}{\partial \varphi}(\sigma, \varphi_0 + 0), \quad \sigma \in (-\infty, -\sigma_0) \cup (\sigma_0, \infty), \tag{6.5}$$

$$\psi(\sigma, \varphi_0 - 0) = \psi(\sigma, \varphi_0 + 0) = 1, \quad \sigma \in (-\sigma_0, \sigma_0), \tag{6.6}$$

where $\sigma_0 = \frac{1}{2} \log(b/a)$. It is readily justified that $g(\tau) \equiv 0$, so the problem reduces to finding the function f that satisfies the dual integral equations

$$\int_0^\infty \sinh(\tau\varphi_0) f(\tau) \cos(\tau\sigma) d\tau = 1, \qquad \sigma \in (0, \sigma_0), \tag{6.7}$$

$$\int_0^\infty \frac{\tau \sinh(\pi\tau)}{\sinh[(\pi-\varphi_0)\tau]} f(\tau)\cos(\tau\sigma)\,d\tau = 0, \qquad \sigma \in (\sigma_0, \infty). \tag{6.8}$$

When the strips are coplanar ($\varphi_0 = \frac{1}{2}\pi$), the potential may be found analytically [54, 55]. (See also Chapter 7.) The parameter $\varphi_1 = \frac{1}{2}\pi - \varphi_0$ measures the deviation of the structure from the confluent geometry of coplanar strips. To quantify this effect, let us solve (6.7) and (6.8) by the *definition method* (Section 1.4). Introduce the auxiliary function g by

$$\int_0^\infty \frac{\tau \sinh(\pi\tau)}{\sinh[(\pi-\varphi_0)\tau]} f(\tau)\cos(\tau\sigma)\,d\tau = \begin{cases} g(\sigma), & \sigma \in (0, \sigma_0), \\ 0, & \sigma \in (\sigma_0, \infty). \end{cases} \tag{6.9}$$

An inverse Fourier cosine transform yields

$$f(\tau) = \frac{2}{\pi}\frac{\sinh[(\pi-\varphi_0)\tau]}{\tau\sinh(\pi\tau)} \int_0^{\sigma_0} g(\sigma')\cos(\tau\sigma')\,d\sigma', \tag{6.10}$$

and inserting this expression in (6.7) leads to the first-kind Fredholm integral equation for g,

$$\int_0^{\sigma_0} g(\sigma') K(\sigma,\sigma')\,d\sigma' = \frac{\pi}{2}, \qquad \sigma \in (0, \sigma_0), \tag{6.11}$$

where the kernel K is defined by

$$K(\sigma,\sigma') = \int_0^\infty \frac{\sinh[(\pi-\varphi_0)\tau]\sinh\tau\varphi_0}{\tau\sinh(\pi\tau)} \cos(\tau\sigma)\cos(\tau\sigma')\,d\tau. \tag{6.12}$$

One can readily transform K to a logarithmic type kernel using the cosine-Fourier transform [14] (Vol. 1), valid for $|\mathrm{Re}\,\alpha| < \pi, |\mathrm{Re}\,\beta| < \pi$,

$$\int_0^\infty \frac{\cosh(\beta y) - \cosh(\alpha y)}{y\sinh(\pi y)} \cos xy dy = \frac{1}{2}\log\left[\frac{\cosh x + \cos\alpha}{\cosh x + \cos\beta}\right]. \tag{6.13}$$

By means of some algebraic manipulation $K(\sigma,\sigma')$ is transformed to

$$K(\sigma,\sigma') = \frac{1}{8}\log\left[\frac{(\cosh\sigma + \cosh\sigma')^2 - 4(\cosh\sigma\cosh\sigma' + \cos^2\varphi_1)\sin^2\varphi_1}{(\cosh\sigma - \cosh\sigma')^2}\right]. \tag{6.14}$$

The kernel that corresponds to the coplanar structure ($\varphi_1 = 0$) is

$$K_0(\sigma,\sigma') = \frac{1}{4}\log\left|\frac{\cosh\sigma + \cosh\sigma'}{\cosh\sigma - \cosh\sigma'}\right|. \tag{6.15}$$

Following the basic idea of the method of regularisation, we now split the kernel into two parts, one of which (K_0) is *singular* and the other (K_1) is *regular*,

$$K(\sigma, \sigma') = K_0(\sigma, \sigma') + K_1(\sigma, \sigma'), \tag{6.16}$$

where

$$\begin{aligned} K_1(\sigma, \sigma') &= K(\sigma, \sigma') - K_0(\sigma, \sigma') \\ &= \frac{1}{8} \log \left\{ 1 - 4 \frac{\cosh \sigma \cosh \sigma' + \cos^2 \varphi_1}{(\cosh \sigma + \cosh \sigma')^2} \sin^2 \varphi_1 \right\}. \end{aligned} \tag{6.17}$$

It is evident that the kernel K_1 is regular in both variables σ and σ' within the interval $[0, \sigma_0]$. The parameter φ_1 measures the deviation of solutions from that for coplanar strips.

In the coplanar case ($\varphi_1 = 0$) the integral equation is

$$\int_0^{\sigma_0} g_0(\sigma') \log \left| \frac{\cosh \sigma + \cosh \sigma'}{\cosh \sigma - \cosh \sigma'} \right| d\sigma' = 2\pi, \qquad \sigma \in (0, \sigma_0) \tag{6.18}$$

where function g_0 refers to the coplanar structure. Integral equations with logarithmic kernels are well studied; many with *canonical* kernels have closed form solutions (see, for example, [48]). In particular, the solution of (6.18) is

$$g_0(\sigma') = \frac{2 \cosh \sigma_0}{K(\operatorname{sech} \sigma_0)} \left(\cosh^2 \sigma_0 - \cosh^2 \sigma \right)^{-\frac{1}{2}}, \tag{6.19}$$

where $K(x)$ denotes the complete elliptic integral of the first kind. The corresponding function f_0 is found from (6.10), using the integral representation (B.100) (see Appendix),

$$f_0(\tau) = \frac{\cosh \sigma_0}{K(\operatorname{sech} \sigma_0)} \frac{P_{-\frac{1}{2}+i\frac{\tau}{2}}(\cosh 2\sigma_0)}{\tau \cosh\left(\frac{\pi}{2}\tau\right)}. \tag{6.20}$$

The line charge density may now be calculated (recall $\rho = (ab)^{\frac{1}{2}} e^{\sigma}$) to be

$$\begin{aligned} l(\rho) &= \frac{1}{4\pi} \left\{ E_\varphi\left(\rho, \frac{\pi}{2} + 0\right) - E_\varphi\left(\rho, \frac{\pi}{2} - 0\right) \right\} \\ &= \frac{1}{2\pi} \int_0^\infty \tau f_0(\tau) \cosh\left(\frac{\pi}{2}\tau\right) \cos \tau\sigma d\tau. \end{aligned} \tag{6.21}$$

After substitution of the expression (6.20) for f_0 in (6.21) we may use the well-known integral [19]

$$\int_0^\infty P_{-\frac{1}{2}+i\tau}(\cosh \alpha) \cos(\tau t) d\tau = \frac{H(\alpha - t)}{\sqrt{2(\cosh \alpha - \cosh t)}}, \tag{6.22}$$

to deduce the line charge density equals (in agreement with [48])

$$l(\rho) = \frac{1}{4\pi} \cdot \frac{b}{K(a/b)} \left\{ \left(\rho^2 - a^2\right)\left(b^2 - \rho^2\right) \right\}^{-\frac{1}{2}}. \tag{6.23}$$

In order to examine the potential distribution for non-coplanar strips, we make extensive use of the Mehler-Fock transform [36, 56].

Theorem 11 *Let f be a real valued function defined on the interval $(1,\infty)$, which is piecewise continuous and of bounded variation on every finite subinterval of $(1,\infty)$. Then providing the integrals*

$$\int_1^a |f(x)| (x-1)^{-\frac{3}{4}}\ dx \quad \text{and} \quad \int_a^\infty |f(x)|\, x^{-\frac{1}{2}} \ln x\ dx$$

are finite for every $a > 1$, the representation

$$f(x) = \int_0^\infty \tau \tanh(\pi\tau) P_{-\frac{1}{2}+i\tau}(x)\, d\tau \int_1^\infty f(\xi) P_{-\frac{1}{2}+i\tau}(\xi)\ d\xi \tag{6.24}$$

is valid at every point $x \in (1,\infty)$ where f is continuous. Thus, if f has transform

$$F(\tau) = \int_1^\infty f(\xi) P_{-\frac{1}{2}+i\tau}(\xi)\ d\xi,$$

the inverse transform is

$$f(x) = \int_0^\infty \tau \tanh(\pi\tau) P_{-\frac{1}{2}+i\tau}(x)\, F(\tau) d\tau.$$

We shall apply the Mehler-Fock transform particularly in the form

$$f(\alpha) = \frac{1}{2} \int_0^\infty \tau \tanh\left(\frac{\pi}{2}\tau\right) P_{-\frac{1}{2}+i\frac{\tau}{2}}(\cosh 2\alpha)\, d\tau \times \int_0^\infty f(\sigma) P_{-\frac{1}{2}+i\frac{\tau}{2}}(\cosh 2\sigma) \sinh(2\sigma)\ d\sigma, \tag{6.25}$$

and use the representations (derived from (B.100) of the Appendix)

$$\begin{aligned} P_{-\frac{1}{2}+i\frac{\tau}{2}}(\cosh 2x) &= \frac{2}{\pi} \int_0^x \frac{\cos \tau t\ dt}{\sqrt{\cosh^2 x - \cosh^2 t}}, \\ &= \frac{2}{\pi} \coth\left(\frac{\pi}{2}\tau\right) \int_x^\infty \frac{\sin \tau t\ dt}{\sqrt{\cosh^2 t - \cosh^2 x}}. \end{aligned} \tag{6.26}$$

Now integrate (6.8) to obtain

$$\int_0^\infty \frac{\sinh(\pi\tau)}{\sinh\left[(\pi - \varphi_0)\tau\right]} f(\tau) \sin \tau\sigma\ d\tau = C, \quad \sigma \in (\sigma_0, \infty), \tag{6.27}$$

where C is a constant of integration to be determined. Rescale the function f so that

$$f(\tau) = \tau \frac{\sinh[(\pi - \varphi_0)\tau]}{\sinh(\pi\tau)} F(\tau). \tag{6.28}$$

After some manipulation, we obtain the dual integral equations

$$\int_0^\infty \tau \tanh\left(\frac{\pi}{2}\tau\right)[1 - N(\tau)] F(\tau) \cos\tau\sigma \, d\tau = 2, \ \sigma \in (0, \sigma_0), \tag{6.29}$$

$$\int_0^\infty \tau F(\tau) \sin\tau\sigma \, d\tau = C, \ \sigma \in (\sigma_0, \infty) \tag{6.30}$$

where

$$N(\tau) = \frac{\sinh^2 \tau\varphi_1}{\sinh^2\left(\frac{\pi}{2}\tau\right)}, \ 0 \le \varphi_1 < \frac{\pi}{2}. \tag{6.31}$$

The function N has the asymptotic behaviour

$$N(\tau) = e^{-(\pi - 2\varphi_1)\tau}(1 + O(e^{-2\varphi_1\tau})), \tag{6.32}$$

and plays the role of the asymptotically small parameter in the regularisation method; its magnitude is determined by the ratio $2\varphi_1/\pi$. Making use of the integral representations (6.26), we may obtain the equivalent form

$$\int_0^\infty \tau \tanh\left(\frac{\pi}{2}\tau\right)[1 - N(\tau)] F(\tau) P_{-\frac{1}{2}+i\frac{\tau}{2}}(\cosh 2\sigma) \, d\tau$$
$$= \frac{4}{\pi} \operatorname{sech}\sigma K(\tanh\sigma), \qquad \sigma \in (0, \sigma_0), \tag{6.33}$$

$$\int_0^\infty \tau \tanh\left(\frac{\pi}{2}\tau\right) F(\tau) P_{-\frac{1}{2}+i\frac{\tau}{2}}(\cosh 2\sigma) \, d\tau$$
$$= \frac{2}{\pi} C \operatorname{sech}\sigma K(\operatorname{sech}\sigma), \qquad \sigma \in (\sigma_0, \infty). \tag{6.34}$$

We rearrange (6.33) and (6.34) so that a suitably chosen singular part may be inverted via the Mehler-Fock transform:

$$\begin{aligned} \Phi(\sigma) &= \int_0^\infty \tau \tanh\left(\frac{\pi}{2}\tau\right) F(\tau) P_{-\frac{1}{2}+i\frac{\tau}{2}}(\cosh 2\sigma) \, d\tau \\ &= \begin{cases} F_1(\sigma), & \sigma \in (0, \sigma_0), \\ F_2(\sigma), & \sigma \in (\sigma_0, \infty), \end{cases} \end{aligned} \tag{6.35}$$

where

$$F_1(\sigma) = \int_0^\infty \tau \tanh\left(\frac{\pi}{2}\tau\right) N(\tau) F(\tau) P_{-\frac{1}{2}+i\frac{\tau}{2}}(\cosh 2\sigma)\, d\tau + \frac{4}{\pi} \operatorname{sech} \sigma K(\tanh \sigma), \quad (6.36)$$

$$F_2(\sigma) = \frac{2}{\pi} C \operatorname{sech} \sigma K(\operatorname{sech} \sigma). \quad (6.37)$$

The as yet unknown constant C is determined by invoking the principle that the potential must satisfy the edge condition. This means that the function Φ defined in (6.35) must be continuous, particularly at σ_0, so

$$C = \frac{\pi}{2} \frac{\cosh \sigma_0}{K(\operatorname{sech} \sigma_0)} \int_0^\infty \tau \tanh\left(\frac{\pi}{2}\tau\right) N(\tau) F(\tau) P_{-\frac{1}{2}+i\frac{\tau}{2}}(\cosh 2\sigma)\, d\tau + 2\frac{K(\tanh \sigma_0)}{K(\operatorname{sech} \sigma_0)}. \quad (6.38)$$

It will be seen later that this value is exactly the capacitance per unit length of the non-coplanar strips. Before applying the Mehler-Fock transform, it is advantageous to insert this value for C (6.37), and to replace the complete elliptic integrals thus occurring in (6.36) and (6.37) by their expression in terms of Legendre functions (see Appendix, (B.88)–(B.92)). The application of the Mehler-Fock transform produces the following second-kind Fredholm integral equation for the function F,

$$F(\tau) - \int_0^\infty F(\mu) K(\mu, \tau) d\mu = G(\tau), \ \tau \in (0, \infty), \quad (6.39)$$

where the kernel is

$$K(\mu, \tau) = \frac{\mu}{4} \tanh\left(\frac{\pi}{2}\mu\right) N(\mu) \int_1^{z_0} P_{-\frac{1}{2}+i\frac{\mu}{2}}(z)\, P_{-\frac{1}{2}+i\frac{\tau}{2}}(z)\, dz + \frac{\mu}{4} \tanh\left(\frac{\pi}{2}\mu\right) N(\mu) \frac{P_{-\frac{1}{2}+i\frac{\mu}{2}}(z_0)}{Q_{-\frac{1}{2}}(z_0)} \int_{z_0}^\infty Q_{-\frac{1}{2}}(z)\, P_{-\frac{1}{2}+i\frac{\tau}{2}}(z)\, dz, \quad (6.40)$$

with $z_0 = \cosh(2\sigma_0)$, and

$$G(\tau) = \frac{1}{2} \int_1^{z_0} P_{-\frac{1}{2}}(z)\, P_{-\frac{1}{2}+i\frac{\tau}{2}}(z)\, dz + \frac{1}{2} \frac{P_{-\frac{1}{2}}(z_0)}{Q_{-\frac{1}{2}}(z_0)} \int_{z_0}^\infty Q_{-\frac{1}{2}}(z)\, P_{-\frac{1}{2}+i\frac{\tau}{2}}(z)\, dz. \quad (6.41)$$

The integrals occurring in the definition of G may be explicitly evaluated by using a differential equation for the Legendre functions P_ν, Q_ν and their Wronskian (see Appendix, (B.63) and (B.69)), so that

$$G(\tau) = \frac{2}{\tau^2} \frac{P_{-\frac{1}{2}+i\frac{\tau}{2}}(z_0)}{Q_{-\frac{1}{2}}(z_0)}. \tag{6.42}$$

In a similar fashion, the explicit closed form for the kernel may be seen to equal

$$K(\mu,\tau) = \frac{\mu}{\tau^2} \tanh\left(\frac{\pi}{2}\mu\right) N(\mu)\left(1 - z_0^2\right) P_{-\frac{1}{2}+i\frac{\tau}{2}}(z_0) P_{-\frac{1}{2}+i\frac{\mu}{2}}(z_0) \times \left\{ q(z_0) - \frac{\mu^2 p_\tau(z_0) - \tau^2 p_\mu(z_0)}{\mu^2 - \tau^2} \right\}, \tag{6.43}$$

where

$$q(z_0) = \frac{Q'_{-\frac{1}{2}}(z_0)}{Q_{-\frac{1}{2}}(z_0)} = \left[\frac{d}{dz} \ln Q_{-\frac{1}{2}}(z)\right]_{z=z_0},$$

$$p_\tau(z_0) = \frac{P'_{-\frac{1}{2}+i\frac{\tau}{2}}(z_0)}{P_{-\frac{1}{2}+i\frac{\tau}{2}}(z_0)} = \left[\frac{d}{dz} \ln P_{-\frac{1}{2}+i\frac{\tau}{2}}(z)\right]_{z=z_0}.$$

Notice that in the limiting case of coplanar strips, the function N vanishes, and the explicit solution is

$$F(\tau) = G(\tau) = \frac{2}{\tau^2} \frac{P_{-\frac{1}{2}+i\frac{\tau}{2}}(z_0)}{Q_{-\frac{1}{2}}(z_0)},$$

which coincides exactly with the result stated in Equation (6.20). When the strips are not coplanar ($0 \leq \varphi_1 < \frac{1}{2}\pi$), the integral Equation (6.39) may be satisfactorily solved by standard numerical methods.

We have implied at various points in this book that the same equations may be solvable in different ways. The integral equation (6.39) should be transformed to some discrete form for this process. At the outset, one may ask if there is a regular basis to construct a satisfactory numerical solution. This question was originally answered affirmatively by C.J. Tranter (see [55]). We describe a similar approach, commencing from (6.7) and (6.8). With the rescaling

$$f^*(\tau) = \frac{\tau \sinh(\pi\tau)}{\sinh\left[(\pi - \varphi_0)\tau\right]} f(\tau), \tag{6.44}$$

we obtain the dual integral equations

$$\int_0^\infty M(\tau) f^*(\tau) \cos(\tau\sigma)\, d\tau = 1, \; \sigma \in (0, \sigma_0), \tag{6.45}$$

$$\int_0^\infty f^*(\tau) \cos(\tau\sigma)\, d\tau = 0, \; \sigma \in (\sigma_0, \infty), \tag{6.46}$$

where

$$M(\tau) = \frac{\sinh(\tau\varphi_0)\sinh[(\pi - \varphi_0)\tau]}{\tau\sinh(\pi\tau)} \to 1, \text{ as } \tau \to \infty. \tag{6.47}$$

Now represent f^* as a Neumann series

$$f^*(\tau) = a_0 J_0(\tau\sigma_0) + 2\sum_{n=1}^{\infty}\sqrt{n}A_n J_{2n}(\tau\sigma_0) \tag{6.48}$$

and substitute in (6.45) and (6.46). After interchanging of integration and summation, and using the discontinuous integral

$$\int_0^{\infty} J_{2n}(\tau\sigma_0)\cos(\tau\sigma)\, d\tau = H(\sigma_0 - \sigma)\left(\sigma_0^2 - \sigma^2\right)^{-\frac{1}{2}} T_{2n}\left(\sqrt{1 - \sigma^2/\sigma_0^2}\right), \tag{6.49}$$

it may be verified that Equation (6.46) is satisfied automatically, whilst Equation (6.45) leads to

$$\frac{1}{2}\sum_{n=1}^{\infty} n^{-\frac{1}{2}} A_n T_{2n}\left(\sqrt{1 - z^2}\right) \tag{6.50}$$

$$= 1 - a_0\int_0^{\infty} M(\tau) J_0(\tau\sigma_0)\cos(\tau\sigma_0 z)\, d\tau \tag{6.51}$$

$$+\sum_{n=1}^{\infty} n^{\frac{1}{2}} A_n \int_0^{\infty} \tau^{-1}\mu(\tau) J_{2n}(\tau\sigma_0)\cos(\tau\sigma_0 z)\, d\tau, \tag{6.52}$$

where $z \in (0,1)$ and $\mu(\tau) = 1 - 2\tau M(\tau) = O\left(e^{-2\tau\varphi_0}\right)$ as $\tau \to \infty$. Employing the orthogonality of the Chebyshev polynomials T_{2n} the Equation (6.52) is easily transformed to the i.s.l.a.e. of the second kind

$$A_m - \sum_{n=1}^{\infty} \alpha_{nm} A_n = -\alpha_m a_0, \tag{6.53}$$

where $m = 1, 2, \dots$ and

$$a_0 = \left(1 + \sum_{n=1}^{\infty} \beta_n A_n\right) / \beta_0^*; \tag{6.54}$$

the coefficients are defined by

$$\begin{aligned}
\alpha_{nm} &= 4(nm)^{\frac{1}{2}} \int_0^{\infty} \tau^{-1}\mu(\tau) J_{2n}(\tau\sigma_0)\, J_{2m}(\tau\sigma_0)\, d\tau, \\
\alpha_m &= 4m^{\frac{1}{2}} \int_0^{\infty} M(\tau) J_0(\tau\sigma_0)\, J_{2m}(\tau\sigma_0)\, d\tau, \\
\beta_n &= n^{\frac{1}{2}} \int_0^{\infty} \tau^{-1}\mu(\tau) J_{2n}(\tau\sigma_0)\, J_0(\tau\sigma_0)\, d\tau, \\
\beta_0^* &= \int_0^{\infty} M(\tau) J_0^2(\tau\sigma_0)\, d\tau.
\end{aligned} \tag{6.55}$$

Making use of (6.14), the integral representation for M is

$$M(\tau) = \frac{1}{2\pi} \int_0^\infty \cos(\tau z) \ln\left[1 + \frac{\sin^2\varphi_0}{\sinh^2 \frac{1}{2} z}\right] dz, \tag{6.56}$$

so that

$$\mu(\tau) = 1 - \frac{\tau}{\pi} \int_0^\infty \cos(\tau z) \ln\left[1 + \frac{\sin^2\varphi_0}{\sinh^2 \frac{1}{2} z}\right] dz. \tag{6.57}$$

An integration by parts shows that

$$\mu(\tau) = \frac{2}{\pi} \int_0^\infty \sin(\tau z) \Phi_1(z)\, dz, \tag{6.58}$$

where

$$\begin{aligned} \Phi_1(z) &= \frac{1}{z} + \frac{1}{2}\frac{d}{dz} \ln\left[1 + \frac{\sin^2\varphi_0}{\sinh^2 \frac{1}{2} z}\right] \\ &= \frac{1}{z} - \frac{1}{2} \frac{\sin^2\varphi_0}{\sinh^2 \frac{1}{2} z + \sin^2\varphi_0} \coth\left(\frac{z}{2}\right). \end{aligned} \tag{6.59}$$

Since $\Phi_1(z) \to 0$ as $z \to 0$, we may integrate by parts again to obtain

$$\mu(\tau) = \frac{2}{\pi\tau} \int_0^\infty \cos(\tau z) \Phi_1'(z)\, dz. \tag{6.60}$$

This process may be reiterated; it is clear that the asymptotics for μ decrease faster than any power, as μ has exponentially decreasing behaviour. The representations (6.58) and (6.60) are satisfactory for numerical calculations of the coefficients $\alpha_{nm}, \alpha_m, \beta_n, \beta_0^*$.

6.2 Electrostatic fields of a charged axisymmetric finite open conical conductor

We have previously exploited solutions of Laplace's equations in spherical coordinates (r, θ, φ), which are discontinuous in the radial variable r, to solve various potential problems such as spherical caps. In this chapter we examine finite open conducting surfaces that are part of the conical surface

$$0 \le r < \infty,\ \theta = \theta_0 = \text{constant},\ 0 \le \varphi \le 2\pi.$$

In this context, it is necessary to construct the total solution of Laplace's equation that is discontinuous in the angular variable θ; it is described in

[23]. It may then be employed to construct solutions satisfying Dirichlet boundary conditions on the conductor surface.

The form of separated solutions to Laplace's equations in spherical coordinates is given by (1.62), (1.63), and (1.64). Since solutions must be periodic in φ, the separation constant μ must be an integer $m \geq 0$, and Φ has the form (a_m, b_m constants),

$$\Phi(\varphi) = a_m \cos m\theta + b_m \sin m\theta. \tag{6.61}$$

An appropriate choice for the separation constant ν appearing in (1.60) is $\nu = -\frac{1}{2} + i\tau$, ($\tau$ real) so that the solution for Θ in (1.60) may be written

$$\Theta(\theta) = C_m(\tau) P^m_{-\frac{1}{2}+i\tau}(\cos\theta) + D_m(\tau) Q^m_{-\frac{1}{2}+i\tau}(\cos\theta), \tag{6.62}$$

where C_m, D_m are to be determined. This choice of ν is dictated by the requirement that the energy integral (1.85) is finite; this may be verified by considering the form of the separated solutions for the radial coordinate listed below (6.65). Alternatively, the solution Θ may be expressed in terms of the pair $P^m_{-\frac{1}{2}+i\tau}(\cos\theta)$ and $P^m_{-\frac{1}{2}+i\tau}(-\cos\theta)$, which are also linearly independent solutions of (1.60); the relation

$$Q^m_{-\frac{1}{2}+i\tau}(x) = -i\frac{\pi}{2}\tanh(\pi\tau) P^m_{-\frac{1}{2}+i\tau}(x) + (-1)^m \frac{\pi}{2}\operatorname{sech}(\pi\tau) P^m_{-\frac{1}{2}+i\tau}(-x) \tag{6.63}$$

is valid for $|x| \leq 1$. Thus, we seek solutions for Θ in the form

$$\Theta(\theta) = C_m(\tau) P^m_{-\frac{1}{2}+i\tau}(\cos\theta) + D_m(\tau) P^m_{-\frac{1}{2}+i\tau}(-\cos\theta). \tag{6.64}$$

The separated solutions (1.62) for the radial coordinate take the form

$$\begin{aligned} R(r) &= r^{-\frac{1}{2}}\left(E(\tau) r^{i\tau} + F(\tau) r^{-i\tau}\right) \\ &= e^{-\frac{\sigma}{2}}\left(e(\tau)\cos(\tau\sigma) + f(\tau)\sin(\tau\sigma)\right), \end{aligned} \tag{6.65}$$

where $\sigma = \ln r$. The separated solution $\psi^\tau_m(r,\theta,\varphi)$ for Laplace's equation corresponding to parameters m and τ is the product of (6.62), (6.64), and (6.65), and the general solution is the superposition

$$\psi(r,\theta,\varphi) = \sum_{m=-\infty}^{\infty} \int_{-\infty}^{\infty} \psi^\tau_m(r,\theta,\varphi)\, d\tau. \tag{6.66}$$

Solutions of Laplace's equations in regions bounded by the conical surface $\theta = \theta_0$ must take account of the singular behaviour of the associated Legendre functions in (6.64) at the singular points $\theta = 0, \pi$. The function $P^m_{-\frac{1}{2}+i\tau}(\cos\theta)$ is bounded at $\theta = 0$, yet unbounded at $\theta = \pi$. Thus *internal* conical harmonics in the region $0 \leq \theta \leq \theta_0$ involve $P^m_{-\frac{1}{2}+i\tau}(\cos\theta)$ (and

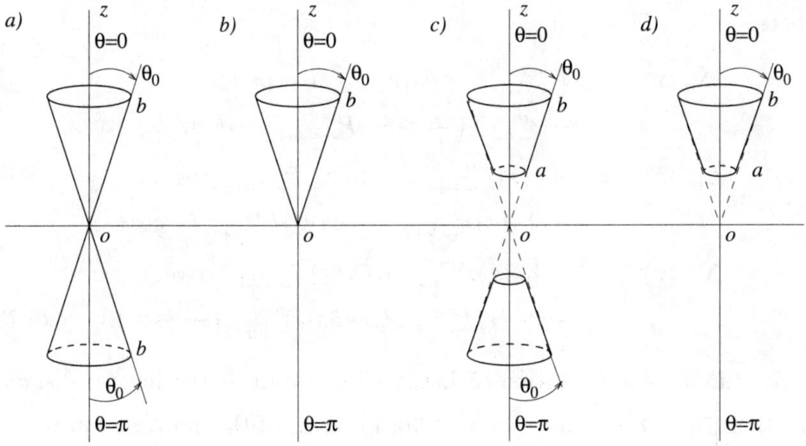

FIGURE 6.2. Various conical structures: (a) the bicone, (b) the cone, (c) a pair of hollow conical frustra, and (d) a hollow conical frustrum of finite length.

$D_m(\tau) = 0$ in (6.64)), whereas *external* conical harmonics in the region $\theta_0 \le \theta \le \pi$ involve $P^m_{-\frac{1}{2}+i\tau}(-\cos\theta)$ (and $C_m(\tau) = 0$). If the conductor is formed from the biconical surface $\theta = \theta_0, \theta = \theta_1$, conical harmonics for the intermediate region $\theta_0 \le \theta \le \theta_1$ employ both terms in (6.64).

Let us consider the Dirichlet boundary value problem for Laplace's equation for an open hollow finite conical conductor of the type shown in Figure 6.2. The boundary condition on the single cone $\theta = \theta_0$, or on the frustrum lying on this surface, is

$$\psi(\sigma, \theta_0 - 0, \varphi) = \psi(\sigma, \theta_0 + 0, \varphi), \quad -\infty < \sigma < \infty, \; 0 \le \varphi \le 2\pi. \tag{6.67}$$

The particular solution (6.64) takes the form

$$\Theta(\theta) = C_m(\tau) \begin{cases} P^m_{-\frac{1}{2}+i\tau}(\cos\theta), & \theta \in (0, \theta_0), \\ P^m_{-\frac{1}{2}+i\tau}(-\cos\theta) P^m_{-\frac{1}{2}+i\tau}(\cos\theta_0) / P^m_{-\frac{1}{2}+i\tau}(\cos\theta_0), & \theta \in (\theta_0, \pi). \end{cases} \tag{6.68}$$

For the finite hollow biconical conductor lying on the bicone $\theta = \theta_0, \theta = \theta_1$, the particular solution (6.64) takes the form

$$\Theta(\theta) = \begin{cases} A_m(\tau) P^m_{-\frac{1}{2}+i\tau}(\cos\theta), & \theta \in (0, \theta_0), \\ \Delta^{-1}_m(\tau) \left[\Delta^{(1)}_m(\tau) P^m_{-\frac{1}{2}+i\tau}(\cos\theta) + \Delta^{(2)}_m(\tau) P^m_{-\frac{1}{2}+i\tau}(-\cos\theta) \right], & \theta \in (\theta_0, \theta_1), \\ D_m(\tau) P^m_{-\frac{1}{2}+i\tau}(-\cos\theta), & \theta \in (\theta_1, \pi), \end{cases} \tag{6.69}$$

where

$$\begin{aligned}\Delta_m(\tau) &= P^m_{-\frac{1}{2}+i\tau}(\cos\theta_0)P^m_{-\frac{1}{2}+i\tau}(-\cos\theta_1)\\&\quad -P^m_{-\frac{1}{2}+i\tau}(-\cos\theta_0)P^m_{-\frac{1}{2}+i\tau}(\cos\theta_1),\\ \Delta^{(1)}_m(\tau) &= A_m(\tau)P^m_{-\frac{1}{2}+i\tau}(-\cos\theta_1)P^m_{-\frac{1}{2}+i\tau}(\cos\theta_0)\\&\quad -D_m(\tau)P^m_{-\frac{1}{2}+i\tau}(-\cos\theta_1)P^m_{-\frac{1}{2}+i\tau}(-\cos\theta_0),\\ \Delta^{(2)}_m(\tau) &= -A_m(\tau)P^m_{-\frac{1}{2}+i\tau}(\cos\theta_0)P^m_{-\frac{1}{2}+i\tau}(\cos\theta_1)\\&\quad +D_m(\tau)P^m_{-\frac{1}{2}+i\tau}(\cos\theta_0)P^m_{-\frac{1}{2}+i\tau}(-\cos\theta_1).\end{aligned} \tag{6.70}$$

The fundamental solution of Laplace's equation is the inverse distance function $\left|\vec{r}-\vec{r'}\right|^{-1}$ given by (1.209); in terms of the notation introduced it takes the form

$$\left|\vec{r}-\vec{r'}\right|^{-1} = e^{-\frac{1}{2}(\sigma+\sigma')}\left\{2\left(\cosh\left(\sigma-\sigma'\right)-\cos\psi\right)\right\}^{-\frac{1}{2}}, \tag{6.71}$$

where $\sigma=\ln r$, $\sigma'=\ln r'$ and

$$\cos\psi = \cos\theta\cos\theta' + \sin\theta\sin\theta'\cos(\phi-\phi').$$

It is representable as the integral transform [23]

$$\left|\vec{r}-\vec{r'}\right|^{-1} = e^{-\frac{1}{2}(\sigma+\sigma')}\int_0^\infty \operatorname{sech}(\pi\tau)\,P_{-\frac{1}{2}+i\tau}(\cos\psi)\cos\tau\left(\sigma-\sigma'\right)d\tau. \tag{6.72}$$

Employing the addition formula for the Legendre function [1]

$$P_\nu(\cos x\cos y \pm \sin x\sin y\cos a) = \sum_{k=0}^{\infty}(\pm 1)^k\cos ka P^k_\nu(\cos x)P^{-k}_\nu(\cos y), \tag{6.73}$$

where $x \geq 0$, $y<\pi$, and $x+y<\pi$, and the relation [1]

$$P^{-k}_\nu(y) = (-1)^k\frac{\Gamma(\nu-k+1)}{\Gamma(\nu+k+1)}P^k_\nu(y),\quad |y|<1, \tag{6.74}$$

we finally obtain the representation which is discontinuous in θ :

$$\left|\vec{r}-\vec{r'}\right|^{-1} = e^{-\frac{1}{2}(\sigma+\sigma')}\sum_{m=0}^{\infty}(2-\delta_{0m})\cos m(\phi-\phi')\times$$
$$\int_0^\infty \operatorname{sech}(\pi\tau)\frac{\Gamma\left(\frac{1}{2}+i\tau-m\right)}{\Gamma\left(\frac{1}{2}+i\tau+m\right)}P(\theta,\theta')\cos\tau\left(\sigma-\sigma'\right)d\tau \tag{6.75}$$

where

$$P(\theta,\theta') = \begin{Bmatrix} P^m_{-\frac{1}{2}+i\tau}(-\cos\theta)P^m_{-\frac{1}{2}+i\tau}(\cos\theta'), & \theta<\theta' \\ P^m_{-\frac{1}{2}+i\tau}(\cos\theta)P^m_{-\frac{1}{2}+i\tau}(-\cos\theta'), & \theta>\theta' \end{Bmatrix}.$$

The function (6.75) describes the potential generated by an elementary point charge located at the point (r',θ',φ'). When it is located on the z-axis at points with $\theta'=0,\pi$, this expression simplifies because $P^m_{-\frac{1}{2}+i\tau}(1) = \delta_{0m}$, to

$$\psi^0(\sigma,\theta) = e^{-\frac{1}{2}(\sigma+\sigma')}\int_0^\infty \operatorname{sech}(\pi\tau)\,P_{-\frac{1}{2}+i\tau}(\mp\cos\theta)\cos\tau\,(\sigma-\sigma')\,d\tau, \tag{6.76}$$

where the minus (respectively plus) sign refers to the choice $\theta'=0$ (respectively π).

Let us consider the simplest problem, the earthed semi-infinite cone $\theta=\theta_0$ in the presence of an elementary positive charge located on the z-axis at $\sigma=\sigma',\theta'=0$. This is a standard internal boundary value problem with Dirichlet boundary conditions given on the conical surface. The total electrostatic potential ψ is sought as the sum

$$\psi = \psi^0 + \psi^1$$

of the primary potential ψ^0 given by (6.76) and an induced potential ψ^1, subject to the boundary condition

$$\psi(\sigma,\theta_0) = 0, \quad -\infty<\sigma<\infty. \tag{6.77}$$

The induced potential ψ^1 is constructed as a superposition of internal conical harmonics

$$\psi^1(\sigma,\theta) = e^{-\frac{1}{2}(\sigma+\sigma')}\int_0^\infty \operatorname{sech}(\pi\tau)\,f(\tau)\,P_{-\frac{1}{2}+i\tau}(\cos\theta)\cos\tau\,(\sigma-\sigma')\,d\tau, \tag{6.78}$$

where the function f is found by the boundary condition to be

$$f(\tau) = -\frac{P_{-\frac{1}{2}+i\tau}(-\cos\theta_0)}{P_{-\frac{1}{2}+i\tau}(\cos\theta_0)}. \tag{6.79}$$

The surface charge density $S(\sigma)$ is easily deduced to be

$$\begin{aligned} S(\sigma) &= \left[-e^{-\sigma}\frac{\partial}{\partial\theta}\psi(\sigma,\theta)\right]_{\theta=\theta_0} \\ &= \frac{2}{\pi}e^{-\frac{1}{2}(3\sigma+\sigma')}\operatorname{cosec}\theta_0\int_0^\infty \frac{\cos\tau\,(\sigma-\sigma')}{P_{-\frac{1}{2}+i\tau}(\cos\theta_0)}d\tau. \end{aligned} \tag{6.80}$$

When $\theta_0 = \frac{1}{2}\pi$, the cone degenerates to the plane $z = 0$. Using the value [36]

$$P_{-\frac{1}{2}+i\tau}(0) = \sqrt{\pi}\left|\Gamma\left(\frac{3}{4} + i\frac{\tau}{2}\right)\right|^{-2}, \tag{6.81}$$

and the tabulated integral [14]

$$\int_0^\infty |\Gamma(a + ix)|^2 \cos xy \, dx = \pi\Gamma(2a)\, 2^{-2a} \operatorname{sech}^{2a}(y/2), \ (a, y > 0), \tag{6.82}$$

we may verify that the surface charge density $S(\rho)$ due to the point charge located on the z-axis at a distance d from the origin of the cylindrical coordinate system is

$$S(\rho) = 2d\left(\rho^2 + d^2\right)^{-\frac{3}{2}};$$

this is in accord with results obtained by elementary methods.

When $\theta_0 \ll 1$, the structure becomes a very sharp hollow cone. In this case it is convenient to use the hypergeometric representation (see Appendix, (B.98))

$$\begin{aligned} P_{-\frac{1}{2}+i\tau}(\cos\theta_0) &= {}_2F_1\left(\frac{1}{2} + i\tau, \frac{1}{2} - i\tau; 1; \sin^2\frac{1}{2}\theta_0\right) \\ &= 1 + \left(\frac{1}{4} + \tau^2\right)\sin^2\frac{1}{2}\theta_0 + O\left(\sin^4\frac{1}{2}\theta_0\right). \end{aligned} \tag{6.83}$$

Inserting this approximation in (6.80) produces the approximation for surface charge density

$$S(\sigma) \simeq \frac{1}{r^{\frac{3}{2}} r'^{\frac{1}{2}}} \operatorname{cosec}\theta_0 \operatorname{cosec}\frac{1}{2}\theta_0 \left(\frac{r}{r'}\right)^{2/\theta_0}, \quad r < r'. \tag{6.84}$$

Thus $S(\sigma) \to 0$ as $r \to 0$.

If the elementary charge is located outside the cone (at $\sigma = \sigma', \theta' = \pi$), then Equation (6.80) is replaced by

$$S(\sigma) = \frac{2}{\pi} e^{-\frac{1}{2}(3\sigma + \sigma')} \operatorname{cosec}\theta_0 \int_0^\infty \frac{\cos\tau(\sigma - \sigma')}{P_{-\frac{1}{2}+i\tau}(-\cos\theta_0)} d\tau. \tag{6.85}$$

For the same limiting structure of the very sharp hollow cone ($\theta_0 \ll 1$), the approximation

$$P_{-\frac{1}{2}+i\tau}(-\cos\theta_0) \simeq \frac{2}{\pi}\cosh(\tau\pi)\ln\left(\frac{2}{\theta_0}\right) \tag{6.86}$$

is relevant and

$$S(\sigma) \simeq \frac{(r+r')^{-1}}{r} \frac{\operatorname{cosec}\theta_0}{\ln(2/\theta_0)}. \tag{6.87}$$

The approximation (6.87) has a singularity at the vertex of the cone. This phenomenon is discussed in some detail by Hobson [23].

Remove the lower and upper parts of the infinite hollow cone to form the finite hollow frustrum

$$a \leq r \leq b, \ \theta = \theta_0,$$

shown in Figure 6.2. It is convenient to introduce the normalised radial coordinate

$$\rho = r/\sqrt{ab},$$

so that the frustrum is given by

$$\rho_0^{-1} \leq r \leq \rho_0, \ \theta = \theta_0,$$

where $\rho_0 = (b/a)^{\frac{1}{2}}$. Also replace the variable $\sigma = \ln r$ previously introduced by $\sigma = \ln \rho$, so the frustrum is described by $\sigma \in (-\sigma_0, \sigma_0)$, where $\sigma_0 = \frac{1}{2}\ln(b/a)$. The partial solutions for the variable σ may still be described by Equation (6.65). Accordingly, let us consider the Dirichlet boundary value problem for the frustrum. The axisymmetric potential $\psi(\sigma, \theta)$ must satisfy the mixed boundary conditions

$$\frac{\partial}{\partial\theta}\psi(\sigma,\theta_0-0) = \frac{\partial}{\partial\theta}\psi(\sigma,\theta_0+0), \ \sigma \in (-\infty,-\sigma_0) \cup (\sigma_0,\infty) \tag{6.88}$$

$$\psi(\sigma,\theta_0-0) = \psi(\sigma,\theta_0+0) = \Phi(\sigma), \ \sigma \in (-\sigma_0,\sigma_0), \tag{6.89}$$

where Φ is the given value of the potential on the conductor. We consider the simplest case where the frustrum is charged to unit potential ($\Phi = 1$). Using the superposition (6.66), the solution for ψ is sought in the form

$$\psi(\sigma,\theta) = e^{-\frac{\sigma}{2}} \int_0^\infty d\tau \left\{f_c(\tau)\cos\tau\sigma + f_s(\tau)\sin\tau\sigma\right\} F(\tau,\theta), \tag{6.90}$$

where

$$F(\tau,\theta) = \begin{cases} P_{-\frac{1}{2}+i\tau}(\cos\theta), & 0 \leq \theta \leq \theta_0, \\ P_{-\frac{1}{2}+i\tau}(\cos\theta_0)P_{-\frac{1}{2}+i\tau}(\cos\theta)/P_{-\frac{1}{2}+i\tau}(-\cos\theta_0), & \theta_0 \leq \theta \leq \pi \end{cases}$$

and f_c, f_s are unknown functions to be determined. Because of the symmetric subdivision of the real line, the integral equations for f_c and f_s can be decoupled to the following sets of dual integral equations:

$$\int_0^\infty f_c(\tau)P_{-\frac{1}{2}+i\tau}(\cos\theta_0)\cos\tau\sigma d\tau = \cosh(\sigma/2), \ \sigma \in (0,\sigma_0), \tag{6.91}$$

$$\int_0^\infty f_c(\tau) \frac{\cosh(\pi\tau)}{P_{-\frac{1}{2}+i\tau}(-\cos\theta_0)} \cos\tau\sigma d\tau = 0, \; \sigma \in (\sigma_0, \infty), \tag{6.92}$$

and

$$\int_0^\infty f_s(\tau) P_{-\frac{1}{2}+i\tau}(\cos\theta_0) \sin\tau\sigma d\tau = \sinh(\sigma/2), \; \sigma \in (0, \sigma_0), \tag{6.93}$$

$$\int_0^\infty f_s(\tau) \frac{\cosh(\pi\tau)}{P_{-\frac{1}{2}+i\tau}(-\cos\theta_0)} \sin\tau\sigma d\tau = 0, \; \sigma \in (\sigma_0, \infty). \tag{6.94}$$

When $\theta_0 = \frac{1}{2}\pi$, these equations describe the nontrivial geometry of the charged annular disc [55].

The first step is to integrate (6.92) and to differentiate (6.93), respectively, obtaining

$$\int_0^\infty f_c(\tau) \frac{\cosh(\pi\tau)}{\tau P_{-\frac{1}{2}+i\tau}(-\cos\theta_0)} \sin\tau\sigma d\tau = C, \; \sigma \in (\sigma_0, \infty), \tag{6.95}$$

where C is a constant of integration to be determined and

$$\int_0^\infty \tau f_s(\tau) P_{-\frac{1}{2}+i\tau}(\cos\theta_0) \cos\tau\sigma d\tau = \frac{1}{2}\cosh(\sigma/2), \; \sigma \in (0, \sigma_0). \tag{6.96}$$

Rescale the unknown functions

$$\begin{aligned} f_c(\tau) &= \tau^2 \operatorname{sech}(\pi\tau) P_{-\frac{1}{2}+i\tau}(-\cos\theta_0) F_c(\tau), & (6.97) \\ f_s(\tau) &= \tau \operatorname{sech}(\pi\tau) P_{-\frac{1}{2}+i\tau}(-\cos\theta_0) F_s(\tau), & (6.98) \end{aligned}$$

so that

$$\int_0^\infty \tau \tanh(\pi\tau) M(\tau;\theta_0) F_c(\tau) \cos\tau\sigma d\tau = \cosh(\sigma/2), \; \sigma \in (0, \sigma_0) \tag{6.99}$$

$$\int_0^\infty \tau F_c(\tau) \sin\tau\sigma d\tau = C, \; \sigma \in (\sigma_0, \infty), \tag{6.100}$$

and

$$\int_0^\infty \tau \tanh(\pi\tau) M(\tau;\theta_0) F_s(\tau) \cos\tau\sigma d\tau = \frac{1}{2}\cosh(\sigma/2), \; \sigma \in (0, \sigma_0) \tag{6.101}$$

$$\int_0^\infty \tau F_s(\tau) \sin\tau\sigma d\tau = 0, \; \sigma \in (\sigma_0, \infty), \tag{6.102}$$

where the function M is defined by

$$M(\tau;\theta_0) = \tau \operatorname{cosech}(\pi\tau) P_{-\frac{1}{2}+i\tau}(\cos\theta_0) P_{-\frac{1}{2}+i\tau}(-\cos\theta_0). \tag{6.103}$$

The asymptotics for the conical functions (see Appendix, (B.101)) show that

$$\lim_{\tau\to\infty} M(\tau;\theta_0) = \frac{1}{\pi}\operatorname{cosec}\theta_0,$$

and the function

$$\begin{aligned} N(\tau;\theta_0) &= 1 - \pi\sin\theta_0 M(\tau;\theta_0) \\ &= 1 - \pi\tau\frac{\sin\theta_0}{\sinh(\pi\tau)} P_{-\frac{1}{2}+i\tau}(\cos\theta_0) P_{-\frac{1}{2}+i\tau}(-\cos\theta_0) \\ &= O(\tau^{-2}) \text{ as } \tau\to\infty. \end{aligned} \tag{6.104}$$

Equations (6.99) and (6.100) may be rearranged in the form

$$\begin{aligned} \Phi_c(\sigma) &= \int_0^\infty \tau\tanh(\pi\tau) F_c(\tau)\, P_{-\frac{1}{2}+i\tau}(\cosh\sigma) d\tau \\ &= \begin{cases} F_1(\sigma), & \sigma\in(0,\sigma_0), \\ F_2(\sigma), & \sigma\in(\sigma_0,\infty), \end{cases} \end{aligned} \tag{6.105}$$

where

$$\begin{aligned} F_1(\sigma) &= \pi\sin\theta_0 + \int_0^\infty \tau\tanh(\pi\tau) F_c(\tau)\, N(\tau;\theta_0) P_{-\frac{1}{2}+i\tau}(\cosh\sigma) d\tau, \\ F_2(\sigma) &= \frac{2}{\pi} C Q_{-\frac{1}{2}}(\cosh\sigma). \end{aligned}$$

Likewise, (6.101) and (6.102) may be rearranged in the form

$$\begin{aligned} \Phi_s(\sigma) &= \int_0^\infty \tau\tanh(\pi\tau) F_s(\tau)\, P_{-\frac{1}{2}+i\tau}(\cosh\sigma) d\tau \\ &= \begin{cases} F_3(\sigma), & \sigma\in(0,\sigma_0), \\ 0, & \sigma\in(\sigma_0,\infty), \end{cases} \end{aligned} \tag{6.106}$$

where

$$F_3(\sigma) = \frac{\pi}{2}\sin\theta_0 + \int_0^\infty \tau\tanh(\pi\tau) F_s(\tau)\, N(\tau;\theta_0) P_{-\frac{1}{2}+i\tau}(\cosh\sigma) d\tau.$$

Note that in deriving (6.105), we also used the relation (B.89) of the Appendix. As shown previously, the value of the constant C is determined by enforcing the requirement of continuity on Φ_c at σ_0, giving

$$\begin{aligned} C = \frac{\pi}{2}\int_0^\infty \tau\tanh(\pi\tau) F_c(\tau)\, N(\tau;\theta_0) \frac{P_{-\frac{1}{2}+i\tau}(\cosh\sigma_0)}{Q_{-\frac{1}{2}}(\cosh\sigma_0)} d\tau \\ + \frac{\pi^2}{2}\frac{\sin\theta_0}{Q_{-\frac{1}{2}}(\cosh\sigma_0)}. \end{aligned} \tag{6.107}$$

The final step uses the inverse of the Mehler-Fock transform to convert both (6.105) and (6.106) to second-kind Fredholm integral equations to be solved for the functions F_c and F_s, respectively, obtaining

$$F_s(\tau) = \int_0^\infty F_s(\xi) K_s(\xi,\tau) d\xi = G_s(\tau) \tag{6.108}$$

and

$$F_c(\tau) = \int_0^\infty F_c(\xi) K_c(\xi,\tau) d\xi = G_c(\tau). \tag{6.109}$$

The inhomogeneous terms are

$$G_s(\tau) = \frac{\pi}{2} \frac{\sin\theta_0}{\frac{1}{4}+\tau^2} \left(1-z_0^2\right) P'_{-\frac{1}{2}+i\tau}(z_0), \tag{6.110}$$

$$G_c(\tau) = \tau^{-2}\pi \sin\theta_0 \left(1-z_0^2\right) P_{-\frac{1}{2}+i\tau}(z_0) \times \left\{ \frac{Q'_{-\frac{1}{2}}(z_0)}{Q_{-\frac{1}{2}}(z_0)} - \frac{1}{1+4\tau^2} \frac{P'_{-\frac{1}{2}+i\tau}(z_0)}{P_{-\frac{1}{2}+i\tau}(z_0)} \right\}, \tag{6.111}$$

and the kernels, respectively, are defined to be

$$K_s(\xi,\tau) = \xi \tanh(\pi\xi) N(\xi;\theta_0) R(\xi,\tau;z_0), \tag{6.112}$$

$$K_c(\xi,\tau) = \xi \tanh(\pi\xi) N(\xi;\theta_0) R(\xi,\tau;z_0) + \xi \tanh(\pi\xi) N(\xi;\theta_0) \frac{P_{-\frac{1}{2}+i\xi}(z_0)}{Q_{-\frac{1}{2}}(z_0)} R^*(\tau;z_0), \tag{6.113}$$

where

$$\begin{aligned} R(\xi,\tau;z_0) &= \int_{-1}^{z_0} P_{-\frac{1}{2}+i\xi}(z) P_{-\frac{1}{2}+i\tau}(z) dz, \\ R^*(\tau;z_0) &= \int_{z_0}^{\infty} Q_{-\frac{1}{2}}(z) P_{-\frac{1}{2}+i\tau}(z) dz. \end{aligned}$$

Both R and R^* may be readily evaluated in closed form (in the same way as Formula (B.97) of the Appendix):

$$R(\xi,\tau;z_0) = \frac{\left(1-z_0^2\right)}{\left(\tau^2-\xi^2\right)} \left\{ P_{-\frac{1}{2}+i\xi}(z_0) P'_{-\frac{1}{2}+i\tau}(z_0) - P'_{-\frac{1}{2}+i\xi}(z_0) P_{-\frac{1}{2}+i\tau}(z_0) \right\}, \tag{6.114}$$

$$R^*(\tau; z_0) = \frac{(1 - z_0^2)}{\tau^2}\left\{P_{-\frac{1}{2}+i\tau}(z_0)Q'_{-\frac{1}{2}}(z_0) - P'_{-\frac{1}{2}+i\tau}(z_0)Q_{-\frac{1}{2}}(z_0)\right\}. \quad (6.115)$$

For an effective numerical solution, the kernels of (6.108) and (6.109) must converge sufficiently fast as $\xi \to \infty$. This depends completely on the asymptotics of the function N. It is convenient to examine the function

$$N^*(\tau; \theta_0) = \tanh(\pi\tau)N(\tau; \theta_0) = (\tanh(\pi\tau) - 1) + N_1^*(\tau; \theta_0), \quad (6.116)$$

where

$$N_1^*(\tau; \theta_0) = 1 - \pi\tau\frac{\sin\theta_0}{\cosh(\pi\tau)}P_{-\frac{1}{2}+i\tau}(\cos\theta_0)P_{-\frac{1}{2}+i\tau}(-\cos\theta_0). \quad (6.117)$$

Since $\tanh(\pi\tau) - 1 = -2e^{-2\pi\tau}\left(1 + O\left(e^{-2\pi\tau}\right)\right)$ as $\tau \to \infty$, it is quite clear that the major contribution to the asymptotic behaviour of N^* is dominated by that of N_1^*. The asymptotic behaviour of N_1^* can be determined by examining an integral representation for the product of conical functions appearing in (6.117). Integrate both sides of (6.75) twice. Assuming $\sigma' = 0, \varphi' = 0, \theta' = \theta_0$, first integrate w.r.t. φ over the interval $[0, 2\pi]$. Multiply the result of the first integration by $\cos\mu\sigma$ and integrate w.r.t. σ over the interval $(0, \infty)$ to obtain

$$\frac{1}{\sqrt{2}}\int_0^\infty d\sigma\cos\tau\sigma\int_0^{2\pi}\{\cosh\sigma - \cos\theta\cos\theta_0 - \sin\theta\sin\theta_0\cos\varphi\}^{-\frac{1}{2}}d\varphi$$
$$= \pi^2\operatorname{sech}(\pi\tau)\begin{cases} P_{-\frac{1}{2}+i\tau}(-\cos\theta)P_{-\frac{1}{2}+i\tau}(\cos\theta_0), & \theta < \theta_0 \\ P_{-\frac{1}{2}+i\tau}(\cos\theta)P_{-\frac{1}{2}+i\tau}(-\cos\theta_0), & \theta > \theta_0. \end{cases} \quad (6.118)$$

Using the tabulated integral [14]

$$\int_0^\pi \frac{dx}{\sqrt{a \pm b\cos x}} = \frac{2}{\sqrt{a+b}}K\left(\sqrt{\frac{2b}{a+b}}\right), \quad a > b > 0, \quad (6.119)$$

and the relation

$$Q_{-\frac{1}{2}}(1 + 2x^2) = \frac{1}{\sqrt{1+x^2}}K\left(\frac{1}{\sqrt{1+x^2}}\right), \quad (6.120)$$

we obtain the desired result

$$\frac{2}{\pi^2}\frac{\cosh(\pi\tau)}{\sqrt{\sin\theta\sin\theta_0}}\int_0^\infty Q_{-\frac{1}{2}}\left(\frac{\cosh z - \cos\theta\cos\theta_0}{\sin\theta\sin\theta_0}\right)\cos\tau z\, dz$$
$$= \begin{cases} P_{-\frac{1}{2}+i\tau}(-\cos\theta)P_{-\frac{1}{2}+i\tau}(\cos\theta_0), & \theta < \theta_0 \\ P_{-\frac{1}{2}+i\tau}(\cos\theta)P_{-\frac{1}{2}+i\tau}(-\cos\theta_0), & \theta > \theta_0. \end{cases} \quad (6.121)$$

Setting $\theta = \theta_0$, the value of $N_1^*(\tau;\theta_0)$ is deduced to be

$$N_1^*(\tau;\theta_0) = 1 - \frac{2\tau}{\pi}\int_0^\infty Q_{-\frac{1}{2}}\left(1 + 2\frac{\sinh^2(z/2)}{\sin^2\theta_0}\right)\cos\tau z\, dz. \tag{6.122}$$

After a single integration by parts this may be written as

$$N_1^*(\tau;\theta_0) = \frac{2}{\pi}\int_0^\infty S_1(z)\sin\tau z\, dz, \tag{6.123}$$

where

$$S_1(z) = \frac{1}{z} + \frac{d}{dz}Q_{-\frac{1}{2}}\left(1 + 2\frac{\sinh^2(z/2)}{\sin^2\theta_0}\right).$$

Since $S_1(z) \backsim z\ln z$ when $z \ll 1$, and $S_1(z) \to 0$ as $z \to 0$, another integration by parts produces

$$N_1^*(\tau;\theta_0) = \frac{2}{\pi\tau}\int_0^\infty S_2(z)\cos\tau z\, dz \tag{6.124}$$

where $S_2(z) = S_1'(z)$; it may now be deduced that $N_1^*(\tau;\theta_0) = O(\tau^{-2})$ as $\tau \to \infty$.

Thus, standard methods for second-kind Fredholm equations may be employed effectively to obtain numerical solutions to (6.108) and (6.109).

6.3 The slotted hollow spindle

The electrostatic potential of several structures related to the *spindle* can be deduced from the solutions already obtained for conical structures. The basic structure of the spindle is the surface of revolution obtained by revolving an arc of a circle about the chord $\overline{OA}$ joining its end points A and O. The vector $\overrightarrow{OA}$ may be chosen to lie along the positive direction of the z-axis; the coordinate origin may be located at O. (See Figure 6.3.)

Under inversion in the sphere of radius $R = \overline{OA}$ centred at A, the image of the spindle is the infinite (right circular) cone with vertex O and axis coinciding with the z-axis. The half-angle α of the cone equals half the angle subtended by the chord $\overline{OA}$ at the centre of its circle.

Bouwkamp's theorem (see Chapter 3) may be used to calculate the capacitance of the spindle (see, for example, [51]). The calculation is equivalent to the calculation of the electrostatic field surrounding the grounded semi-infinite cone in the presence of a unit negative charge located at A; in the usual spherical coordinates the charge is located at $(r,\theta,\phi) = (R,\pi,0)$.

The potential ψ^0 due to this charge is given by (6.76), with $\sigma = \ln r$ and $\sigma' = \ln R$,

$$\psi^0(\sigma,\theta) = -e^{-\frac{1}{2}(\sigma+\sigma')}\int_0^\infty \operatorname{sech}(\pi\tau)\, P_{-\frac{1}{2}+i\tau}(\cos\theta)\cos\tau(\sigma-\sigma')\, d\tau. \tag{6.125}$$

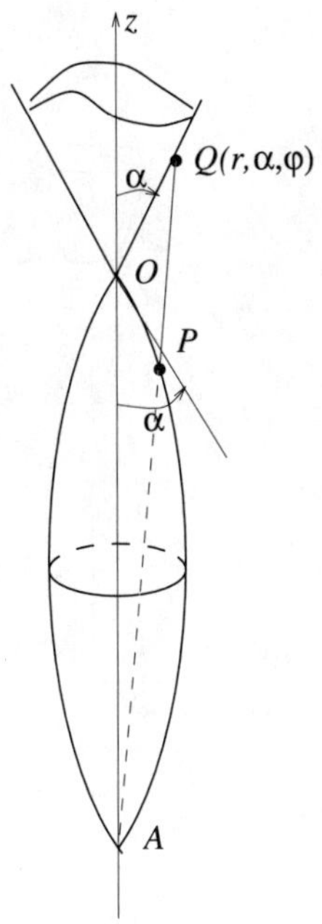

FIGURE 6.3. The spindle. The semi-infinite cone is its image under an inversion with centre A.

The induced potential has the form

$$\psi^1(\sigma,\theta) = e^{-\frac{1}{2}(\sigma+\sigma')}\int_0^\infty \operatorname{sech}(\pi\tau)\, f(\tau) P_{-\frac{1}{2}+i\tau}(-\cos\theta)\cos\tau\,(\sigma-\sigma')\, d\tau, \tag{6.126}$$

where the function f is to be determined. The total potential vanishes on the grounded conical surface,

$$\psi^0(\sigma,\alpha) + \psi^1(\sigma,\alpha) = 0,\ \sigma \in (-\infty,\infty), \tag{6.127}$$

so that

$$f(\tau) = \frac{P^m_{-\frac{1}{2}+i\tau}(\cos\alpha)}{P^m_{-\frac{1}{2}+i\tau}(-\cos\alpha)}. \tag{6.128}$$

The capacitance C of the spindle is deduced from the value of the induced potential at the point of inversion A,

$$C = R^2\psi^1(\sigma',\pi) = R\int_0^\infty \operatorname{sech}(\pi\tau)\frac{P_{-\frac{1}{2}+i\tau}(\cos\alpha)}{P_{-\frac{1}{2}+i\tau}(-\cos\alpha)} d\tau. \tag{6.129}$$

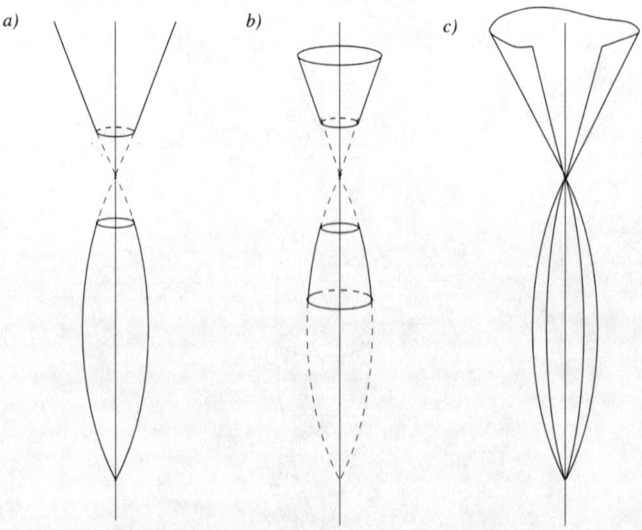

FIGURE 6.4. The spindle with various apertures. The conical structures that are their images under inversion are also shown. (a) An axisymmetric circular hole, (b) a pair of axisymmetric circular holes, and (c) a nonsymmetric azimuthal slot.

When $\alpha = \frac{1}{2}\pi$, the spindle degenerates to a sphere of radius $a = \frac{1}{2}R$, and the value of the capacitance given by (6.129) coincides with the well-known capacitance C_0 of the sphere: $C_0 = a$. It will be convenient to normalise the capacitance given by (6.151) against C_0. When $\alpha \ll 1$,

$$P^m_{-\frac{1}{2}+i\tau}(\cos\alpha) \simeq 1, \; P^m_{-\frac{1}{2}+i\tau}(-\cos\alpha) \simeq \frac{2}{\pi}\cosh(\pi\tau)\ln\left(\frac{2}{\alpha}\right), \tag{6.130}$$

so that the normalised capacitance is approximately

$$C/C_0 \simeq 1/\ln\left(\frac{2}{\alpha}\right), \quad \alpha \ll 1. \tag{6.131}$$

It should be noted that the calculations above are valid when $0 < \alpha \leq \frac{1}{2}\pi$. When $\alpha > \frac{1}{2}\pi$, the image of the spindle under inversion is a spherical shell with circular apertures centred at its poles.

Some structures formed by removing part of the spindle surface are shown in Figure 6.4. Under inversion the spindle with a symmetrically placed circular aperture is equivalent to the semi-infinite frustrum, whilst the spindle with two symmetrically placed circular apertures is equivalent to the finite conical frustrum. Perhaps the most interesting open spindle-shaped conductor is obtained by introducing an azimuthal slot. Under inversion its image is the semi-infinite cone with an azimuthal sector removed. The introduction of this aperture breaks the axial symmetry present in all the conical structures considered above. Together with the toroid with azimuthal cuts analysed in Chapter 5, this structure allows us to illustrate a very significant extension of analytic and semi-analytic techniques to the determination of the three dimensional potential distribution surrounding *nonsymmetric* open conducting surfaces.

Consider, therefore, the problem of determining the electrostatic field surrounding a charged hollow spindle with an azimuthal slot. The equivalent problem is to find the electrostatic field induced on the grounded semi-infinite cone with an azimuthal (or sectoral) slot by a unit negative charge, located at the inversion centre. Let $2\varphi_0$ be the angular width of the sectoral slot.

The free-space potential ψ^0 is given by (6.125). Based on previous results, the induced potential ψ^1 may be represented as

$$\psi^1(\sigma,\theta,\varphi) = e^{-\frac{1}{2}(\sigma+\sigma')} \int_0^\infty F_\tau(\theta,\varphi)\cos\tau(\sigma-\sigma')\,d\tau, \tag{6.132}$$

where

$$F_\tau(\theta,\varphi) = \operatorname{sech}(\pi\tau)\sum_{m=0}^\infty (2-\delta_{0m})\,f_m(\tau)\cos(m\varphi)H(\tau,\theta), \tag{6.133}$$

with

$$H(\tau,\theta) = \begin{cases} P^m_{-\frac{1}{2}+i\tau}(\cos\theta), & \theta<\theta_0, \\ P^m_{-\frac{1}{2}+i\tau}(\cos\theta_0)P^m_{-\frac{1}{2}+i\tau}(-\cos\theta)/P^m_{-\frac{1}{2}+i\tau}(-\cos\theta_0), & \theta>\theta_0, \end{cases}$$

and the functions f_m $(m = 0,1,2,\ldots)$ are unknowns to be found. The free-space potential may also be written in the analogous form

$$\psi^0(\sigma,\theta) = e^{-\frac{1}{2}(\sigma+\sigma')} \int_0^\infty F^0_\tau(\theta)\cos\tau(\sigma-\sigma')\,d\tau, \tag{6.134}$$

where

$$F^0_\tau(\theta) = -\operatorname{sech}(\pi\tau)P_{-\frac{1}{2}+i\tau}(\cos\theta).$$

For *all* $\sigma\in(-\infty,\infty)$, the following boundary conditions apply to the total potential $\psi=\psi^0+\psi^1$,

$$\begin{aligned} \frac{\partial}{\partial\theta}\psi(\sigma,\theta_0-0,\varphi) &= \frac{\partial}{\partial\theta}\psi(\sigma,\theta_1+0,\varphi), \quad \varphi\in(0,\varphi_0), & (6.135)\\ \psi(\sigma,\theta_0-0,\varphi) &= \psi(\sigma,\theta_1+0,\varphi)=0, \ \ \varphi\in(\varphi_0,\pi). & (6.136) \end{aligned}$$

Because these boundary conditions apply for the complete interval $(-\infty,\infty)$, we may apply a Fourier transform to express them in terms of F_τ and its derivative,

$$\frac{\partial}{\partial\theta}F_\tau(\theta_0-0,\varphi) = \frac{\partial}{\partial\theta}F_\tau(\theta_0+0,\varphi), \quad \varphi\in(0,\varphi_0)\,, \tag{6.137}$$

$$F_\tau(\theta_0-0,\varphi) = F_\tau(\theta_0+0,\varphi) = -F_\tau^0(\theta_0), \varphi\in(\varphi_0,\pi)\,. \tag{6.138}$$

Enforcement of these conditions produces the following dual series equations

$$\sum_{m=0}^{\infty}\frac{(-1)^m(2-\delta_{0m})}{P^m_{-\frac{1}{2}+i\tau}(-\cos\theta_0)}f_m(\tau)\frac{\Gamma\left(\frac{1}{2}+i\tau+m\right)}{\Gamma\left(\frac{1}{2}+i\tau-m\right)}\cos(m\varphi)=0,$$

$$\varphi\in(0,\varphi_0)\,, \tag{6.139}$$

$$\sum_{m=0}^{\infty}(2-\delta_{0m})\,f_m(\tau)P^m_{-\frac{1}{2}+i\tau}(\cos\theta_0)\cos(m\varphi)=P_{-\frac{1}{2}+i\tau}(\cos\theta_0),$$

$$\varphi\in(\varphi_0,\pi)\,, \tag{6.140}$$

where the value of the Wronskian W of $P^m_{-\frac{1}{2}+i\tau}(x)$ and $P^m_{-\frac{1}{2}+i\tau}(-x)$ has been employed. We introduce the functions

$$F_m(\tau)=\frac{(-1)^m}{m}\frac{\Gamma\left(\frac{1}{2}+i\tau+m\right)}{\Gamma\left(\frac{1}{2}+i\tau-m\right)}\frac{f_m(\tau)}{P^m_{-\frac{1}{2}+i\tau}(-\cos\theta_0)}, \tag{6.141}$$

and separate in (6.139) and (6.140) the terms with index $m=0$ to obtain

$$\sum_{m=1}^{\infty}mF_m(\tau)\cos(m\varphi)=-\frac{f_0(\tau)}{2P_{-\frac{1}{2}+i\tau}(-\cos\theta_0)},$$

$$\varphi\in(0,\varphi_0)\,, \tag{6.142}$$

$$\sum_{m=1}^{\infty}G_m(\tau,\theta_0)F_m(\tau)\cos(m\varphi)=\frac{1}{2}\left[1-f_0(\tau)\right]P_{-\frac{1}{2}+i\tau}(\cos\theta_0),$$

$$\varphi\in(\varphi_0,\pi)\,, \tag{6.143}$$

where

$$G_m(\tau,\theta_0)=(-1)^m\,m\frac{\Gamma\left(\frac{1}{2}+i\tau-m\right)}{\Gamma\left(\frac{1}{2}+i\tau+m\right)}P^m_{-\frac{1}{2}+i\tau}(\cos\theta_0)P^m_{-\frac{1}{2}+i\tau}(-\cos\theta_0). \tag{6.144}$$

We now investigate the asymptotic behaviour of the function $G_m(\tau,\theta_0)$ as $m\to\infty$. For these purposes τ is fixed. From the definition of the associated Legendre functions (B.102),

$$G_m(\tau,\theta_0) = m\frac{\cosh(\pi\tau)}{\pi}\frac{\Gamma\left(\frac{1}{2}+i\tau+m\right)\Gamma\left(\frac{1}{2}-i\tau+m\right)}{\Gamma^2(m+1)} \times$$
$${}_2F_1\left(\frac{1}{2}-i\tau,\frac{1}{2}+i\tau;m+1;\sin^2\frac{\theta_0}{2}\right) \times$$
$${}_2F_1\left(\frac{1}{2}-i\tau,\frac{1}{2}+i\tau;m+1;\cos^2\frac{\theta_0}{2}\right). \qquad (6.145)$$

Rearrange the Gamma function factors as

$$\frac{\Gamma\left(\frac{1}{2}+i\tau+m\right)\Gamma\left(\frac{1}{2}-i\tau+m\right)}{\Gamma^2(m+1)}$$
$$= \frac{\Gamma^2\left(m+\frac{1}{2}\right)}{\Gamma^2(m+1)}\frac{\left|\Gamma\left(\frac{1}{2}+i\tau+m\right)\right|^2}{\Gamma^2\left(m+\frac{1}{2}\right)}$$
$$= \frac{\Gamma^2\left(m+\frac{1}{2}\right)}{\Gamma^2(m+1)}\prod_{n=0}^{\infty}\left[1+\frac{\tau^2}{\left(n+m+\frac{1}{2}\right)^2}\right]^{-1}. \qquad (6.146)$$

From Field's formula (see Appendix, (B.7)) we may deduce that, as $m \to \infty$,

$$\frac{\Gamma^2\left(m+\frac{1}{2}\right)}{\Gamma^2(m+1)} = \frac{1}{m}\left(1-\frac{1}{4m}+O(m^{-2})\right). \qquad (6.147)$$

Moreover, it is easy to make the estimate

$$\prod_{n=0}^{\infty}\left[1+\frac{\tau^2}{\left(n+m+\frac{1}{2}\right)^2}\right]^{-1} = 1-\tau^2\sum_{n=0}^{\infty}\frac{1}{\left(n+m+\frac{1}{2}\right)^2}+O(m^{-2})$$
$$= 1-\frac{\tau^2}{m}+O(m^{-2}), \qquad (6.148)$$

as $m \to \infty$. Finally, from the definition of the Gaussian hypergeometric series it is easily verified that the product of the hypergeometric factors occurring in (6.145) is

$$1+\left(\frac{1}{4}+\tau^2\right)m^{-1}+O(m^{-2}) \qquad (6.149)$$

as $m \to \infty$. Combining these estimates shows that

$$G_m(\tau,\theta_0) = \pi^{-1}\cosh(\pi\tau)\left(1+O(m^{-2})\right), \qquad (6.150)$$

as $m \to \infty$.

We therefore introduce the parameter

$$\varepsilon_m(\tau) = 1 - \pi \operatorname{sech}(\pi\tau) G_m(\tau, \theta_0) = O(m^{-2}), \tag{6.151}$$

and rewrite the dual series (6.142) and (6.143) in the form

$$\sum_{m=1}^{\infty} m F_m(\tau) \cos(m\varphi) = -\frac{f_0(\tau)}{2P_{-\frac{1}{2}+i\tau}(-\cos\theta_0)}, \ \varphi \in (0, \varphi_0), \tag{6.152}$$

$$\begin{aligned}\sum_{m=1}^{\infty} F_m(\tau) \cos(m\varphi) &= \frac{\pi}{2} \operatorname{sech}(\pi\tau) \left[1 - f_0(\tau)\right] P_{-\frac{1}{2}+i\tau}(\cos\theta_0) \\ &+ \sum_{m=1}^{\infty} \varepsilon_m(\tau) F_m(\tau) \cos(m\varphi), \varphi \in (\varphi_0, \pi).\end{aligned} \tag{6.153}$$

When the slot in the spindle closes ($\varphi_0 \to 0$), it may be verified that

$$f_0(\tau) = 1, \ f_m(\tau) = 0 \ (m > 0)$$

and the solution reduces to that which was previously obtained (see (6.126) and (6.128)). It is clear that the dual series (6.152) and (6.153) may be solved by the standard technique for trigonometric kernels outlined in Section 2.2. It is conveniently done by substituting $\varphi = \pi - \vartheta$ and replacing $F_m(\tau) = (-1)^m F_m^*(\tau)$. Then $\vartheta_0 = \pi - \varphi_0$ is the angular half-width of the conductor surface (rather than the slot).

The solution may now be deduced from the dual series (2.39), (2.40) and their solution (2.61), (2.62) with the following identification of values:

$$\begin{aligned} m &= n, \ F_m^*(\tau) = x_n, \ f_0(\tau) = x_0, \ q_n = \varepsilon_n(\tau); \\ a &= \left\{2P_{-\frac{1}{2}+i\tau}(-\cos\theta_0)\right\}^{-1}, \ b = \frac{\pi}{2} \operatorname{sech}(\pi\tau) P_{-\frac{1}{2}+i\tau}(\cos\theta_0); \\ g_0 &= \frac{\pi}{2} \operatorname{sech}(\pi\tau) P_{-\frac{1}{2}+i\tau}(\cos\theta_0); \end{aligned} \tag{6.154}$$

the remaining parameters (g_n, r_n, f_n, f_0) all vanish.

The capacitance of the slotted spindle may now be deduced. According to Bouwkamp's theorem, it is

$$C = e^{2\sigma'} \psi(\sigma', \pi, 0) = R \int_0^{\infty} \frac{P_{-\frac{1}{2}+i\tau}(\cos\theta_0)}{\cosh(\pi\tau) P_{-\frac{1}{2}+i\tau}(-\cos\theta_0)} f_0(\tau) d\tau. \tag{6.155}$$

Some further details about the calculation of this value are provided in the next section where the slotted charged sphere is considered.

6.4 A spherical shell with an azimuthal slot

As remarked in the previous section, when $\theta_0 = \frac{1}{2}\pi$, the slotted spindle degenerates to a spherical shell with an azimuthal slot. The image under the inversion described in that section is not a cone with a sectoral slot but is, more simply, a plane with a sectoral cut of half-width φ_0. (See Figure 6.5.) A case of particular interest is the hemispherical shell and its image, the half-plane (occurring when $\varphi_0 = \frac{1}{2}\pi$). The capacitance of the hemisphere was computed in Section 1.4 to be

$$C = a\left(\frac{1}{2} + \frac{1}{\pi}\right).$$

It provides a benchmark value for spherical shells with sectoral slots of arbitrary angle. When $\theta_0 = \frac{1}{2}\pi$, the parameter $\varepsilon_m(\tau)$ introduced in (6.151) may be written in the form

$$\varepsilon_m(\tau) = 1 - \frac{1}{2}m\frac{\left|\Gamma\left(\frac{1}{4} + \frac{1}{2}i\tau + \frac{1}{2}m\right)\right|^2}{\left|\Gamma\left(\frac{3}{4} + \frac{1}{2}i\tau + \frac{1}{2}m\right)\right|^2}. \tag{6.156}$$

Although the parameter has a simpler form than when $0 < \theta_0 < \frac{1}{2}\pi$, it is still not possible to solve the associated potential problem in a closed form. However, it is possible to obtain some analytical approximations in two limiting cases: the narrow cut ($\varphi_0 \ll 1$) and the narrow sectoral conductor ($\vartheta_0 = \pi - \varphi_0 \ll 1$). The problem has some similarities with the azimuthally slotted degenerate torus treated in Chapter 5, and so some repetitious details will be suppressed.

Setting $\theta_0 = \frac{1}{2}\pi$, it follows from (6.155) that the capacitance for a spherical shell with an azimuthal slot is

$$C = R\int_0^\infty \operatorname{sech}(\pi\tau) f_0(\tau) d\tau. \tag{6.157}$$

Making use of the identification (6.154), we may recognise that

$$f_0(\tau) = \left\{b(\tau) - a(\tau)\ln\left[\frac{1-t_0}{2}\right]\right\}^{-1} \times \left\{b(\tau) + \frac{1+t_0}{2}\sum_{n=1}^{\infty}\left(\frac{2}{n}\right)^{\frac{1}{2}} F_n^*(\tau)\varepsilon_n(\tau)\hat{P}_{n-1}^{(0,1)}(t_0)\right\} \tag{6.158}$$

where the parametric dependence of $a = a(\tau)$ and $b = b(\tau)$ in (6.154) is made explicit; $t_0 = \cos\vartheta_0$.

When $\varphi_0 = \pi - \vartheta_0 \ll 1$, it is readily observed from (6.158) that

$$f_0(\tau) = b(\tau)\left\{b(\tau) - 2a(\tau)\ln\cos\frac{\varphi_0}{2}\right\}^{-1} + O(\varphi_0^2). \tag{6.159}$$

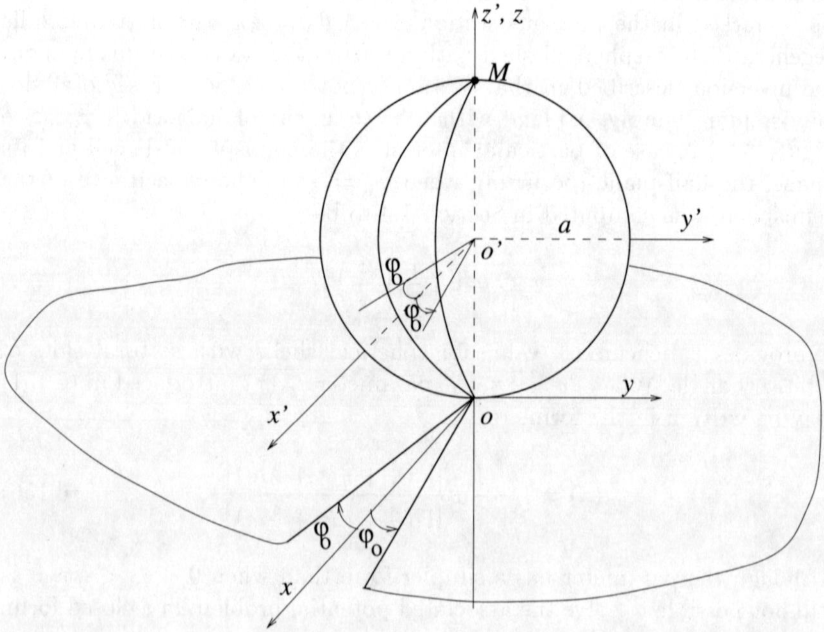

FIGURE 6.5. A spherical shell with an azimuthal slot; its image under inversion is the xOy plane with a sectoral slot removed.

When the spindle closes, the function $f_0(\tau)$ becomes 1; thus if we define

$$\varepsilon = -\ln\cos\frac{\varphi_0}{2},$$

then

$$f_0(\tau) = 1 - \frac{2a(\tau)\varepsilon}{\{b(\tau) + 2a(\tau)\varepsilon\}} + O(\varphi_0^2) = 1 - 2\frac{a(\tau)}{b(\tau)}\varepsilon\left\{1 + O(\varepsilon)\right\} \quad (6.160)$$

when the slot is *logarithmically narrow* ($\varepsilon \ll 1$).

When the conductor is a narrow sector ($\vartheta_0 \ll 1$), we may deduce from (6.158) that

$$f_0(\tau) = \left\{b(\tau) - 2a(\tau)\ln\sin\frac{\vartheta_0}{2}\right\}^{-1}\left\{b(\tau) + \sum_{n=1}^{\infty} F_n^*(\tau)\varepsilon_n(\tau) + O(\vartheta_0^2)\right\}, \quad (6.161)$$

where the functions F_n^* may be approximated as the solution of the infinite system (with a confluent matrix)

$$F_m^*(\tau) - \frac{2a(\tau)}{b(\tau) - 2a(\tau)\ln\sin\frac{1}{2}\vartheta_0}\frac{1}{m}\sum_{n=1}^{\infty} F_n^*(\tau)\varepsilon_n(\tau) = \frac{1}{m}\frac{2a(\tau)b(\tau)}{b(\tau) - 2a(\tau)\ln\sin\frac{1}{2}\vartheta_0}. \quad (6.162)$$

This system can be solved by multiplying both sides of (6.162) by $\varepsilon_m(\tau)$ and summing over m. Thus

$$\sum_{n=1}^{\infty} F_n^*(\tau)\varepsilon_n(\tau) = \frac{2a(\tau)b(\tau)A(\tau)}{b(\tau) - 2a(\tau)\left\{A(\tau) + \ln\sin\frac{1}{2}\vartheta_0\right\}}, \quad (6.163)$$

where

$$A(\tau) = \sum_{m=1}^{\infty} \frac{\varepsilon_m(\tau)}{m}. \quad (6.164)$$

Insertion of (6.163) in (6.161) shows that

$$f_0(\tau) = \frac{b(\tau)}{b(\tau) - 2a(\tau)\left\{A(\tau) + \ln\sin\frac{1}{2}\vartheta_0\right\}}\left\{1 + O(\vartheta_0^2)\right\}. \quad (6.165)$$

If we introduce the parameter

$$\varepsilon' = -\left\{\ln\sin\frac{1}{2}\vartheta_0\right\}^{-1},$$

then

$$f_0(\tau) = \frac{b(\tau)}{2a(\tau)}\varepsilon'\left\{1 + O(\varepsilon')\right\}, \quad (6.166)$$

when the sector is logarithmically narrow ($\varepsilon' \ll 1$). Thus the capacitance of the logarithmically narrow slot is

$$C_1 = C_0 - \frac{2}{\pi}\varepsilon R\int_0^{\infty}\left\{P_{-\frac{1}{2}+i\tau}(-\cos\theta_0)\right\}^{-2} d\tau + O(\varepsilon^2), \quad (6.167)$$

where C_0 is the capacitance of the corresponding closed spindle (see (6.129)), and the capacitance of the logarithmically narrow sector is

$$C_2 = \frac{\pi}{2}\varepsilon' R\int_0^{\infty} \operatorname{sech}^2(\pi\tau)\left\{P_{-\frac{1}{2}+i\tau}(\cos\theta_0)\right\}^2 d\tau + O(\varepsilon'^2). \quad (6.168)$$

When $\theta_0 = \frac{1}{2}\pi$, tabulated values of the integrals occurring in (6.167) and (6.168) are $\frac{1}{16}\pi^2$ and $\frac{1}{2}$, respectively (see [15]), so that

$$C_1/C_0 = 1 - \frac{\pi}{4}\varepsilon + O(\varepsilon^2),\ C_2/C_0 = \frac{\pi}{2}\varepsilon' + O(\varepsilon'^2). \tag{6.169}$$

Now consider the needle-shaped spindle ($\theta_0 \ll 1$) with a logarithmically narrow slot ($\varepsilon \ll 1$) and, in addition, suppose that $\varphi_0 \ll \theta_0$. The approximation for the capacitance is

$$\frac{C_1}{C_0} \simeq \frac{1}{\ln(2/\theta_0)}\left\{1 - \frac{\varepsilon}{\ln(2/\theta_0)}\right\}. \tag{6.170}$$

When $\theta_0 \ll 1$ and $\vartheta_0 \ll 1$, the structure very nearly becomes a straight finite strip with some variable width and its capacitance is approximately

$$\frac{C_2}{C_0} \simeq \varepsilon'. \tag{6.171}$$

Comparing (6.169) and (6.171), we may recognise the difference of a factor of $\frac{1}{2}\pi$ in capacitance between the spherically curved crescent-shaped strip (6.169) and its flat analogy (6.171). On the other hand, the characteristic factor of $\{\ln(2/\theta_0)\}^{-1}$ present in (6.170) is notably absent in (6.171). It is therefore important to recognise that these approximations are not uniformly valid in the problem parameters, and that the regime of their validity is best delineated by numerical methods; nonetheless, the approximations are useful at the extreme limit of the parameter range.

7
Two-dimensional Potential Theory

Historically, two-dimensional potential problems have been studied more extensively than have three-dimensional problems. Apart from the apparent simplicity of lower dimension, the main reasons are that powerful methods, based upon conformal mapping techniques and the well-developed theory of analytic functions, are available in the plane; these provide rather clear procedures to facilitate the solution of mixed boundary value problems in potential theory.

Basically, analytic function theory techniques reduce the potential problem to the well-known Riemann-Hilbert problem of the determination of an analytical function on some contour bounding a domain [45]; various concrete applications of this technique can be found in [18] and [53]. Applications of the *conformal mapping method* are so numerous that classic texts on electromagnetic theory invariably describe and solve a variety of electrostatic problems with this technique (see, for example, [54, 66]).

Despite the lower dimension, it should be observed that boundary value problems in two-dimensional potential theory involve an additional abstraction compared to that for three-dimensional bodies of finite extent, even for open surfaces with sharp edges. Whilst it is reasonable to imagine an extremely long, but at the same time finite conductor charged to some potential, its extension to infinity, at the same constant potential as for the finite conductor, raises some questions about the physical reality or relevance of the model. A physicist might reasonably question the source of infinite energy needed to charge this infinitely long conductor.

It is not surprising, then, that two-dimensional potential problems, even properly stated, require some *nonphysical* behaviour of the potential function at infinity. This manifests itself as a logarithmic dependence on distance from the conductors, so the potential is unbounded at infinity. Although strange from the physical perspective, the mathematical issue simply concerns the choice of the class of functions required for a satisfactory two-dimensional potential theory. Generally speaking, if the conductor is modelled as an infinitely long object of constant cross-section, the basic postulates of potential theory force a logarithmic increase to solutions at large distances from the conductor.

Some simple illustrative examples will indicate distinctive features of two-dimensional potentials. The electrostatic potential $\psi(\vec{r})$, due to some electrified conductor held at unit potential in two dimensions, is defined by the single-layer potential [66]

$$\psi(\vec{r}) = -\frac{1}{2\pi}\int_L \log\left|\vec{r}-\vec{r'}\right| \sigma\left(\vec{r'}\right) dl \tag{7.1}$$

where $G_2(\vec{r},\vec{r}') = -(1/2\pi)\log|\vec{r}-\vec{r}'|$ is two-dimensional Green's function, σ is the linear charge density on the cross-sectional contour L, dl is the element of the contour integral, and $\vec{r}, \vec{r}'$ are position vectors of observation points and points on L, respectively.

We will consider a variety of canonical structures that are infinite cylinders of constant cross-section, into which apertures are introduced to produce longitudinally slotted cylinders (the edges of the slots are parallel to the cylindrical axis).

By way of introduction, we consider the circular arc (Section 7.1), and then circular cylinders with multiple slots (Section 7.2), various configurations of thin strips (Section 7.3), and elliptic cylinders with multiple slots (Section 7.4). In Section 7.5, a singly-slotted cylinder with arbitrary cross-section is considered. Although this structure is noncanonical, our purpose is to demonstrate how to regularise the integral equations of potential theory in a rather more general setting than the simpler canonical structures discussed in the earlier sections. The process transforms the integral equations to a second-kind system of equations with its attendant benefits: a well-conditioned system of equations for numerical solution after truncation.

7.1 The circular arc

Consider an infinitely long, singly-slotted circular cylinder whose cross-section is an arc L of a circle of radius a (see Figure 7.1). Polar coordinates (r, φ), where $r = \rho/a$, are convenient for this configuration. Assume that the right half of the arc (given by $\varphi \in (0, \varphi_0)$) is charged to unit potential, but the left half (given by $\varphi \in (-\varphi_0, 0)$) is charged either to unit positive or negative value, i.e., $\psi(1,\varphi) = (-1)^l$, $(l = 0, 1)$. If σ^l denotes the charge

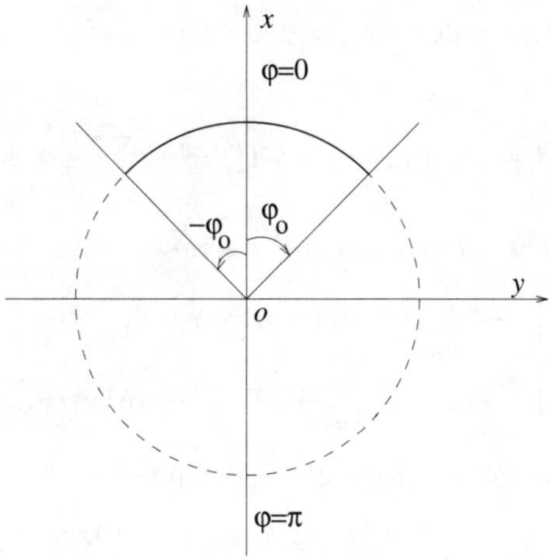

FIGURE 7.1. The circular arc.

distribution on L, it is evident that $\sigma^l(-\varphi) = (-1)^l \sigma^l(\varphi)$. (When $l = 1$, an infinitesimally small insulating gap is placed at $\varphi = 0$.) Then the potential can be represented as

$$\begin{aligned}\psi^l(r,\varphi) &= -\frac{1}{4\pi}\int_{-\varphi_0}^{\varphi_0} \log\left|1 - 2r\cos(\varphi - \varphi') + r^2\right| \sigma^l(\varphi')\, d\varphi' \\ &= -\frac{1}{4\pi}\int_0^{\varphi_0} K^l(r,\varphi,\varphi')\sigma^l(\varphi')\, d\varphi', \qquad (7.2)\end{aligned}$$

where

$$K^l(r,\varphi,\varphi') = \log\left|1 - 2r\cos(\varphi - \varphi') + r^2\right| + (-1)^l \log\left|1 - 2r\cos(\varphi + \varphi') + r^2\right|. \qquad (7.3)$$

When $l = 0$, it can be readily shown that

$$\psi^l(r,\varphi) - \frac{1}{2\pi} q \log\left(r^{-1}\right)$$

is a regular harmonic function, as $r \to \infty$, where

$$q = 2\int_0^{\varphi_0} \sigma^0(\varphi')\, d\varphi'$$

is the (total) charge per unit length; when $l = 1, \psi^l(r,\varphi)$ behaves as a regular harmonic function, as $r \to \infty$.

First consider the uniformly charged strip $(l=0)$. Using the expansion [19],

$$\log(1-2t\cos(\varphi-\varphi')+t^2) = -2\sum_{k=1}^{\infty}\frac{t^k}{k}\cos kx, \tag{7.4}$$

the kernel of 7.2 has the cosine Fourier series

$$K^0(r,\varphi,\varphi') = \begin{cases} -4\sum_{n=1}^{\infty} n^{-1}r^n\cos n\varphi\cos n\varphi', & r<1, \\ 4\log r - 4\sum_{n=1}^{\infty} n^{-1}r^{-n}\cos n\varphi\cos n\varphi', & r>1. \end{cases} \tag{7.5}$$

Extend the domain of definition of σ^0, defining

$$\sigma^0_{tot}(\varphi') = \begin{cases} \sigma^0(\varphi'), & \varphi'\in(0,\varphi_0) \\ 0, & \varphi'\in(\varphi_0,\pi) \end{cases}; \tag{7.6}$$

this even function has a Fourier series expansion

$$\sigma^0_{tot} = \sum_{m=0}^{\infty}(2-\delta^0_m)\,x_m\cos m\varphi', \tag{7.7}$$

with unknown Fourier coefficients x_m to be determined. Substitute these expansions into (7.2) to obtain

$$\psi(r,\varphi) = \begin{cases} \sum_{n=1}^{\infty} n^{-1}x_n r^n\cos n\varphi, & r<1 \\ -x_0\log r + \sum_{n=1}^{\infty} n^{-1}x_n r^{-n}\cos n\varphi, & r>1 \end{cases}. \tag{7.8}$$

This representation can also be obtained by the method of separation of variables applied directly to Laplace's equation.

The boundary condition at $r=1$, $\varphi\in(0,\varphi_0)$ is

$$\psi(1+0,\varphi) = \psi(1-0,\varphi) = 1; \tag{7.9}$$

on the slot $r=1$, $\varphi\in(\varphi_0,\pi)$, the boundary condition, which follows directly from the definition (7.6), is

$$\sigma^0_{tot}(\varphi) = \left.\frac{\partial\psi(r,\varphi)}{\partial r}\right|_{r=1-0}^{r=1+0} = 0. \tag{7.10}$$

Enforcement of these boundary conditions produces the following dual series equations:

$$\sum_{n=1}^{\infty}\frac{1}{n}x_n\cos n\varphi = 1, \qquad \varphi\in(0,\varphi_0) \tag{7.11}$$

$$\sum_{n=1}^{\infty}x_n\cos n\varphi = -\frac{1}{2}x_0, \qquad \varphi\in(\varphi_0,\pi). \tag{7.12}$$

The method developed in Section 2.2 shows that the closed form solution is

$$\begin{aligned} x_0 &= -\left\{\log\left(\sin\frac{\varphi_0}{2}\right)\right\}^{-1}, \\ x_m &= -\frac{1}{2}\left\{\log\left(\sin\frac{\varphi_0}{2}\right)\right\}^{-1}(1+\cos\varphi_0)\,P_{m-1}^{(0,1)}(\cos\varphi_0) \\ &= -\frac{1}{2}\left\{\log\left(\sin\frac{\varphi_0}{2}\right)\right\}^{-1}\{P_m(\cos\varphi_0)+P_{m-1}(\cos\varphi_0)\}, \end{aligned} \tag{7.13}$$

when $m > 0$. The capacitance of the cylindrically shaped strip (per unit length) is thus

$$C = 2\int_0^{\pi}\sigma_{tot}^0(\varphi')\,d\varphi' = 2\pi x_0 = -2\pi\left\{\log\left(\sin\frac{\varphi_0}{2}\right)\right\}^{-1}. \tag{7.14}$$

On the interval $[0,\varphi_0]$, the line charge density equals

$$\sigma_{tot}^0(\varphi) = \frac{1}{4\pi}\left.\frac{\partial\psi(r,\varphi)}{\partial r}\right|_{r=1-0}^{r=1+0} = \frac{1}{4\pi}\sum_{n=0}^{\infty}\left(2-\delta_n^0\right)x_n\cos m\varphi, \tag{7.15}$$

and its value is easily deduced from the discontinuous series (1.109) to be

$$\sigma_{tot}^0(\varphi) = -\frac{1}{2\sqrt{2}\pi}\cos\frac{\varphi}{2}\left\{\log\left(\sin\frac{\varphi_0}{2}\right)\right\}^{-1}(\cos\varphi-\cos\varphi_0)^{-\frac{1}{2}},\ \varphi<\varphi_0; \tag{7.16}$$

it vanishes when $\varphi > \varphi_0$.

When the circular arc comprises oppositely charged halves (Figure 7.1), the potential is bounded; there is no logarithmic term. Physically, the structure is a two-dimensional dipole. Set $l = 1$ in (7.2) and again use expansion (7.4) to obtain

$$K^1(r,\varphi;\varphi') = -4\sum_{n=1}^{\infty}n^{-1}\sin n\varphi\sin n\varphi'\begin{cases} r^n, & r<1. \\ r^{-n}, & r>1. \end{cases} \tag{7.17}$$

As before, introduce the extended or *total* line charge density

$$\sigma_{tot}^1(\varphi') = \begin{cases} \sigma^0(\varphi'), & \varphi'\in(0,\varphi_0) \\ 0, & \varphi'\in(\varphi_0,\pi) \end{cases}, \tag{7.18}$$

and represent this odd function as a Fourier sine series

$$\sigma_{tot}^1(\varphi') = \sum_{m=1}^{\infty}y_m\sin m\varphi'. \tag{7.19}$$

Substitute (7.17) and (7.19) into (7.2) to obtain

$$\psi^1(r,\varphi) = \sum_{n=1}^{\infty} n^{-1} y_n \sin n\varphi \begin{cases} r^n, & r < 1. \\ r^{-n}, & r > 1. \end{cases} \tag{7.20}$$

Enforcing the mixed boundary conditions on the arc $r = 1$ produces the dual series equations for the unknown coefficients y_n:

$$\sum_{n=1}^{\infty} n^{-1} y_n \sin n\varphi = 1, \quad \varphi \in (0, \varphi_0) \tag{7.21}$$

$$\sum_{n=1}^{\infty} y_n \sin n\varphi = 0, \quad \varphi \in (\varphi_0, \pi). \tag{7.22}$$

The solution of these equations (see Section 2.2) is (with $z_0 = \cos\varphi_0$),

$$y_n = \frac{\sqrt{2}}{\pi} n \int_{z_0}^{1} \frac{P_{n-1}^{(0,1)}(t)}{(1-t)^{\frac{1}{2}}} dt = \frac{\sqrt{2}}{\pi} n \int_{z_0}^{1} \frac{P_n(t) + P_{n-1}(t)}{(1+t)(1-t)^{\frac{1}{2}}} dt. \tag{7.23}$$

The format of this solution (7.23) has some rather satisfactory features. For example, one may conveniently calculate the distribution of the potential on the circle $r = 1$ to be

$$\psi(1,\varphi) = 2\pi^{-1} \arctan\left[\sqrt{2}\sin\frac{1}{2}\varphi_0 \cos\frac{1}{2}\varphi \{\cos\varphi_0 - \cos\varphi\}^{-\frac{1}{2}}\right] \tag{7.24}$$

when $\varphi > \varphi_0$; when $0 < \varphi < \varphi_0$, $\psi(1,\varphi) = 1$.

7.2 Axially slotted open circular cylinders

In this section, slotted circular cylinders with multiple apertures are considered. A restricted selection of electrostatic problems that are distinctive of this geometry are examined. Our first calculation is of the electrostatic field due to a pair of charged circular arcs, asymmetrically placed as shown in Figure 7.2. The second calculation is of the field generated by the *quadrupole lens* also shown in Figure 7.2; for the sake of simplicity, when the arcs are all positively charged we restrict attention to the symmetrical case ($\varphi_1 = \pi - \varphi_0$).

As in the previous section, the conductors lie on the contour of the unit circle and are charged to potentials $V_1 = 1$ and $V_2 = (-1)^l$ as shown in Figure 7.2; the index $l = 0$ or 1. The potential associated with the pair of charged circular arcs, at potentials V_1 (defined by $\varphi \in (\varphi_0, \varphi_1)$) and V_2 (defined by $\varphi \in (-\varphi_1, -\varphi_0)$) is

$$\psi^l(r,\varphi) = -\frac{1}{4\pi} \int_{\varphi_0}^{\varphi_1} K_1^l(r,\varphi,\varphi')\, \sigma^l(\varphi')\, d\varphi', \tag{7.25}$$

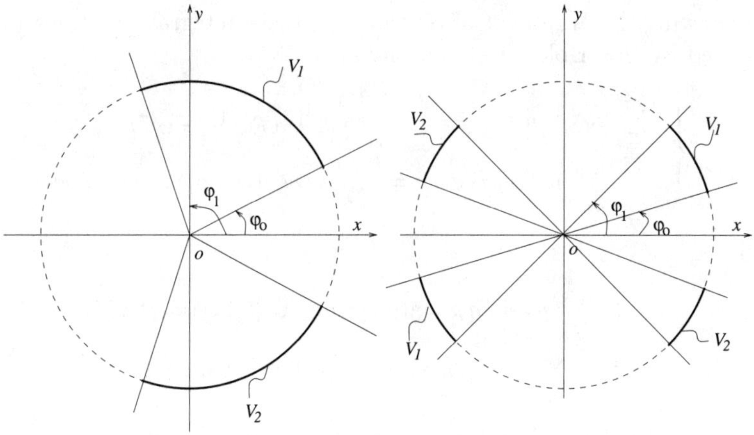

FIGURE 7.2. The circular arc (left) and quadrupole (right).

where $K_1^l = K^l$ is defined by (7.3). For the two-dimensional quadrupole lens in which the pair of arcs defined by $\varphi \in (\varphi_0, \varphi_1) \cup (-(\pi - \varphi_0), -(\pi - \varphi_1))$ is held at potential V_1, and the pair of arcs defined by $\varphi \in (-\varphi_0, -\varphi_1) \cup (\pi - \varphi_0, \pi - \varphi_1)$ is held at potential V_2, the potential is

$$\psi^l(r, \varphi) = -\frac{1}{4\pi} \int_{\varphi_0}^{\varphi_1} K_2(r, \varphi, \varphi') \sigma^l(\varphi') \, d\varphi' \tag{7.26}$$

where

$$K_2(r, \varphi, \varphi') = \log\left[\left(r^2+1\right)^2 - 4r^2\cos^2(\varphi - \varphi')\right] + (-1)^l \log\left[\left(r^2+1\right)^2 - 4r^2\cos(\varphi + \varphi')\right].$$

First consider the pair of charged arcs. Enforcement of the boundary condition

$$\psi^l(1+0, \varphi) = \psi^l(1-0, \varphi) = 1, \qquad \varphi \in (\varphi_0, \varphi_1),$$

(and the corresponding condition on the arc at potential V_2) produces a first-kind Fredholm integral equation for the unknown charge density σ^l. On equicharged arcs ($l = 0$), the density σ^0 satisfies

$$-\frac{1}{2\pi} \int_{\varphi_0}^{\varphi_1} \sigma^0(\varphi') \ln 4 \left| \sin^2 \frac{1}{2}\varphi - \sin^2 \frac{1}{2}\varphi' \right| d\varphi' = 1, \qquad \varphi \in (\varphi_0, \varphi_1), \tag{7.27}$$

whereas on oppositely charged arcs ($l = 1$), the density σ^1 satisfies

$$\frac{1}{2\pi} \int_{\varphi_0}^{\varphi_1} \sigma^1(\varphi') \log \left| \frac{\tan \frac{1}{2}\varphi + \tan \frac{1}{2}\varphi'}{\tan \frac{1}{2}\varphi - \tan \frac{1}{2}\varphi'} \right| d\varphi' = 1, \qquad \varphi \in (\varphi_0, \varphi_1). \tag{7.28}$$

Following the argument of Section 7.1, these integral equations may be replaced by the triple series equations

$$\left\{ \begin{array}{ll} \sum_{n=1}^{\infty} x_n \cos n\varphi = -\tfrac{1}{2} x_0, & \varphi \in (0, \varphi_0) \cup (\varphi_1, \pi), \\ \sum_{n=1}^{\infty} n^{-1} x_n \cos n\varphi = 1, & \varphi \in (\varphi_0, \varphi_1), \end{array} \right. \tag{7.29}$$

and

$$\left\{ \begin{array}{ll} \sum_{n=1}^{\infty} y_n \sin n\varphi = 0, & \varphi \in (0, \varphi_0) \cup (\varphi_1, \pi), \\ \sum_{n=1}^{\infty} n^{-1} y_n \sin n\varphi = 1, & \varphi \in (\varphi_0, \varphi_1), \end{array} \right. \tag{7.30}$$

where the densities σ^0 and σ^1 are respectively expanded in cosine and sine Fourier series,

$$\sigma^0(\varphi') = \sum_{m=0}^{\infty} \left(2 - \delta_m^0\right) x_m \cos m\varphi', \tag{7.31}$$

$$\sigma^1(\varphi') = \sum_{n=1}^{\infty} y_n \sin n\varphi', \tag{7.32}$$

with unknown coefficients $\{x_n\}_{n=0}^{\infty}$ and $\{y_n\}_{n=1}^{\infty}$.

The symmetric situation ($\varphi_1 = \pi - \varphi_0$) is quickly solved. The odd index coefficients all vanish, and the Equations (7.29) reduce to the following dual series equations for the even index coefficients x_{2n},

$$\begin{array}{ll} \sum_{n=1}^{\infty} x_{2n} \cos n\vartheta = -\tfrac{1}{2} x_0, & \vartheta \in (0, \vartheta_0), \\ \sum_{n=1}^{\infty} n^{-1} x_{2n} \cos n\vartheta = 2, & \vartheta \in (\vartheta_0, \pi), \end{array} \tag{7.33}$$

where $\vartheta = 2\varphi$ and $\vartheta_0 = 2\varphi_0$. The substitution $\vartheta = \pi - \theta$ transforms (7.33) to

$$\begin{array}{ll} \sum_{n=1}^{\infty} n^{-1} X_n \cos n\theta = 2, & \theta \in (0, \theta_0), \\ \sum_{n=1}^{\infty} X_n \cos n\theta = -\tfrac{1}{2} x_0, & \theta \in (\theta_0, \pi), \end{array} \tag{7.34}$$

where $X_n = (-1)^n x_{2n}$, $\theta_0 = \pi - \vartheta_0 = \pi - 2\varphi_0$, and $\theta = \pi - 2\varphi$. Comparing Equations (7.11) and (7.12) with (7.34), the solution of this symmetric case is

$$\begin{array}{c} x_0 = -2 \left\{\log\left(\cos\varphi_0\right)\right\}^{-1}, \\ x_{2n} = -\left\{\log\left(\cos\varphi_0\right)\right\}^{-1} \left\{P_n\left(\cos\varphi_0\right) - P_{n-1}\left(\cos\varphi_0\right)\right\}. \end{array} \tag{7.35}$$

Now consider (7.30) in the more general case, in which the parameters φ_0 and φ_1 are unrelated, taking arbitrary values in $(0, \pi)$, with $\varphi_0 < \varphi_1$. The *standard* form of triple series equations involving the trigonometrical kernels $\{\sin n\varphi\}_{n=1}^{\infty}$ is obtained by setting $y_n = na_n$; Equations (7.30) become

$$\begin{aligned} \sum_{n=1}^{\infty} na_n \sin n\varphi &= 0, & \varphi &\in (0, \varphi_0) \cup (\varphi_1, \pi), \\ \sum_{n=1}^{\infty} a_n \sin n\varphi &= 1, & \varphi &\in (\varphi_0, \varphi_1). \end{aligned} \tag{7.36}$$

From the results of Section 2.7, Equations (7.36) are equivalent to the symmetric triple equations

$$\begin{aligned} \sum_{n=1}^{\infty} nb_n \sin n\vartheta &= 0, & \vartheta &\in (0, \vartheta_0) \cup (\pi - \vartheta_0, \pi), \\ \sum_{n=1}^{\infty} b_n \sin n\vartheta &= \left(\tan \tfrac{1}{2}\varphi_0 \tan \tfrac{1}{2}\varphi_1\right)^{\frac{1}{2}}, & \vartheta &\in (\vartheta_0, \pi - \vartheta_0), \end{aligned} \tag{7.37}$$

where $\tan \frac{1}{2}\vartheta_0 = \tan \frac{1}{2}\varphi_0 \cot \frac{1}{2}\varphi_1$.

In turn these equations may be reduced to the following dual series equations for the odd index Fourier coefficients b_{2n+1},

$$\begin{aligned} \sum_{n=0}^{\infty} \left(n + \tfrac{1}{2}\right) b_{2n+1} \sin\left(n + \tfrac{1}{2}\right)\theta &= 0, & \theta &\in (0, \theta_0) \\ \sum_{n=0}^{\infty} b_{2n+1} \sin\left(n + \tfrac{1}{2}\right)\theta &= \left(\tan \tfrac{1}{2}\varphi_0 \tan \tfrac{1}{2}\varphi_1\right)^{\frac{1}{2}}, & \theta &\in (\theta_0, \pi) \end{aligned} \tag{7.38}$$

where $\theta = 2\vartheta$, and $\theta_0 = 2\vartheta_0$; all the even index coefficients b_{2n} vanish.

It should be noted that original coefficients $\{a_n\}_{n=1}^{\infty}$ are related to $\{b_n\}_{n=1}^{\infty}$ by (2.264). By means of the Abel integral transform, we deduce from Equations (7.38) that

$$\sum_{n=0}^{\infty} b_{2n+1} P_n(z) = \frac{2}{\pi} \left(\tan \frac{\varphi_0}{2} \tan \frac{\varphi_1}{2}\right)^{\frac{1}{2}} \begin{cases} F_1(z), & z \in (-1, z_0), \\ F_2(z), & z \in (z_0, 1), \end{cases} \tag{7.39}$$

where $z = \cos\theta$ and $z_0 = \cos\theta_0$, and

$$\begin{aligned} F_1(z) &= K\left(\sqrt{\frac{1+z}{2}}\right), \\ F_2(z) &= K\left(\sqrt{\frac{1-z}{2}}\right) K\left(\sqrt{\frac{1+z_0}{2}}\right) / K\left(\sqrt{\frac{1-z_0}{2}}\right). \end{aligned}$$

(K is the complete elliptic integral of the first kind.) Orthogonality of the Legendre polynomials on $[-1,1]$ instantly implies

$$b_{2n+1} = \frac{2}{\pi}\left(\tan\frac{\varphi_0}{2}\tan\frac{\varphi_1}{2}\right)^{\frac{1}{2}}\left(n+\frac{1}{2}\right)\times$$
$$\left\{\int_{-1}^{z_0} F_1(z)P_n(z)\,dz + \int_{z_0}^{1} F_2(z)P_n(z)\,dz\right\}. \quad (7.40)$$

The integrals occurring in (7.40) are readily calculated, if one recalls the relationship between the complete elliptic integrals and the Legendre functions

$$P_{-\frac{1}{2}}(z) = \frac{2}{\pi}K\left(\sqrt{\frac{1-z}{2}}\right), Q_{-\frac{1}{2}}(z) = K\left(\sqrt{\frac{1+z}{2}}\right). \quad (7.41)$$

An integration by parts and use of the differential equation for the Legendre functions produces the compact result

$$b_{2n+1} = \left(\tan\frac{\varphi_0}{2}\tan\frac{\varphi_1}{2}\right)^{\frac{1}{2}}\left\{\left(n+\frac{1}{2}\right)K\left(\sqrt{\frac{1-z_0}{2}}\right)\right\}^{-1} P_n(z_0). \quad (7.42)$$

The capacitance of these oppositely charged circular arcs equals

$$C = \frac{1}{4\pi}\sum_{n=0}^{\infty} a_{2n+1} = \frac{1}{4\pi}\left(\tan\frac{\varphi_0}{2}\tan\frac{\varphi_1}{2}\right)^{\frac{1}{2}}\sum_{n=0}^{\infty} b_{2n+1}$$
$$= \frac{1}{2\pi}K\left(\sqrt{\frac{1+z_0}{2}}\right)/K\left(\sqrt{\frac{1-z_0}{2}}\right). \quad (7.43)$$

In terms of the original parameters this capacitance is

$$C = \frac{1}{2\pi}K\left(\frac{1-q}{1+q}\right)/K\left(\frac{2\sqrt{q}}{1+q}\right) = \frac{1}{2\pi}K(t)/K\left(\sqrt{1-t^2}\right), \quad (7.44)$$

where $q = \tan\frac{1}{2}\varphi_0 \cot\frac{1}{2}\varphi_1$, and $t = \sin\frac{1}{2}(\varphi_1-\varphi_0)/\sin\frac{1}{2}(\varphi_1+\varphi_0)$.

The calculation of the line charge density σ based on the evident relationship between the transformed series (see Section 2.7) is

$$\sigma(\varphi) = \frac{1}{4\pi}\sum_{n=1}^{\infty} n a_n \sin n\varphi$$
$$= \frac{1}{4\pi}\left(\sin^2\frac{\varphi}{2} + \tan\frac{\varphi_0}{2}\tan\frac{\varphi_1}{2}\cos^2\frac{\varphi}{2}\right)^{-1}\sum_{n=1}^{\infty} n b_n \sin n\vartheta \quad (7.45)$$

where

$$\tan\frac{1}{2}\vartheta = \tan\frac{\varphi}{2}\left\{\tan\frac{\varphi_0}{2}\tan\frac{\varphi_1}{2}\right\}^{-\frac{1}{2}}.$$

Suppressing the details of an uncomplicated but bulky transformation, we deduce the final formula for the charge density on the asymmetric disposed arcs to be

$$\sigma(\varphi) = \frac{1}{K\left(\sqrt{1-t^2}\right)}\frac{\sin\frac{1}{2}(\varphi_1+\varphi_0)}{\sqrt{(\cos\varphi_0-\cos\varphi)(\cos\varphi-\cos\varphi_1)}} \tag{7.46}$$

where the parameter t was defined above.

Finally, consider the quadrupole lens charged so that the potentials $V_1 = -V_2 = 1$ $(l = 1)$. From (7.26) the electrostatic potential is

$$\psi^1(r,\varphi) = -\frac{1}{4\pi}\int_{\varphi_0}^{\varphi_1}\log\left[\frac{1-2r^2\cos(2\varphi-2\varphi')+r^4}{1-2r^2\cos(2\varphi+2\varphi')+r^4}\right]\sigma^1(\varphi')\,d\varphi'. \tag{7.47}$$

We expand the kernel of Equation (7.47) as

$$-4\sum_{n=1}^{\infty} n^{-1}\sin 2n\varphi\sin 2n\varphi'\begin{cases} r^{2n}, & r<1 \\ r^{-2n}, & r>1 \end{cases} \tag{7.48}$$

and line charge density as a Fourier sine series

$$\sigma(\varphi') = \sum_{n=1}^{\infty} y_{2n}\sin 2n\varphi', \tag{7.49}$$

so that (cf. (7.20))

$$\psi^1(r,\varphi) = \sum_{n=1}^{\infty} n^{-1}y_{2n}\sin 2n\varphi\begin{cases} r^{2n}, & r<1 \\ r^{-2n}, & r>1 \end{cases}. \tag{7.50}$$

By the same argument as above, we obtain triple series equations for the coefficients $\{y_{2n}\}_{n=1}^{\infty}$,

$$\begin{aligned} \sum_{n=1}^{\infty} y_{2n}\sin 2n\phi &= 0, & \phi &\in (0,\phi_0)\cup(\phi_1,\pi), \\ \sum_{n=1}^{\infty} n^{-1}y_{2n}\sin 2n\phi &= 1, & \phi &\in (\phi_0,\phi_1), \end{aligned} \tag{7.51}$$

where $\phi = 2\varphi$, $\phi_0 = 2\varphi_0$, and $\phi_1 = 2\varphi_1$.

Equations (7.30) and (7.51) are the same, so that the solution of (7.51) is given by (7.42) with replacement of the parameters ϕ_0 and ϕ_1 by $2\varphi_0$ and $2\varphi_1$ respectively. With this replacement, Formulae (7.44) and (7.46) hold for the quadrupole lens.

In principle, more complicated configurations of cylindrical strips lying on the contour of a circle may be tackled by this approach. The resulting series equations are naturally more complex, but considerable simplification occurs if the components are symmetrically located.

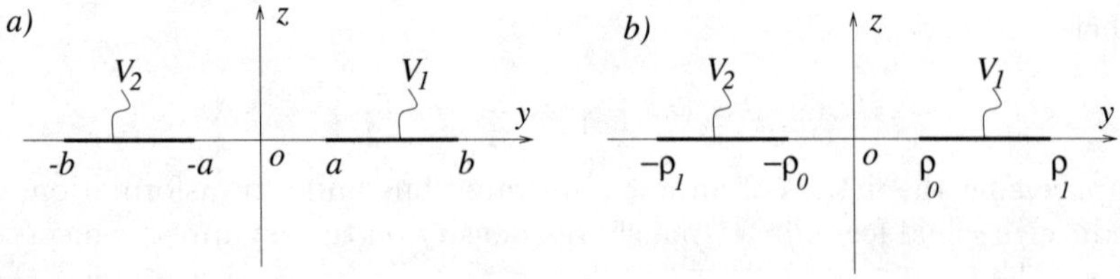

FIGURE 7.3. Pairs of charged thin strips.

7.3 Electrostatic potential of systems of charged thin strips

In many respects, potential problems for flat strips are similar to those for cylindrically-shaped strips. A notable difference is the extraction of *zero terms* in the functional equations with continuous spectrum (integral equations), which is analogous to the extraction of zero-order Fourier coefficients in series equations.

Let us consider the canonical example of the pair of charged coplanar flat strips shown in Figure 7.3; the strips occupy the regions $a \leq y' \leq b$, $-b \leq y' \leq -a$ and are charged to potentials $V_1 = 1, V_2 \, (-1)^l$ respectively, where $l = 0$ or 1.

It is convenient to solve this problem in rescaled Cartesian coordinates (ρ, z) derived from standard coordinates (y', z') by $\rho = y' / (ab)^{\frac{1}{2}}$; $z = z' / (ab)^{\frac{1}{2}}$. Thus the strips occupy the regions $\rho_0 \leq \rho \leq \rho_1, -\rho_1 \leq \rho \leq -\rho_0$, where $\rho_0 = (a/b)^{\frac{1}{2}}$ and $\rho_1 = \rho_0^{-1} = (b/a)^{\frac{1}{2}}$.

The total potential ψ is the sum of single-layer potentials ψ_1, ψ_2 derived from the right-half plane and left-half plane strips, respectively:

$$\psi(\rho, z) = \psi_1(\rho, z) + \psi_2(\rho, z), \tag{7.52}$$

where

$$\psi_1(\rho, z) = -\frac{1}{4\pi} \int_{\rho_0}^{\rho_1} \ln\left[(\rho - \rho')^2 + z^2\right] \sigma_1(\rho') \, d\rho', \tag{7.53}$$

$$\psi_2(\rho, z) = -\frac{1}{4\pi} \int_{-\rho_1}^{-\rho_0} \ln\left[(\rho - \rho')^2 + z^2\right] \sigma_2(\rho') \, d\rho'. \tag{7.54}$$

The symmetry of the problem instantly implies

$$\begin{aligned}
\sigma_1(\rho') &= \sigma_2(-\rho') \overset{def}{=} \sigma^0(\rho') \quad \text{if } l = 0, \\
\sigma_1(\rho') &= -\sigma_2(-\rho') \overset{def}{=} \sigma^1(\rho') \quad \text{if } l = 1,
\end{aligned}$$

so that a single representation for the potential is

$$\psi^l(\rho, z) = -\frac{1}{4\pi}\int_{\rho_0}^{\rho_1} K(\rho, z; \rho', z')\sigma^l(\rho')\,d\rho' \tag{7.55}$$

where the kernel

$$K(\rho, z; \rho') = \log\left[\sqrt{(\rho-\rho')^2 + z^2}\right] + (-1)^l \log\left[\sqrt{(\rho+\rho')^2 + z^2}\right].$$

A first-kind Fredholm integral equation for the line charge density σ^l is obtained by enforcement of the boundary condition on the strips,

$$\psi^l(\rho, +0) = \psi^l(\rho, -0) = 1, \qquad \rho \in (\rho_0, \rho_1), \tag{7.56}$$

yielding

$$-\frac{1}{2\pi}\int_{\rho_0}^{\rho_1} \sigma^0(\rho') \log\left|\rho^2 - \rho'^2\right| d\rho' = 1, \qquad \rho \in (\rho_0, \rho_1), \tag{7.57}$$

and

$$\frac{1}{2\pi}\int_{\rho_0}^{\rho_1} \sigma^1(\rho') \log\left|\frac{\rho+\rho'}{\rho-\rho'}\right| d\rho' = 1, \qquad \rho \in (\rho_0, \rho_1). \tag{7.58}$$

We now use familiar mathematical tools to reduce both equations to triple integral equations for some unknown Fourier coefficients. First, represent the logarithmic kernels by their Fourier transforms

$$\begin{aligned} \log\left|\rho^2 - \rho'^2\right| &= 2\log\rho + \log\left|1 - \frac{\rho'^2}{\rho^2}\right| \\ &= 2\log\rho + 2\int_0^\infty \frac{1-\cos\xi\rho'}{\xi}\cos\xi\rho\, d\xi, \end{aligned} \tag{7.59}$$

and

$$\log\left|\frac{\rho+\rho'}{\rho-\rho'}\right| = 2\int_0^\infty \frac{\sin\xi\rho'}{\xi}\sin\xi\rho\, d\xi. \tag{7.60}$$

Then extend the domain of σ^l by introducing the functions σ^l_{tot} with their associated Fourier transforms,

$$\sigma^0_{tot} = \left\{\begin{array}{ll} 0, & \rho \in (0, \rho_0) \cup (\rho_1, \infty) \\ \sigma^0(\rho), & \rho \in (\rho_0, \rho_1) \end{array}\right\} = \int_0^\infty g(\lambda)\cos(\lambda\rho)\, d\lambda, \tag{7.61}$$

$$\sigma^1_{tot} = \left\{\begin{array}{ll} 0, & \rho \in (0, \rho_0) \cup (\rho_1, \infty) \\ \sigma^1(\rho), & \rho \in (\rho_0, \rho_1) \end{array}\right\} = \int_0^\infty f(\lambda)\sin(\lambda\rho)\, d\lambda. \tag{7.62}$$

Equations (7.57) and (7.58) can be replaced by the equivalent integral equations on an extended range of integration,

$$-\frac{1}{2\pi}\int_0^\infty \sigma_{tot}^0(\rho')\log\left|\rho^2-\rho'^2\right|d\rho' = 1, \qquad \rho\in(\rho_0,\rho_1) \tag{7.63}$$

and

$$\frac{1}{2\pi}\int_0^\infty \sigma_{tot}^1(\rho')\log\left|\frac{\rho+\rho'}{\rho-\rho'}\right|d\rho' = 1, \qquad \rho\in(\rho_0,\rho_1). \tag{7.64}$$

Substitution of the Fourier transforms for the kernels (7.59) and (7.60), and unknown functions σ^l ((7.61) and(7.62)), produces the integral equations

$$\int_0^\infty \xi^{-1}\left[g(\xi)-g(0)\right]\cos(\xi\rho)\,d\xi = 2+g(0)\log\rho,\ \rho\in(\rho_0,\rho_1), \tag{7.65}$$

$$\int_0^\infty \xi^{-1}f(\xi)\sin(\xi\rho)\,d\xi = 2, \qquad \rho\in(\rho_0,\rho_1). \tag{7.66}$$

Equations (7.61) and (7.62) provide the complementary part of the triple integral equations for unknown functions g and f, respectively: for oppositely charged flat strips ($l=1$),

$$\begin{aligned}\int_0^\infty f^*(\xi)\sin(\xi\rho)\,d\xi &= 0, \quad \rho\in(0,\rho_0)\cup(\rho_1,\infty),\\ \int_0^\infty \xi^{-1}f^*(\xi)\sin(\xi\rho)\,d\xi &= 1, \quad \rho\in(\rho_0,\rho_1),\end{aligned} \tag{7.67}$$

whereas for positively charged strips ($l=0$),

$$\begin{aligned}\int_0^\infty g^*(\xi)\cos(\xi\rho)\,d\xi &= 0, \quad \rho\in(0,\rho_0)\cup(\rho_1,\infty),\\ \int_0^\infty \xi^{-1}\left[g^*(\xi)-g^*(0)\right]\cos(\xi\rho)\,d\xi &= 1+g^*(0)\log\rho, \quad \rho\in(\rho_0,\rho_1),\end{aligned} \tag{7.68}$$

where $g^*(\xi)=\frac{1}{2}g(\xi)$, and $f^*(\xi)=\frac{1}{2}f(\xi)$.

Equations (7.67) are easily reduced to those solved in the previous section by means of the transform $\varphi = 2\arctan(\rho)$ (see (7.36)), and we deduce

$$\sigma^1(y) = b\left\{K\left(\frac{2q^{\frac{1}{2}}}{1+q}\right)\right\}^{-1}\frac{1+q}{\sqrt{(y^2-a^2)(b^2-y^2)}}, \qquad a<y<b, \tag{7.69}$$

where $q=r_0^2=a/b$. It is interesting to compare this result with that for the cylindrically-shaped strips, given by (7.46). Apart from a factor of $\frac{1}{4}\pi$, which is due to a different definition of σ^1, the result also coincides with that of Sneddon [55], once the complete elliptic integral identity [1]

$$K\left(\frac{2\sqrt{q}}{1+q}\right) = (1+q)\,K(q) \tag{7.70}$$

is taken into account. As might be expected, the expression for the capacitance per unit length for flat strips has a similar format to that for cylindrically-shaped strips (7.44) :

$$C = 2\frac{K(\kappa)}{K'(\kappa)}, \tag{7.71}$$

where $\kappa = (1-q)/(1+q)$.

The solution of the Equations (7.68) may be approached in many ways. One approach is to reduce (7.64) to triple integral equations with sine function kernels, and then to use the relationship between integral and series equations. A second way is to find the relationship between integral and series equations involving the cosine functions. Both approaches require rather bulky transforms. However, a simpler way exploits the well-known mathematical device employed in [55]. First, rescale the standard coordinates (y', z'), setting $r = y'/b, z = z'/b$, so that Equations (7.68) become

$$\int_0^\infty g(\lambda)\cos(\lambda r)\,d\lambda = 0, \;\; r \in (0, r_0) \cup (1, \infty), \tag{7.72}$$

$$\int_0^\infty \lambda^{-1}\left[g(\lambda) - g(0)\right]\cos(\lambda r)\,d\lambda = 2 + g(0)\log r, \;\; r \in (r_0, 1), \tag{7.73}$$

where $r_0 = a/b$. The mathematical device is a variant of the substitution method, and assumes that the unknown function g has an expansion in a Neumann series

$$g(\lambda) = \sum_{n=0}^{\infty} a_n J_{2n}(\lambda), \tag{7.74}$$

where $\{a_n\}_{n=0}^{\infty}$ are the unknown coefficients to be determined. The well-known discontinuous integral [19]

$$\int_0^\infty J_{2n}(\xi)\cos(\xi r)\,d\xi = \left(1-r^2\right)^{-\frac{1}{2}} T_{2n}\left(\sqrt{1-r^2}\right) H(1-r) \tag{7.75}$$

shows that the integral Equation (7.72) is satisfied automatically, for $r \in (1, \infty)$. Substitution of (7.74) into (7.73), and use of another identity [19] (valid when $n > 0$),

$$\int_0^\infty \xi^{-1} J_{2n}(\xi)\cos(\xi r)\,d\xi = \frac{1}{2n} T_{2n}\left(\sqrt{1-r^2}\right), \qquad r < 1,$$

leads to the following dual series equations for the unknown coefficients a_n:

$$\begin{cases} \sum_{n=0}^{\infty} a_n T_{2n}\left(\sqrt{1-r^2}\right) = 0, & r \in (0, r_0), \\ \sum_{n=1}^{\infty} n^{-1} a_n T_{2n}\left(\sqrt{1-r^2}\right) = 4 + 2a_0 \log r, & r \in (r_0, 1). \end{cases} \tag{7.76}$$

In deducing (7.76), we have used the obvious relationship $a_0 = g(0)$.

The remaining steps are now obvious; the substitution $\cos \frac{1}{2}\varphi = \sqrt{1-r_0^2}$ converts these equations to trigonometric form

$$\sum_{n=1}^{\infty} a_n \cos n\varphi = -A_0, \quad \varphi \in (0, \varphi_0), \tag{7.77}$$

$$\sum_{n=1}^{\infty} n^{-1} a_n \cos n\varphi = 4 + 2a_0 \log\left(\sin \frac{\varphi}{2}\right), \quad \varphi \in (\varphi_0, \pi),$$

where $\cos \frac{1}{2}\varphi_0 = \sqrt{1-r_0^2}$. Following the general scheme outlined in Section 2.2 we obtain

$$\sum_{n=1}^{\infty} n^{-1} a_n P_n^{(-1,0)}(\cos\varphi) = \begin{cases} -2a_0 \log\left(\cos \frac{1}{2}\varphi\right), & \varphi \in (0, \varphi_0) \\ 4 + 2a_0 \log\left[\frac{1}{2}\left(1 + \sin \frac{\varphi}{2}\right)\right], & \varphi \in (\varphi_0, \pi). \end{cases} \tag{7.78}$$

A standard continuity argument establishes that

$$a_0 = -2\left\{\log\left[\frac{1}{2}\left(1 + \sin\frac{\varphi_0}{2}\right)\cos\frac{\varphi_0}{2}\right]\right\}^{-1}, \tag{7.79}$$

so the final solution for the coefficients is

$$a_m = -2a_0 P_m^{(-1,0)}(\cos\varphi_0) + a_0 m \int_{-1}^{x_0} \frac{P_m^{(-1,0)}(x)\,dx}{\left(1 + \sqrt{\frac{1}{2}(1-x)}\right)\sqrt{\frac{1}{2}(1-x)}}, \tag{7.80}$$

where $x_0 = \cos\varphi_0$. The capacitance of the two strips is $C = \frac{\pi}{2} g(0) = \frac{\pi}{2} a_0$, so that

$$C = -\pi\left\{\log\left[\frac{1}{2}\left(1 + \sin\frac{\varphi_0}{2}\right)\cos\frac{\varphi_0}{2}\right]\right\}^{-1}$$
$$= -\pi\left\{\log\left[\frac{1}{2}(1 + r_0)\sqrt{1 - r_0^2}\right]\right\}^{-1}, \tag{7.81}$$

where we recall that $r_0 = a/b$.

Finally let us consider the quadrupole lens system of four charged electrodes, each of which is a flat strip (see Figure 7.4). Use coordinates r, z

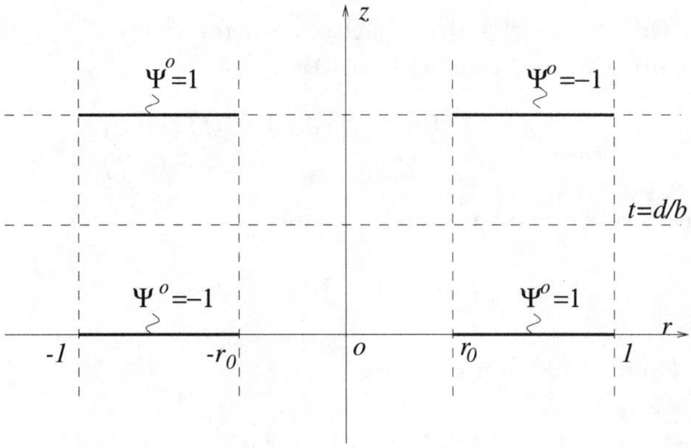

FIGURE 7.4. The charged thin-strip quadrupole.

so that the strips are separated by a distance $2t = 2d/b$ and comprise the four segments specified by $r \in (-1, -r_0) \cup (r_0, 1)$, $z = \pm 1$. The segments in the first and third quadrants are positively charged to unit potential, whilst the remaining strips are negatively charged (to unit potential). The potential may be constructed as a sum of two dipole-like potentials in two ways. Group upper and lower pairs of strips, as dipole-like structures, so that

$$\psi(r,z) = \psi_{up}(r,z) + \psi_{low}(r,z) \tag{7.82}$$

where

$$\psi_{up}(r,z) = \frac{1}{4\pi}\int_{r_0}^{1} \log\left[\frac{(r+r')^2 + \left(z-\frac{t}{2}\right)^2}{(r-r')^2 + \left(z-\frac{t}{2}\right)^2}\right]\sigma(r')\,dr', \tag{7.83}$$

$$\psi_{low}(r,z) = -\frac{1}{4\pi}\int_{r_0}^{1} \log\left[\frac{(r+r')^2 + \left(z+\frac{t}{2}\right)^2}{(r-r')^2 + \left(z+\frac{t}{2}\right)^2}\right]\sigma(r')\,dr', \tag{7.84}$$

σ being the line charge density on that electrode in the first quadrant $(r > 0,\ z > 0)$. A variant grouping of the electrodes is vertical; however, both representations provide the same quadrupole potential distribution.

We now construct the Fourier integral representation of the function ψ, in the three domains $z > \frac{1}{2}t$, $|z| < \frac{1}{2}t$, and $z < -\frac{1}{2}t$. First, use the Fourier transform representation [19] of the logarithmic function,

$$\frac{1}{4}\log\frac{(a+b)^2 + p^2}{(a-b)^2 + p^2} = \int_0^\infty \xi^{-1} e^{-\xi p}\sin(\xi a)\sin(\xi b)\,d\xi, \tag{7.85}$$

valid when $\operatorname{Re} p > |\operatorname{Im} a| + |\operatorname{Im} b|$. Also, as before, extend the domain of σ to obtain a function σ_{tot}, defined on $(0, \infty)$ by

$$\sigma_{tot}(r') = \begin{cases} 0, & r' \in (0, r_0) \cup (1, \infty), \\ \sigma(r'), & r' \in (0, 1), \end{cases} \tag{7.86}$$

with Fourier sine transform representation

$$\sigma_{tot}(r') = \int_0^\infty f(\lambda) \sin \lambda r' d\lambda. \tag{7.87}$$

The desired representation is

$$\psi(r, z) = \int_0^\infty \lambda^{-1} f(\lambda) \sinh\left(\lambda \frac{t}{2}\right) e^{-\lambda z} \sin(\lambda r)\, d\lambda, \quad z > \frac{t}{2}, \tag{7.88}$$

$$\psi(r, z) = \int_0^\infty \lambda^{-1} f(\lambda) \sinh(\lambda z)\, e^{-\lambda \frac{t}{2}} \sin(\lambda r)\, d\lambda, \quad |z| < \frac{t}{2}, \tag{7.89}$$

$$\psi(r, z) = -\int_0^\infty \lambda^{-1} f(\lambda) \sinh\left(\lambda \frac{t}{2}\right) e^{\lambda z} \sin(\lambda r)\, d\lambda, \; z < \frac{-t}{2}. \tag{7.90}$$

It is evident that the electrostatic potential defined by (7.88)–(7.90) is continuous, including across the interfaces $|z| = t/2$, so the following triple integral equations for the unknown function f hold:

$$\int_0^\infty f(\lambda) \sin \lambda r d\lambda = 0, \; r \in (0, r_0) \cup (1, \infty), \tag{7.91}$$

$$\int_0^\infty \lambda^{-1} \left(1 - e^{-\lambda t}\right) f(\lambda) \sin \lambda r d\lambda = -2, \; r \in (r_0, 1). \tag{7.92}$$

The value of the discontinuous integral [19]

$$\int_0^\infty J_{2n+1}(\lambda) \sin \lambda r d\lambda = r\left(1 - r^2\right)^{-\frac{1}{2}} U_{2n}\left(\sqrt{1 - r^2}\right) H(1 - r) \tag{7.93}$$

suggests the following Neumann series representation for f,

$$f(\lambda) = \sum_{n=0}^\infty b_n J_{2n+1}(\lambda); \tag{7.94}$$

it satisfies (7.91) automatically when $r > 1$. The remaining two integral equations are transformed to the following dual series equations for the unknown coefficients b_n,

$$\sum_{n=0}^\infty b_n \sin(2n + 1)\theta = 0, \quad \theta \in (0, \theta_0), \tag{7.95}$$

$$\sum_{n=0}^\infty (2n + 1)^{-1} b_n \sin(2n + 1)\theta = -2 + F(\theta), \theta \in \left(\theta_0, \frac{\pi}{2}\right), \tag{7.96}$$

where $\cos\theta = \sqrt{1-r^2}$, $\cos\theta_0 = \sqrt{1-r_0^2}$,

$$F(\theta) = \sum_{n=0}^{\infty} b_n \alpha_n(\theta) \tag{7.97}$$

and

$$\alpha_n(\theta) = \int_0^{\infty} \lambda^{-1} e^{-\lambda t} J_{2n+1}(\lambda) \sin(\lambda \sin\theta)\, d\lambda.$$

Using the expansion [1]

$$\sin(\lambda \sin\theta) = 2\sum_{k=0}^{\infty} J_{2k+1}(\lambda) \sin(2k+1)\theta, \tag{7.98}$$

we may write

$$\begin{aligned} \alpha_n(\theta) &= 2\sum_{k=0}^{\infty} \beta_{nk} \sin(2k+1)\theta, \\ F(\theta) &= 2\sum_{n=0}^{\infty} b_n \sum_{k=0}^{\infty} \beta_{nk} \sin(2k+1)\theta, \end{aligned} \tag{7.99}$$

where

$$\beta_{nk} = \int_0^{\infty} \lambda^{-1} e^{-\lambda t} J_{2n+1}(\lambda)\, J_{2k+1}(\lambda)\, d\lambda. \tag{7.100}$$

After the trivial substitution $\varphi = 2\theta$ (and $\varphi_0 = 2\theta_0$) one obtains the following dual series equations on the standard domain $(0,\pi)$,

$$\sum_{n=0}^{\infty} b_n \sin\left(n+\frac{1}{2}\right)\varphi = 0, \qquad \varphi \in (0,\varphi_0), \tag{7.101}$$

$$\sum_{n=0}^{\infty} \left(n+\frac{1}{2}\right)^{-1} b_n \sin\left(n+\frac{1}{2}\right)\varphi = -4 + 4\sum_{n=0}^{\infty} b_n \sum_{k=0}^{\infty} \beta_{nk} \sin\left(k+\frac{1}{2}\right)\varphi, \qquad \varphi \in (\varphi_0,\pi). \tag{7.102}$$

It should be noted that in the limiting case when $t \to 0$, the conjunction of oppositely charged strips eliminates sources to produce electrostatic field. In this case $\beta_{nk} = \frac{1}{4}\left(k+\frac{1}{2}\right)^{-1} \delta_{nk}$.

Equations (7.101) and (7.102) have a clear physical interpretation. The left-hand side of these equations represents field terms for a single dipole pair of oppositely charged strips. The mutual coupling between the two

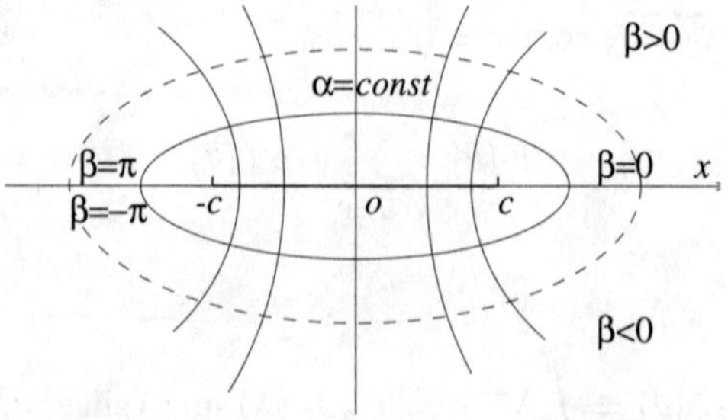

FIGURE 7.5. The elliptic cylinder coordinate system.

dipoles is reflected in the presence of coupling terms on the right-hand side of (7.102). The coefficients β_{nk} measure the strength of this coupling. The situation simplifies when two pairs of strips are well separated, so that $t \gg 1$. The coefficients β_{nk} may be expanded as a rapidly convergent power series in t^{-1}, equalling

$$\sum_{m=0}^{\infty} \frac{(-1)^m \Gamma(2n+2k+2m+3)\,\Gamma(2n+2k+2m+2)\,(2t)^{-2n-2k-2m-2}}{\Gamma(m+1)\,\Gamma(2k+m+2)\,\Gamma(2m+n+2)\,\Gamma(2n+2k+m+3)}. \tag{7.103}$$

The expansion is valid for $t \leq \frac{1}{2}$, and converges rapidly for large t.

7.4 Axially-slotted elliptic cylinders

In this section, we consider cylinders of elliptic cross-section with one or two apertures; these structures are analogues of the slotted circular cylinders considered in previous sections. Elliptic cylinder coordinates (α, β, z) were defined in Section 1.1.6 (see Figure 7.5); briefly, in terms of Cartesian coordinates, the coordinates satisfy

$$x = c\cosh\alpha\cos\beta,\ \ y = c\sinh\alpha\sin\beta,\ \ z = z, \tag{7.104}$$

where the range of parameters is $0 \leq \alpha \leq \infty$, and $-\pi \leq \beta \leq \pi$, the coordinate surfaces $\alpha = \alpha_0 = constant$ form a family of confocal elliptic cylinders with semifocal distance $c = \frac{d}{2}$, and the z-independent solutions ψ of Laplace's equation satisfy

$$\Delta\psi(\alpha,\beta) = \frac{1}{c^2\left(\cosh^2\alpha - \cos^2\beta\right)}\left[\frac{\partial^2\psi}{\partial\alpha^2} + \frac{\partial^2\psi}{\partial\beta^2}\right] = 0. \tag{7.105}$$

The geometry of various slotted elliptic cylinders to be considered are shown in Figure 7.6. The first two (Figures 7.6(a) and 7.6(b)) are portions

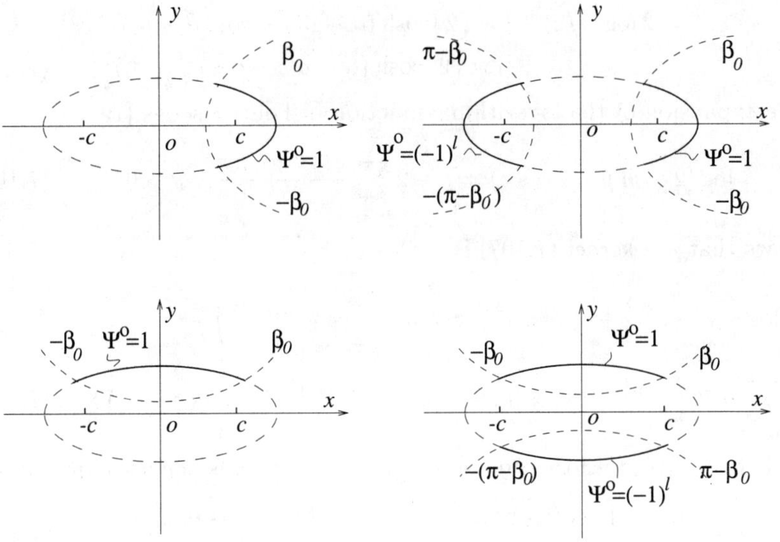

FIGURE 7.6. Various configurations of charged elliptic strips.

of coordinate surfaces. The last two are better described as portions of coordinate surfaces in a variant of elliptic cylinder coordinates to be described later in this section.

The general solution of the Laplace equation, given by a single-layer potential representation, is used to derive the basic series equations for these structures, shown in Figure 7.6 (a)–(d).

First, consider the electrostatic field surrounding a single elliptic arc charged to unit potential (Figure 7.6 (a)). The potential has a single-layer representation

$$\psi(\alpha,\beta) = -\frac{1}{4\pi}\int_{-\beta_0}^{\beta_0} \log\left\{(x-x')^2 + (y-y')^2\right\}\sigma(\beta')\,dl_\beta, \qquad (7.106)$$

where it is to be understood that (x,y) and (x',y') depend on (α,β) and (α',β') according to (7.104), $dl_\beta = c\sqrt{\cosh^2\alpha' - \cos^2\beta'}d\beta'$ is the length differential on the elliptic contour, and $\sigma = \sigma(\beta')$ is the line charge density.

Simple algebra transforms the kernel in (7.106) to the form

$$\begin{aligned}\log\left\{(x-x')^2+(y-y')^2\right\} &= 2\log(c/2)+\log\left\{2\left[\cosh(\alpha+\alpha')-\cos(\beta+\beta')\right]\right\}\\ &\quad+\log\left\{2\left[\cosh(\alpha-\alpha')-\cos(\beta-\beta')\right]\right\}. \qquad (7.107)\end{aligned}$$

The expansion of the logarithmic function in Fourier series [19]

$$\log(2\cosh y-2\cos x)=y-2\sum_{n=1}^{\infty}\frac{e^{-ny}\cos nx}{n},\quad y>0 \qquad (7.108)$$

shows that the kernel (7.107) is

$$\begin{aligned}\log\left\{(x-x')^2+(y-y')^2\right\} &= 2\log\left(\frac{1}{2}ce^{\max(\alpha,\alpha')}\right)-\\ 2\sum_{n=1}^{\infty}n^{-1}&\left\{e^{-n|\alpha-\alpha'|}\cos n(\beta-\beta')+e^{-n(\alpha+\alpha')}\cos n(\beta+\beta')\right\}. \qquad (7.109)\end{aligned}$$

As before, introduce the function σ_{tot}, which extends the domain of σ via

$$\sigma_{tot}(\beta')=\begin{cases}c\sqrt{\cosh^2\alpha'-\cos^2\beta'}\,\sigma(\beta'), & \beta'\in(0,\beta_0)\\ 0, & \beta'\in(\beta_0,\pi)\end{cases} \qquad (7.110)$$

with the understanding $\sigma_{tot}(-\beta')=\sigma_{tot}(\beta')$; assume that σ_{tot} can be represented as a Fourier cosine series

$$\sigma_{tot}(\beta')=\sum_{m=0}^{\infty}(2-\delta_m^0)\,x_m\cos m\beta'. \qquad (7.111)$$

Thus, (7.106) has the equivalent representation

$$\psi(\alpha,\beta)=-\frac{1}{4\pi}\int_0^{\pi}K(\alpha,\beta;\alpha',\beta')\,\sigma_{tot}(\beta')\,d\beta', \qquad (7.112)$$

where

$$\begin{aligned}K(\alpha,\beta;\alpha',\beta') &= -4\sum_{n=1}^{\infty}n^{-1}\left\{e^{-n|\alpha-\alpha'|}+e^{-n(\alpha+\alpha')}\right\}\cos n\beta\cos n\beta'\\ &\quad+4\log\left(\frac{1}{2}ce^{\max(\alpha,\alpha')}\right). \qquad (7.113)\end{aligned}$$

By substitution of (7.111) and (7.113) into (7.112), the form of the potential function ψ in terms of unknown coefficients x_n is

$$\begin{aligned}\psi(\alpha,\beta) &= \sum_{n=1}^{\infty}n^{-1}x_n\left[e^{-n|\alpha-\alpha'|}+e^{-n(\alpha+\alpha')}\right]\cos n\beta\\ &\quad-x_0\log\left(\frac{1}{2}ce^{\max(\alpha,\alpha')}\right). \qquad (7.114)\end{aligned}$$

The major and minor semi-axes are $a = c\cosh\alpha'$ and $b = c\sinh\alpha'$, so that $\frac{1}{2}ce^{\alpha'} = \frac{1}{2}(a+b)$. When the elliptic arc degenerates to a circular arc $(b \to a)$, representation (7.114) transforms to (7.8). The only difference is the reference point, from which the potential is calculated. In order to make both representations compatible, redefine $\psi(\alpha,\beta)$ as

$$\psi(\alpha,\beta) = -x_0(\alpha-\alpha') + \sum_{n=1}^{\infty} n^{-1}x_n\left[e^{-n|\alpha-\alpha'|} + e^{-n(\alpha+\alpha')}\right]\cos n\beta. \tag{7.115}$$

Now use the obvious mixed boundary conditions to obtain the following dual series equations for the unknown Fourier coefficients x_n :

$$\begin{aligned}\sum_{n=1}^{\infty} n^{-1}x_n\left(1+e^{-n2\alpha'}\right)\cos n\beta &= 1, \qquad \beta\in(0,\beta_0),\\ \sum_{n=1}^{\infty} x_n\cos n\beta &= -\frac{1}{2}x_0, \quad \beta\in(\beta_0,\pi).\end{aligned} \tag{7.116}$$

It is instructive to compare (7.116) with its circular analogue (7.11)–(7.12). Formally, the difference is the appearance of a new term $(e^{-n2\alpha'})$, which is a measure of deviation between elliptic and circular strips. Equation (7.116) is transformed in the usual way to the following second-kind Fredholm matrix equation,

$$X_m - \sum_{n=1}^{\infty} X_n\kappa_{nm} = \gamma_m, \tag{7.117}$$

where $m = 1,2,\ldots$, $X_m = (2/m)^{\frac{1}{2}}x_m$,

$$\kappa_{nm} = -e^{-2n\alpha'}\{\log[(1-z_0)/2]\}^{-1}\frac{\hat{P}_n^{(0,-1)}(z_0)\,\hat{P}_m^{(0,-1)}(z_0)}{nm} + e^{-2n\alpha'}\hat{Q}_{nm}^{(-1,0)}(z_0) \tag{7.118}$$

$$\gamma_m = -2\{\log[(1-z_0)/2]\}^{-1}\frac{\hat{P}_m^{(0,-1)}(z_0)}{m}, \tag{7.119}$$

and $z_0 = \cos\beta_0$. Furthermore, for these values of parameters, the normalised Jacobi polynomials $\hat{P}_n^{(0,-1)}$ and $\hat{P}_n^{(-1,0)}$ are defined by

$$\hat{P}_n^{(0,-1)}(x) = (2n)^{\frac{1}{2}}P_n^{(0,-1)}(x), \quad \hat{P}_n^{(-1,0)}(x) = (2n)^{\frac{1}{2}}P_n^{(-1,0)}(x),$$

and the incomplete scalar products are

$$\hat{Q}_{nm}^{(-1,0)}(z_0) = \int_{z_0}^{1}(1-z)^{-1}\hat{P}_n^{(-1,0)}(z)\,\hat{P}_m^{(-1,0)}(z)\,dz.$$

This second-kind system may be solved numerically in the usual way, employing a truncation method that is rapidly convergent.

The field of the slotted elliptic cylinder shown in Figure 7.6(b), in which the slots are symmetrically located and both charged to unit positive potential, may be derived from the solution obtained above for the single elliptic strip. Taking into consideration the charge on both strips, the Fourier series representation for the potential takes the form

$$\psi(\alpha,\beta) = \sum_{n=1}^{\infty} n^{-1} x_{2n} \left[e^{-2n|\alpha-\alpha'|} + e^{-2n(\alpha+\alpha')} \right] \cos 2n\beta - \begin{Bmatrix} 2x_0(\alpha-\alpha'), & (\alpha>\alpha_1) \\ 0, & (\alpha<\alpha_1) \end{Bmatrix}. \tag{7.120}$$

Satisfaction of the boundary conditions produces the dual series equations

$$\sum_{n=1}^{\infty} n^{-1} x_{2n} \left[1 + e^{-4n\alpha'} \right] \cos n\vartheta = 1, \quad \vartheta \in (0,\vartheta_0) \tag{7.121}$$

$$\sum_{n=1}^{\infty} x_{2n} \cos n\vartheta = -\frac{1}{2} x_0, \quad \vartheta \in (\vartheta_0,\pi) \tag{7.122}$$

where $\vartheta = 2\beta$ and $\vartheta_0 = 2\beta_0$.

The solution is readily derived from (7.117) with the following changes: in the matrix elements κ_{nm}, the factor $e^{-2n\alpha'}$ is replaced by $e^{-4n\alpha'}$, the parameter z_0 is replaced by $2z_0^2 - 1 = \cos 2\beta_0$, and the unknown x_m is replaced by x_{2m}. This completes the solution for the slotted elliptic cylinder with identically charged components.

The field of the slotted elliptic cylinder shown in Figure 7.6(b) in which the slots are symmetrically located, but are oppositely charged (each to unit potential), may be derived from the representation

$$\psi(\alpha,\beta) = \sum_{n=0}^{\infty} \frac{x_{2n+1}}{(n+\frac{1}{2})} \left[e^{-(2n+1)|\alpha-\alpha'|} + e^{-(2n+1)(\alpha+\alpha')} \right] \cos(2n+1)\beta, \tag{7.123}$$

where the coefficients x_{2n+1} satisfy the dual series equations

$$\sum_{n=0}^{\infty} \frac{x_{2n+1}}{(n+\frac{1}{2})} \left[1 + e^{-(4n+2)\alpha'} \right] \cos\left(n+\frac{1}{2}\right)\vartheta = 1, \quad \vartheta \in (0,\vartheta_0), \tag{7.124}$$

$$\sum_{n=0}^{\infty} x_{2n+1} \cos\left(n+\frac{1}{2}\right)\vartheta = 0, \quad \vartheta \in (\vartheta_0,\pi), \tag{7.125}$$

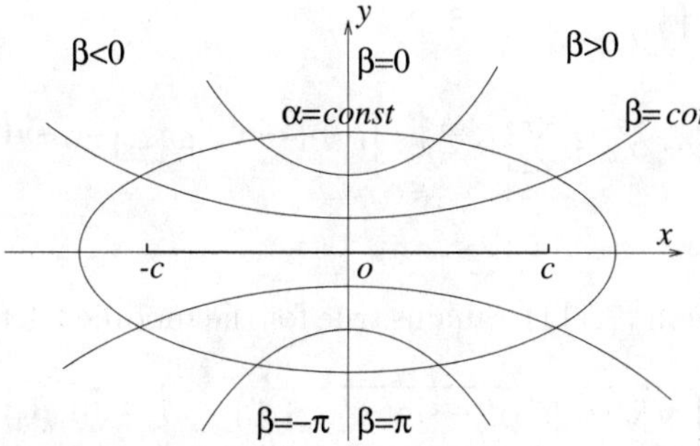

FIGURE 7.7. The oblate elliptic cylinder coordinate system.

and the variable ϑ and parameter ϑ_0 are the same as in Equations (7.121). It can be readily shown that by replacing ϑ by $\pi - \theta$ and identifying b_n with $(-1)^n x_{2n+1}$, Equations (7.124) reduce to equations of the same type as (7.94) and (7.95); however, the term on the right-hand side of (7.124) has a much simpler analytical structure.

Suppressing the intermediate steps, the final form of the system is

$$x_{2m+1} + \sum_{n=0}^{\infty} x_{2n+1} e^{-(4n+2)\alpha'} R_{nm}(z_0) = C_m, \tag{7.126}$$

where $m = 0, 1, 2, \ldots,$

$$C_m = P_m(z_0) / Q_{-\frac{1}{2}}(z_0), \tag{7.127}$$

$$R_{nm}(z_0) = \frac{\left(m+\frac{1}{2}\right)^2}{n+\frac{1}{2}} \left\{ \frac{P_n(z_0)}{Q_{-\frac{1}{2}}(z_0)} \int_{-1}^{z_0} Q_{-\frac{1}{2}}(z) P_m(z)\, dz + Q_{nm}^{(0,0)}(z_0) \right\}, \tag{7.128}$$

and $z_0 = \cos\vartheta_0 = \cos 2\beta_0$. The integrals appearing in (7.128) may be easily evaluated (see Appendix, (B.97)).

The configurations of the charged elliptic strips shown in Figures 7.6 (c) and 7.6 (d) are best described by *oblate* elliptic cylinder coordinates (see Figure 7.7); this system is obtained by replacing the parameter β by $\frac{\pi}{2} - \beta$ in the *prolate* variant of elliptic cylinder coordinates defined at the beginning of this section. Thus,

$$x = \cosh\alpha \sin\beta, \; y = c \sinh\alpha \sin\beta, \; z = z,$$

where the range of parameters is $0 \leq \alpha < \infty$, $-\pi \leq \beta \leq \pi$, and the Laplacian for z-independent potentials is

$$\Delta\psi(\alpha,\beta) = \frac{1}{c^2\left(\cosh^2\alpha - \sin^2\beta\right)} \left[\frac{\partial^2\psi}{\partial\alpha^2} + \frac{\partial^2\psi}{\partial\beta^2}\right]. \tag{7.129}$$

By simple algebra, one may verify that representation of the potential is given by (cf. (7.115))

$$\psi(\alpha,\beta) = -x_0(\alpha-\alpha') + \sum_{n=1}^{\infty} n^{-1}x_n \left[e^{-n|\alpha-\alpha'|} + (-1)^n e^{-n(\alpha+\alpha')}\right] \cos 2n\beta \tag{7.130}$$

where the expansion (7.111) remains true for the modified definition for σ_{tot},

$$\sigma_{tot}(\beta') = \begin{cases} c\sqrt{\cosh^2\alpha' - \sin^2\beta'}\;\sigma(\beta'), & \beta' \in (0,\beta_0), \\ 0, & \beta' \in (\beta_0,\pi). \end{cases} \tag{7.131}$$

The potential function for the elliptic strips shown in Figures 7.6(c) and 7.6(d) may be readily derived from the solutions already obtained in this section with a few simple modifications. For the single elliptic strip (Figure 7.6(c)), multiply the matrix elements κ_{nm} (7.118) by a factor of $(-1)^n$. For the pair of symmetrically located strips (Figure 7.6(d)) both positively charged to unit potential, no changes are needed; however, if the pair of symmetrically located strips (Figure 7.6(d)) are oppositely charged, change the sign of the term containing $e^{-(4n+2)\alpha'}$, replacing it by $-e^{-(4n+2)\alpha'}$. The line charge density σ can now be calculated using definition (7.131).

7.5 Slotted cylinders of arbitrary profile

The study of the slotted elliptical cylinder suggests that the idea of regularisation might beneficially be extended to determine the potential of more general two-dimensional, thin, charged conductors. In examining the elliptic cylinder, we analytically inverted that part of the series equations (see (7.116)) that definitely corresponds to a circular profile. From the perspective of the method of regularisation a *singular part* of the operator associated with the series equations formulation was inverted. The remaining contributions (visible as the terms proportional to $e^{-n2\alpha'}$) are regular (analytic) perturbation terms that measure the deviation of the elliptic profile from the circular.

The purpose of this section is to show how a regularisation approach may be extended to open (slotted) hollow cylinders with arbitrarily profiled cross-section. Although we do not aim to compute the electrostatic fields of all possible configurations, nevertheless we wish to demonstrate how the methods developed for canonical conductors work in the wider context. In particular, the regularised system of equations for an open cylinder of arbitrary cross-section with one slot or aperture will be obtained. This approach has been developed by Tuchkin [52, 65] in the context of a rigorous treatment of diffraction by open thin cylinders of arbitrary cross-section.

The starting point is the construction of the solution to the Dirichlet boundary problem for Laplace equation on an arc of a hollow circular cylinder of unit radius. In cylindrical polar coordinates (r, φ) the electrostatic potential ψ produced by such a thin strip with (as yet unknown) charge density σ is given by the single-layer potential of the type (7.2),

$$\psi(r,\varphi) = -\frac{1}{4\pi}\int_{-\varphi_0}^{\varphi_0} \log\left|1 - 2r\cos(\varphi-\varphi') + r^2\right| \sigma(\varphi')d\varphi'. \tag{7.132}$$

If the conductor is charged to potential $\psi_0(\varphi)$ (as a function of position), enforcement of the boundary condition

$$\psi(1,\varphi) = \psi_0(\varphi), \varphi \in [-\varphi_0, \varphi_0], \tag{7.133}$$

produces the first-kind Fredholm equation

$$-\frac{1}{2\pi}\int_{-\varphi_0}^{\varphi_0} \log\left|2\sin\frac{\varphi-\varphi'}{2}\right| \sigma(\varphi')d\varphi' = \psi_0(\varphi), \quad \varphi \in [-\varphi_0, \varphi_0]. \tag{7.134}$$

Extend the domain of definition of the line charge density σ to a function

$$\sigma^*(\varphi') = \begin{cases} \sigma(\varphi'), & \varphi' \in [-\varphi_0, \varphi_0] \\ 0, & \varphi' \in [-\pi, \pi]\setminus[-\varphi_0, \varphi_0] \end{cases}, \tag{7.135}$$

so that the function σ^* satisfies

$$-\frac{1}{2\pi}\int_{-\pi}^{\pi} \log\left|2\sin\frac{\varphi-\varphi'}{2}\right| \sigma^*(\varphi')d\varphi' = \psi_0(\varphi), \quad \varphi \in [-\varphi_0, \varphi_0]. \tag{7.136}$$

The logarithmic kernel has a Fourier series expansion

$$\log\left|2\sin\frac{\varphi-\varphi'}{2}\right| = -\frac{1}{2}\sum_{n=-\infty}^{n=-\infty}{}' \frac{1}{|n|} e^{in(\varphi-\varphi')}, \tag{7.137}$$

where prime indicates omission of the zero index ($n = 0$) term. Assume that both σ^* and ψ_0 have Fourier series expansions

$$\sigma^*(\varphi') = \sum_{n=-\infty}^{\infty} x_n e^{in\varphi'}, \tag{7.138}$$

$$2\psi_0(\varphi) = \sum_{n=-\infty}^{\infty} f_n e^{in\varphi}, \tag{7.139}$$

where $\{x_n\}_{n=-\infty}^{\infty}$ are unknown coefficients to be determined, but the coefficients $\{f_n\}_{n=-\infty}^{\infty}$ are known. Substitution of these expansions produces the following dual series equations,

$$\sum_{n=-\infty}^{\infty}{}' \frac{1}{|n|} x_n e^{in\varphi} = \sum_{n=-\infty}^{\infty} f_n e^{in\varphi}, \varphi \in [-\varphi_0, \varphi_0], \tag{7.140}$$

$$\sum_{n=-\infty}^{\infty} x_n e^{in\varphi} = 0, \varphi \in [-\pi, \pi]\setminus[-\varphi_0, \varphi_0], \tag{7.141}$$

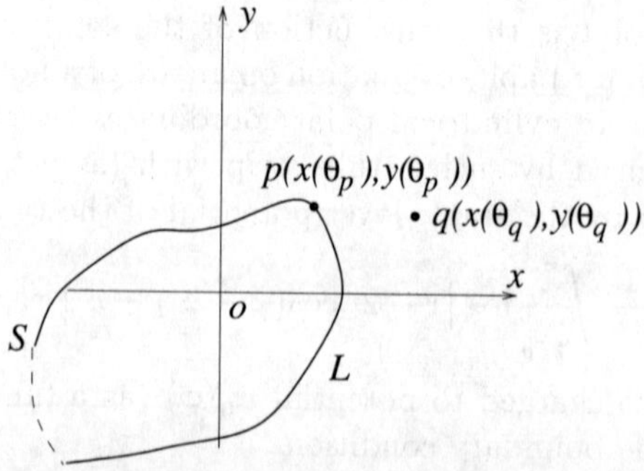

FIGURE 7.8. Cross-section of the arbitrarily shaped, infinitely thin, slotted conductor.

where the prime in Equation (7.140) means that the index zero term is omitted from the summation.

As shown in Section 2.2, the *canonical* equations (7.140) and (7.141) are solvable analytically, and thus provide a starting point for the generalisation to slotted cylinders of arbitrary cross-section.

Let L denote the two-dimensional cross-section in the xy plane of the arbitrarily shaped, infinitely thin, slotted conductor (see Figure 7.8). It will be assumed to be sufficiently smooth; the precise degree of smoothness will become apparent below. Let $p = p(x, y)$ denote a point on the contour L. If the conductor is charged to the potential $\psi_0(p)$ (at each $p \in L$), the Dirichlet boundary conditions to be enforced at each point $p \in L$ are

$$\psi(p - 0) = \psi(p + 0) = \psi_0(p). \tag{7.142}$$

The potential $\psi(q)$ of the electrostatic field produced by this conductor has a single-layer potential representation (7.1) in terms of the surface charge density σ,

$$\psi(q) = -\frac{1}{2\pi}\int_L \log(|p - q|)\sigma(p)dl_p, \qquad q \in R^2, \tag{7.143}$$

where dl_p is the differential of arc length at the point $p \in L$, q is a point at which the electrostatic potential is considered, and $R = |p - q|$ is the distance between the point p on the conductor and the observation point q .

Applying the boundary condition (7.142) to Equation (7.143) yields the integral equation

$$\frac{1}{2\pi}\int_L \log(|p - q|)\sigma(p)dl_p = -\psi_0(q), \qquad q \in L \tag{7.144}$$

for the unknown surface charge density σ. Once this is found, the electrostatic potential at any point q can be found from (7.143), and all the relevant physical quantities such as charge and capacitance are easily calculated.

Our reformulation of the integral Equation (7.144) begins by regarding the open contour L as part of a larger *closed* structure S, which is parametrised by the functions $x(\theta), y(\theta)$ where $\theta \in [-\pi, \pi]$; the parametrising functions are periodic so that $x(-\pi) = x(\pi), y(-\pi) = y(\pi)$. The contour L is parametrised by the subinterval $[-\theta_0, \theta_0]$,

$$L = \{(x(\theta), y(\theta)), \theta \in [-\theta_0, \theta_0]\},$$

and the aperture is created by the removal from S of the segment

$$L' = \{(x(\theta), y(\theta)), \theta \in [-\pi, -\theta_0] \cup [\theta_0, \pi]\}.$$

In order to employ the regularisation procedure to be described, the parametrisation of the contour S must be continuous and twice differentiable at each point p of S. Moreover, the computational effectiveness of the numerical algorithm derived from the regularised system increases as the degree of contour smoothness (differentiability) increases.

With this parametrisation, the differential of arc length is

$$l(\tau) = \sqrt{(x'(\tau))^2 + (y'(\tau))^2},$$

and the integral Equation (7.144) takes the form

$$\frac{1}{2\pi}\int_{-\theta_0}^{\theta_0} \log(R)\sigma_0(\tau)d\tau = -\psi_0(\theta), \qquad \theta \in [-\theta_0, \theta_0], \tag{7.145}$$

where $R(\theta,\tau) = \sqrt{[x(\theta) - x(\tau)]^2 + [y(\theta) - y(\tau)]^2}$ is the distance between points of the contour parametrised by θ and τ, $\sigma_0(\tau) = \sigma(x(\tau), y(\tau))\, l(\tau)$, and $\psi_0(\theta) = -\psi_0(x(\theta), y(\theta))$.

Introduce the new unknown function z, extending the domain of σ_0 and defined by

$$z(\tau) = \begin{cases} \sigma_0(\tau), & \tau \in [-\theta_0, \theta_0], \\ 0, & \tau \in [-\pi, -\theta_0] \cup [\theta_0, \pi]. \end{cases} \tag{7.146}$$

Transform (7.145) to an integral equation for this new unknown over the full interval $[-\pi, \pi]$ of the angular coordinate θ:

$$\frac{1}{2\pi}\int_{-\pi}^{\pi} \log(R)z(\tau)d\tau = -\Psi_0(\theta), \qquad \theta \in [-\theta_0, \theta_0]. \tag{7.147}$$

Equation (7.147), together with the requirement that z vanishes outside the interval $[-\theta_0, \theta_0]$, is completely equivalent to Equation (7.145).

We now convert Equation (7.147) to a dual series with a trigonometric kernel. The function z is represented by its Fourier series, whilst the kernel of (7.147) is expanded as a double Fourier series. The semi-inversion and regularisation of dual series with trigonometric functions kernels described in Chapter 2 is the key technical idea upon which this method relies.

The first stage is to obtain the integral equation in the equivalent form of a dual series equation with exponential functions $e^{in\theta}$. Split the kernel of the integral Equation (7.147) into singular and regular parts:

$$\log(R(\theta,\tau)) = \ln(2\left|\sin\frac{\theta-\tau}{2}\right|) + H(\theta,\tau). \tag{7.148}$$

The singular part of the kernel (7.148) has the expansion (7.137). Our assumptions about the surface S imply that $H(\theta,\tau)$ is smooth and continuously differentiable with respect to θ and τ; this allows its expansion in a double Fourier series,

$$H(\theta,\tau) = \sum_{p=-\infty}^{\infty}\sum_{n=-\infty}^{\infty} h_{np}e^{i(n\theta+p\tau)}, \qquad \theta,\tau\in[-\pi,\pi], \tag{7.149}$$

where

$$\sum_{p=-\infty}^{\infty}\sum_{n=-\infty}^{\infty}(1+|p|^2)(1+|n|^2)\left|h_{np}\right|^2 < \infty.$$

Here the coefficients h_{np} are given by

$$h_{np} = \frac{1}{4\pi^2}\int_{-\pi}^{\pi}\int_{-\pi}^{\pi} H(\theta,\tau)e^{-i(n\theta+p\tau)}d\theta d\tau. \tag{7.150}$$

Represent the conductor potential function ψ_0 and the unknown z in their Fourier series:

$$2\psi_0(\theta) = \sum_{n=-\infty}^{\infty} g_n e^{in\theta}, \qquad \theta\in[-\pi,\pi] \tag{7.151}$$

$$z(\tau) = \sum_{n=-\infty}^{\infty} \varsigma_n e^{in\tau}, \qquad \tau\in[-\pi,\pi]. \tag{7.152}$$

Inserting (7.137) and (7.148)–(7.152) into Equation (7.147) and recalling that $z(\tau)$ vanishes outside the interval $[-\theta_0,\theta_0]$, we obtain the following dual series equations with exponential kernels:

$$\sum_{n=-\infty}^{\infty}{}' |n|^{-1}\varsigma_n e^{in\theta} - 2\sum_{n=-\infty}^{\infty} e^{in\theta}\sum_{p=-\infty}^{\infty} h_{n,-p}\varsigma_p = \sum_{n=-\infty}^{\infty} g_n e^{in\theta}, \quad \theta \in [-\theta_0,\theta_0], \tag{7.153}$$

$$\sum_{n=-\infty}^{\infty}\varsigma_n e^{in\theta} = 0, \qquad \theta\in[-\pi,-\theta_0]\cup[\theta_0,\pi]. \tag{7.154}$$

Thus the integral Equation (7.147) is converted to equivalent dual series equations defined on two subintervals of $[-\pi, \pi]$, with unknowns $\{\varsigma_n\}_{n=-\infty}^{\infty}$ to be found.

Following the procedure of Section 2.2, we convert this system one with real trigonometric kernels. Introduce the new unknowns

$$x_n = (\zeta_n + \zeta_{-n})/|n|, \quad y_n = (\zeta_n - \zeta_{-n})/|n|, \tag{7.155}$$

where $n = 1, 2, \ldots$. Set

$$g_n^+ = g_n + g_{-n}, \quad g_n^- = g_n - g_{-n}, \quad (n = 1, 2, ...), \tag{7.156}$$

and define the matrices from the coefficients $\{h_{np}\}_{n,p=-\infty}^{\infty}$ (7.150) by

$$\begin{aligned}
k_{np}^{(++)} &= \left[(h_{n,p} + h_{n,-p}) + (h_{-n,p} + h_{-n,-p})\right]/(2 + 2\delta_{n0}), & n, p \geq 0;\\
k_{np}^{(+-)} &= \left[(h_{n,p} - h_{n,-p}) + (h_{-n,p} - h_{-n,-p})\right]/(2 + 2\delta_{n0}), & n \geq 0, p \geq 1;\\
k_{np}^{(-+)} &= \left[(h_{n,p} + h_{n,-p}) - (h_{-n,p} + h_{-n,-p})\right]/2, & n \geq 1, p \geq 0;\\
k_{np}^{(--)} &= \left[(h_{n,p} - h_{n,-p}) - (h_{-n,p} - h_{-n,-p})\right]/2, & n, p \geq 1.
\end{aligned} \tag{7.157}$$

We may therefore reduce the system of Equations (7.154) to two coupled systems of dual series equations with trigonometric function kernels,

$$\begin{aligned}
\sum_{n=1}^{\infty} x_n \cos n\theta &= a_0 + \sum_{n=1}^{\infty} a_n \cos n\theta, & \theta \in [0, \theta_0],\\
\sum_{n=1}^{\infty} n x_n \cos n\theta &= -\zeta_0, & \theta \in [\theta_0, \pi],
\end{aligned} \tag{7.158}$$

and

$$\begin{aligned}
\sum_{n=1}^{\infty} y_n \sin n\theta &= \sum_{n=1}^{\infty} c_n \sin n\theta, & \theta \in [0, \theta_0],\\
\sum_{n=1}^{\infty} n y_n \sin n\theta &= 0, & \theta \in [\theta_0, \pi],
\end{aligned} \tag{7.159}$$

where

$$\begin{aligned}
a_0 &= g_0 + 2k_{00}^{(++)}\zeta_0 + 2\sum_{p=1}^{\infty} p(k_{0p}^{(++)} x_p - k_{0p}^{(+-)} y_p),\\
a_n &= g_n^+ + 2k_{n0}^{(++)}\zeta_0 + 2\sum_{p=1}^{\infty} p(k_{np}^{(++)} x_p - k_{np}^{(+-)} y_p),\\
c_n &= g_n^- + 2k_{n0}^{(-+)}\zeta_0 - 2\sum_{p=1}^{\infty} p(k_{np}^{(--)} y_p - k_{np}^{(-+)} x_p).
\end{aligned} \tag{7.160}$$

These equations are now in standard form to apply the results of Section 2.2, and we write down the regularised system of linear equations obtained by this process. It produces two coupled matrix equations (of second kind) with the rescaled unknowns

$$X_n = x_n\sqrt{2n},\ Y_n = y_n\sqrt{2n},\ X_0 = 2\zeta_0. \tag{7.161}$$

Setting $t_0 = \cos\theta_0$, the systems are

$$Y_m + \sum_{p=1}^{\infty}\sqrt{2p}\sum_{n=1}^{\infty}\sqrt{2n}\left[Y_p k_{np}^{(--)} - X_p k_{np}^{(-+)}\right]\hat{Q}_{n-1,m-1}^{(0,1)}(t_0)$$
$$= \sum_{n=1}^{\infty}\sqrt{2n}(X_0 k_{np}^{(-+)} + g_n^-)\hat{Q}_{n-1,m-1}^{(0,1)}(t_0), \tag{7.162}$$

and

$$X_m - \sum_{p=1}^{\infty}\sqrt{2p}\sum_{n=1}^{\infty}\sqrt{2n}\left[X_p k_{np}^{(++)} - Y_p k_{np}^{(+-)}\right]\hat{Q}_{n-1,m-1}^{(1,0)}(t_0)$$
$$= \sum_{n=1}^{\infty}\sqrt{2n}(X_0 k_{n0}^{(++)} + g_n^+)\hat{Q}_{n-1,m-1}^{(1,0)}(t_0) + X_0(1+t_0)\frac{1}{m}\hat{P}_{m-1}^{(0,1)}(t_0), \tag{7.163}$$

where $m = 1, 2, \ldots$; an additional equation, which is to be solved together with the Equations (7.162) and (7.163), is

$$\sum_{p=1}^{\infty}\sqrt{2p}(X_p k_{0p}^{(++)} - Y_p k_{0p}^{(+-)}) +$$
$$\frac{(1+t_0)}{2}\sum_{p=1}^{\infty}\sqrt{2p}\sum_{n=1}^{\infty}\sqrt{\frac{2}{n}}\left[X_p k_{np}^{(++)} - Y_p k_{np}^{(+-)}\right]\hat{P}_{n-1}^{(0,1)}(t_0)$$
$$= -\frac{(1+t_0)}{2}\sum_{n=1}^{\infty}\sqrt{\frac{2}{n}}(X_0 k_{n0}^{(++)} + g_n^+)\hat{P}_{n-1}^{(0,1)}(t_0)$$
$$- g_0 - X_0\left[k_{00}^{(++)} + \frac{1}{2}\ln(\frac{(1-t_0)}{2})\right]. \tag{7.164}$$

Here

$$\hat{Q}_{n,m}^{(0,1)}(t_0) = \int_{t_0}^{1}(1+t)\hat{P}_n^{(0,1)}(t)\hat{P}_m^{(0,1)}(t)dt$$

is the usual normalised incomplete scalar product.

This regularised system of equations is a coupled Fredholm matrix system of second kind, which may be satisfactorily solved by the usual process of truncation. In addition to the standard considerations about truncation number, some attention must be paid to the rate of convergence of the double Fourier series representation (7.149) of the distance between points on the cylinder profile. With this proviso, the regularisation approach and the resulting system of equations provides a satisfactory basis for numerical computations of the electrostatic fields surrounding open (singly-slotted) hollow cylinders with arbitrarily profiled cross-sections.

8
More Complicated Structures

In this chapter, we consider a class of structures which, from a technical point of view, is more complicated than those classes examined in previous chapters. The class comprises plates, some of simple geometric or canonical shape, and others with a greater degree of complexity. Complexity is a relative notion. The determination of the potential for an electrified circular disc is not complicated, and its solution has been known for a long time [8]; however, the analogous problem for an electrified elliptic plate seems to be more complex, and its rigorous solution has been obtained only comparatively recently [5]. In the same way, the potential associated with a charged thin spherical shell, with an elliptic hole, or with the charged spherically conformal elliptic plate, provides problems of equal complexity. Rather more complex are problems generated by crossed plates, or by polygon plates, etc. In this hierarchy, arbitrarily-shaped flat plates present the most complex problem structures for analytical methods.

In this chapter we outline how the integral methods may be used for a unified treatment of determining the potential for all these charged structures, from the electrified disc to arbitrarily-shaped charged flat plates. The circular and elliptic discs are considered in Sections 8.1 and 8.2, respectively; this forms the basis for calculating the capacitance of a spherically-curved elliptic plate. Plates that are regular polygons are examined in Section 8.3. The finite rectangular strip is considered in Section 8.4; considerable manipulation is required to demonstrate that the regularised system is indeed dominantly diagonal. In the final section (8.5) we calculate the capacitance of a *coupled* pair of charged conductors, the spherical cap and the circular disc. This example is interesting because the components are

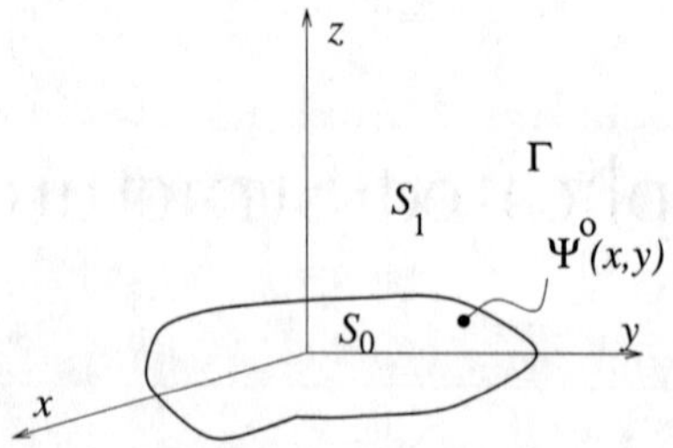

FIGURE 8.1. The flat plate S_0 with arbitrarily-shaped boundary Γ. It lies on the xOy plane and has complement S_1.

parts of coordinate surfaces belonging to *different* coordinate systems, and the resultant equations are particular cases of the *integro-series equations* briefly described in Section 2.9.

8.1 Rigorous solution methods for charged flat plates

In this section we examine the canonical problem of an electrified circular disc; it provides a starting point for the generalisation of integral methods to more general structures.

Consider an arbitrarily-shaped flat plate occupying the finite surface region S_0 in the plane $z = 0$ (see Figure 8.1); let S_1 be the (unbounded) complementary part of this plane and Γ be its boundary contour. The potential generated by the structure may be represented in the form of the single-layer potential,

$$\psi(x,y,z) = -\frac{1}{4\pi}\iint_{S_0} \frac{\sigma(x',y')\,dx'dy'}{\sqrt{(x-x')^2+(y-y')^2+z^2}}, \qquad (x,y,z)\in R^3\backslash\Gamma \tag{8.1}$$

where σ is surface charge density induced when the structure is immersed in a known potential field ψ^0; enforcement of the boundary condition

$$\psi = \psi^0(\vec{r}), \quad \vec{r}\in S_0$$

provides an integral equation determining σ.

Recall that the inverse distance

$$\left|\vec{r}-\vec{r'}\right|^{-1} = \left\{(x-x')^2+(y-y')^2+(z-z')^2\right\}^{-\frac{1}{2}} \tag{8.2}$$

is a fundamental solution of the Laplace equation in R^3. In order to represent ψ as a double Fourier transform, we employ the representation (1.206) of the inverse distance that is discontinuous in z,

$$\left\{(x-x')^2+(y-y')^2+z^2\right\}^{-\frac{1}{2}}$$
$$=\frac{1}{2\pi}\int_{-\infty}^{\infty}d\nu\cos\nu\left(x-x'\right)\int_{-\infty}^{\infty}d\mu\frac{\cos\mu\left(y-y'\right)}{\sqrt{\nu^2+\mu^2}}e^{-\sqrt{\nu^2+\mu^2}|z|}. \quad (8.3)$$

Then extend to domain of definition of σ to the whole of the plane $z=0$ via

$$\sigma_t\left(x',y'\right)=\begin{cases}\sigma\left(x',y'\right), & (x',y')\in S_0,\\ 0, & (x',y')\in S_1.\end{cases} \quad (8.4)$$

For the most general structures, the double Fourier transform of the function σ_t may be expressed in terms of four unknown functions f, g, h, and t,

$$\sigma_t\left(x,y\right)=\int_{-\infty}^{\infty}d\nu\cos\nu x\int_{-\infty}^{\infty}d\mu\left\{f\left(\nu,\mu\right)\cos\mu y+h\left(\nu,\mu\right)\sin\mu y\right\}$$
$$+\int_{-\infty}^{\infty}d\nu\sin\nu x\int_{-\infty}^{\infty}d\mu\left\{g\left(\nu,\mu\right)\cos\mu y+t\left(\nu,\mu\right)\sin\mu y\right\}. \quad (8.5)$$

The boundary condition

$$\psi\left(x,y,+0\right)=\psi\left(x,y,-0\right)=\psi^0\left(x,y\right), \qquad (x,y)\in S_0 \quad (8.6)$$

now provides an integral equation for the extended surface charge density σ_t,

$$-\frac{1}{4\pi}\int_{-\infty}^{\infty}\int_{-\infty}^{\infty}\frac{\sigma_t\left(x',y'\right)}{\sqrt{(x-x')^2+(y-y')^2}}dx'dy'=\psi^0\left(x,y\right), \quad (x,y)\in S_0. \quad (8.7)$$

Substitution of the double Fourier transforms (8.5) and (8.3) for the functions σ_t and inverse distance, respectively, together with the recognition that σ_t vanishes on S_1, produces the dual integral equations

$$-\frac{\pi}{4}\int_{-\infty}^{\infty}d\nu\cos\nu x\int_{-\infty}^{\infty}\frac{d\mu}{\left(\nu^2+\mu^2\right)^{\frac{1}{2}}}\left\{f\left(\nu,\mu\right)\cos\mu y+h\left(\nu,\mu\right)\sin\mu y\right\}$$
$$-\frac{\pi}{4}\int_{-\infty}^{\infty}d\nu\sin\nu x\int_{-\infty}^{\infty}\frac{d\mu}{\left(\nu^2+\mu^2\right)^{\frac{1}{2}}}\left\{g\left(\nu,\mu\right)\cos\mu y+t\left(\nu,\mu\right)\sin\mu y\right\}$$
$$=\psi^0\left(x,y\right), \qquad (x,y)\in S_0, \quad (8.8)$$

and

$$\int_{-\infty}^{\infty} d\nu \cos\nu x \int_{-\infty}^{\infty} d\mu \left\{ f(\nu,\mu)\cos\mu y + h(\nu,\mu)\sin\mu y \right\}$$
$$+ \int_{-\infty}^{\infty} d\nu \sin\nu x \int_{-\infty}^{\infty} d\mu \left\{ g(\nu,\mu)\cos\mu y + t(\nu,\mu)\sin\mu y \right\}$$
$$= 0, \qquad (x,y) \in S_1. \quad (8.9)$$

Equations (8.8) and (8.9) describe the most general electrostatic field for an arbitrarily-shaped charged flat plate. To consider the special case of a charged circular disc, introduce the parametrisation by $x = \rho\cos\phi$, $y = \rho\sin\phi$, so that the bounding contour Γ is $\rho = 1$. Then, setting $\tau = \sqrt{\nu^2 + \mu^2}$, use the expansions in series (derived from the generating series, see Appendix, (B.139)–(B.142)),

$$\cos(\nu x)\cos(\mu y) = \cos(\nu\rho\cos\phi)\cos(\mu\rho\sin\phi)$$
$$= \sum_{m=0}^{\infty} \left(2 - \delta_m^0\right) J_{2m}(\tau\rho)\, T_{2m}\left(\mu\tau^{-1}\right) T_{2m}(\cos\phi), \quad (8.10)$$

$$\cos(\nu x)\sin(\mu y) = \cos(\nu\rho\cos\phi)\sin(\mu\rho\sin\phi)$$
$$= 2\nu\tau^{-1}\sin\phi \sum_{m=0}^{\infty} J_{2m+1}(\tau\rho)\, U_{2m}\left(\mu\tau^{-1}\right) U_{2m}(\cos\phi), \quad (8.11)$$

$$\sin(\nu x)\cos(\mu y) = \sin(\nu\rho\cos\phi)\cos(\mu\rho\sin\phi)$$
$$= 2\nu\tau^{-1} \sum_{m=0}^{\infty} J_{2m+1}(\tau\rho)\, U_{2m}\left(\mu\tau^{-1}\right) T_{2m+1}(\cos\phi), \quad (8.12)$$

$$\sin(\nu x)\sin(\mu y) = \sin(\nu\rho\cos\phi)\sin(\mu\rho\sin\phi)$$
$$= 2\nu\tau^{-1}\sin\phi \sum_{m=0}^{\infty} J_{2m+2}(\tau\rho)\, U_{2m+1}\left(\mu\tau^{-1}\right) U_{2m+1}(\cos\phi). \quad (8.13)$$

If the given potential $\psi^0(x,y)$ is representable as a trigonometric series (or equivalently as a series in the Chebyshev polynomials $T_m(\cos\phi)$ and $U_m(\cos\phi)$), then using the orthogonality of the even or odd Chebyshev polynomials on $\left(0, \frac{\pi}{2}\right)$ as appropriate, one may deduce dual integral equations, involving the Bessel function kernels of the form $J_m\left(\sqrt{\nu^2+\mu^2}\rho\right)$ for the unknowns f, g, h, and t.

In the simplest case, suppose that the circular disc is raised to unit potential so that $\psi^\circ(x,y) = 1$ on S_0. The bivariate dual integral equations become

$$\int_0^\infty d\nu \int_0^\infty d\mu F(\nu,\mu) J_0\left(\sqrt{\nu^2+\mu^2}\rho\right) = 1, \quad 0 \le \rho < 1 \tag{8.14}$$

$$\int_0^\infty d\nu \int_0^\infty d\mu \sqrt{\nu^2+\mu^2} F(\nu,\mu) J_0\left(\sqrt{\nu^2+\mu^2}\rho\right) = 0, \quad \rho > 1 \tag{8.15}$$

where the as yet unknown function F represents the electrostatic potential by

$$\psi(x,y) = \int_0^\infty d\nu \cos \nu x \int_0^\infty d\mu F(\nu,\mu) e^{-\sqrt{\nu^2+\mu^2}|z|} \cos(\mu y). \tag{8.16}$$

For a circular disc, it is obvious that F depends only upon $\tau = \sqrt{\nu^2+\mu^2}$, so that $F(\nu,\mu) = F\left(\sqrt{\nu^2+\mu^2}\right) = F(\tau)$, and the dual integral equations become

$$\int_0^\infty d\tau.\tau F(\tau) J_0(\tau\rho) = \frac{2}{\pi}, \quad 0 \le \rho < 1, \tag{8.17}$$

$$\int_0^\infty d\tau.\tau^2 F(\tau) J_0(\tau\rho) = 0, \quad \rho > 1. \tag{8.18}$$

The solution is given by (see Section 2.6) $F(\tau) = 4\pi^{-2}\tau^{-2}\sin\tau$, so

$$F(\nu,\mu) = \frac{4}{\pi^2}\left(\nu^2+\mu^2\right)^{-1} \sin\left(\sqrt{\nu^2+\mu^2}\right). \tag{8.19}$$

The *substitution method* provides an alternative and very useful method for solving Equations (8.14) and (8.15). Seek the solution F as an expansion in the Neumann series

$$F(\nu,\mu) = \left(\nu^2+\mu^2\right)^{-\frac{3}{4}} \sum_{k=0}^\infty x_k J_{2k+\frac{1}{2}}\left(\sqrt{\nu^2+\mu^2}\right) \tag{8.20}$$

where the coefficients x_k are to be found. Insertion of (8.20) into (8.14) and (8.15) yields (using again the substitution $\tau = \sqrt{\nu^2+\mu^2}$)

$$\sum_{k=0}^\infty x_k \int_0^\infty \tau^{-\frac{1}{2}} J_0(\tau\rho) J_{2k+\frac{1}{2}}(\tau)\, d\tau = \frac{2}{\pi}, \quad 0 \le \rho < 1, \tag{8.21}$$

$$\sum_{k=0}^\infty x_k \int_0^\infty \tau^{\frac{1}{2}} J_0(\tau\rho) J_{2k+\frac{1}{2}}(\tau)\, d\tau = 0, \quad \rho > 1. \tag{8.22}$$

The integrals occurring in (8.21) and (8.22) have the values [14]

$$\int_0^\infty \tau^{-\frac{1}{2}} J_0(\tau\rho) J_{2k+\frac{1}{2}}(\tau)\, d\tau = 2^{-\frac{1}{2}} \frac{\Gamma\left(k+\frac{1}{2}\right)}{\Gamma(k+1)} P_{2k}\left(\sqrt{1-\rho^2}\right), \quad (8.23)$$

when $0 \le \rho < 1$, and

$$\int_0^\infty \tau^{\frac{1}{2}} J_0(\tau\rho) J_{2k+\frac{1}{2}}(\tau)\, d\tau = 2^{\frac{1}{2}} \frac{\Gamma(k+1)}{\Gamma\left(k+\frac{1}{2}\right)} \frac{P_{2k}\left(\sqrt{1-\rho^2}\right)}{\sqrt{1-\rho^2}} H(1-\rho). \quad (8.24)$$

When $\rho > 1$, the integrals occurring in (8.22) therefore vanish identically for each k, so that the equation is satisfied automatically; when $0 \le \rho < 1$, Equation (8.21) becomes

$$\sum_{k=0}^\infty \frac{\Gamma\left(k+\frac{1}{2}\right)}{\Gamma(k+1)} x_k P_{2k}\left(\sqrt{1-\rho^2}\right) = \frac{2\sqrt{2}}{\pi}. \quad (8.25)$$

Because the even order Legendre polynomials are orthogonal on $(0,1)$,

$$\int_0^1 \rho\left(1-\rho^2\right)^{-\frac{1}{2}} P_{2k}\left(\sqrt{1-\rho^2}\right) P_{2n}\left(\sqrt{1-\rho^2}\right) d\rho = (4n+1)^{-1} \delta_{kn}, \quad (8.26)$$

we may deduce

$$x_n = (2/\pi)^{\frac{3}{2}} \delta_{n0} \quad (n = 1, 2, \ldots). \quad (8.27)$$

Thus

$$F(\tau) = \tau^{-\frac{3}{2}} (2/\pi)^{\frac{3}{2}} J_{\frac{1}{2}}(\tau) = 4\pi^{-2}\tau^{-2} \sin\tau,$$

in agreement with the previously obtained result (8.19).

8.2 The charged elliptic plate

As well as its own intrinsic interest, the calculation of electrostatic potential due to a charged elliptic plate demonstrates basic steps of a more general method to calculate the potential of a flat charge plate of arbitrary shape. The fundamental idea is to use a parametrisation that reduces the original problem to *disc-like* equations with *disc-like* solutions.

Guided by the results of the previous section, let us consider the problem in Cartesian coordinates (see Figure 8.2). When the plate is charged to unit potential ($\psi^0 = 1$ on S_0), the form of the potential to be found is also

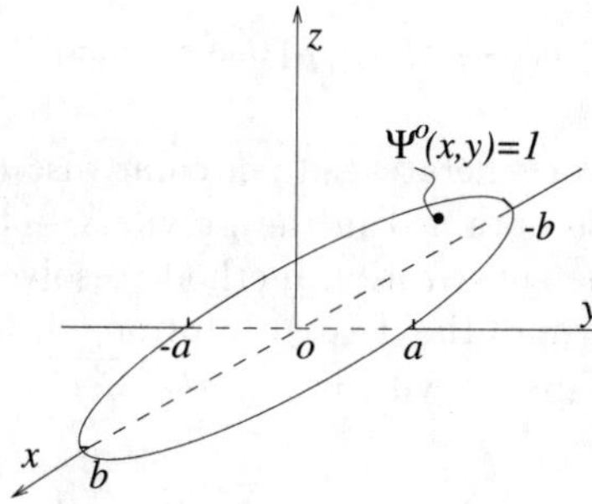

FIGURE 8.2. The charged elliptic disc.

given by (8.16). It should be noted that this simpler form is the result of symmetry. If a and b denote the minor and major semi-axes, respectively, introduce the coordinates

$$x = b\rho\cos\phi,\ \ y = a\rho\sin\phi \tag{8.28}$$

so that the boundary of the elliptic plate Γ is given by $\rho = 1$. Let $q = a/b$, so that $q \le 1$.

Use the boundary conditions (8.6) to obtain the dual integral equations for the unknown function $F(\nu,\mu)$, valid for $\phi \in \left(0, \frac{1}{2}\pi\right)$,

$$\int_0^\infty d\nu \cos(\nu b\rho\cos\phi) \int_0^\infty d\mu F(\nu,\mu)\cos(\mu a\rho\sin\phi) = 1, \qquad 0 \le \rho < 1, \tag{8.29}$$

$$\int_0^\infty d\nu \cos(\nu b\rho\cos\phi) \int_0^\infty d\mu \sqrt{\nu^2+\mu^2} F(\nu,\mu)\cos(\mu a\rho\sin\phi) = 0, \quad \rho > 1. \tag{8.30}$$

Again, use the series expansion (cf.(8.10)) involving even Chebyshev polynomials $T_{2m}(\cos\phi)$ with $\tau = \sqrt{\nu^2 + q^2\mu^2}$,

$$\cos(\nu b\rho\cos\phi)\cos(\mu a\rho\sin\phi)$$
$$= \sum_{m=0}^{\infty} \left(2-\delta_m^0\right) J_{2m}(b\rho\tau)\, T_{2m}\left(q\mu\tau^{-1}\right) T_{2m}(\cos\phi), \tag{8.31}$$

to reduce (8.29) and (8.30) to the equivalent dual integral equations involving the Bessel function kernel of form $J_0(\tau b\rho)$,

$$\int_0^\infty d\nu \int_0^\infty d\mu F(\nu,\mu) J_0(\tau b\rho) = 1, \qquad 0 \le \rho < 1, \tag{8.32}$$

$$\int_0^\infty d\nu \int_0^\infty d\mu \sqrt{\nu^2+\mu^2} F(\nu,\mu)\, J_0(\tau b \rho) = 0, \qquad \rho > 1. \tag{8.33}$$

When elliptic plate degenerates into circular disc ($q = 1, b = 1$), equations identical to those obtained in the previous section are obtained. As before, we may use the substitution method to solve these disc-like equations. The modified form of the desired solution (cf. (8.20)) that takes into account the elliptic shape is (with $\tau = \sqrt{\nu^2 + q^2\mu^2}$)

$$F(\nu,\mu) = \left(\nu^2+\mu^2\right)^{-\frac{1}{2}} \tau^{-\frac{1}{2}} \sum_{k=0}^{\infty} x_k J_{2k+\frac{1}{2}}(\tau b) \tag{8.34}$$

where the coefficients x_k are to be found. Insertion of this representation into Equations (8.32) and (8.33) produces the dual equations

$$\varkappa(q) \sum_{k=0}^{\infty} x_k \int_0^\infty \tau^{-\frac{1}{2}} J_0(\tau b\rho)\, J_{2k+\frac{1}{2}}(\tau b)\, d\tau = 1, \; 0 \le \rho < 1, \tag{8.35}$$

$$\sum_{k=0}^{\infty} x_k \int_0^\infty \tau^{\frac{1}{2}} J_0(\tau b\rho)\, J_{2k+\frac{1}{2}}(\tau b)\, d\tau = 0, \; \rho > 1, \tag{8.36}$$

where

$$\varkappa(q) = \int_0^1 \frac{dt}{\sqrt{(1-t^2)\left[1-(1-q^2)\,t^2\right]}} = K\left(\sqrt{1-q^2}\right) \tag{8.37}$$

is a complete elliptic integral of first kind.

As before, (8.24) shows that the second Equation (8.36) is automatically satisfied; from (8.23) one may transform (8.35) to

$$\sum_{k=0}^{\infty} \frac{\Gamma\left(k+\frac{1}{2}\right)}{\Gamma(k+1)} x_k P_{2k}\left(\sqrt{1-\rho^2}\right) = (2b)^{\frac{1}{2}} / K\left(\sqrt{1-q^2}\right), \qquad 0 \le \rho < 1. \tag{8.38}$$

This has the closed form solution

$$x_k = \delta_{0k}\,(2b)^{\frac{1}{2}} / K\left(\sqrt{1-q^2}\right) \qquad (k = 0, 1, 2, \ldots). \tag{8.39}$$

The solution for the unknown function $F(\nu,\mu)$ is deduced from (8.34) to be

$$F(\nu,\mu) = \frac{2}{\pi} \frac{\sin\left(\sqrt{\nu^2+q^2\mu^2}\,b\right)}{\sqrt{(\nu^2+q^2\mu^2)(\nu^2+\mu^2)}} \cdot \frac{1}{K\left(\sqrt{1-q^2}\right)}. \tag{8.40}$$

When the elliptic disc is circular ($q = 1$) the solution (8.40) coincides with (8.19) on the assumption that $b = 1$.

We may now calculate the capacitance C of the elliptic plates. At unit potential, the value of C numerically coincides with the total charge Q accumulated on the elliptic plate. This may be calculated by integration over the surface of the surface charge density, which equals the jump in normal component of the electrostatic field across the plate,

$$\sigma(x,y) = \frac{1}{4\pi}\left\{E_z(x,y,-0) - E_z(x,y,+0)\right\}.$$

The capacitance is readily found to be

$$C = b/K\left(\sqrt{1-q^2}\right). \tag{8.41}$$

8.2.1 *The spherically-curved elliptic plate*

The method of inversion allows us to calculate the capacitance of a curved elliptic plate. From the perspective of inversion, we are naturally led to consider the spherically-curved elliptic plate, conformal with the surface of a sphere, shown in Figure 8.3. Let M be the centre of inversion of a sphere of radius $2R$; consider the plane tangent to this sphere at the point O' antipodal to M. Under inversion, the image of this tangent plane is a sphere of radius R and centre O located at the midpoint of the segment MO'. The image of an ellipse lying in the tangent plane and centred at O' is spherically conformal; it is an elliptically-shaped region of the spherical surface. The image of the tangent plane with the elliptic disc removed is a spherical shell with an elliptic aperture. Introduce axes as shown in Figure 8.3: the z-axis coincides with OM, and the usual spherical polars (r,θ,ϕ) and cylindrical polars (ρ,ϕ,z) are centred at O. The elliptic disc lies in the plane $z=-R$. The map given by

$$\rho = 2R\tan\frac{1}{2}\theta$$

corresponds to inversion in the sphere of radius $2R$ centred at M, followed by the antipodal map $(r,\theta,\phi)\longmapsto(r,\pi-\theta,2\pi-\phi)$; it is the image of the elliptic disc under this map that is shown in Figure 8.3.

The boundary of the elliptic plate is specified

$$\rho(\varphi) = b/\sqrt{1+\kappa^2\sin^2\varphi},$$

where $\kappa = q^{-1}\sqrt{1-q^2}$, $q = a/b$; thus, the boundary of the spherically conformal elliptic region is given by

$$\theta(\varphi) = 2\arctan\left\{\frac{b}{2R}\left(1+\kappa^2\sin^2\varphi\right)^{-\frac{1}{2}}\right\}. \tag{8.42}$$

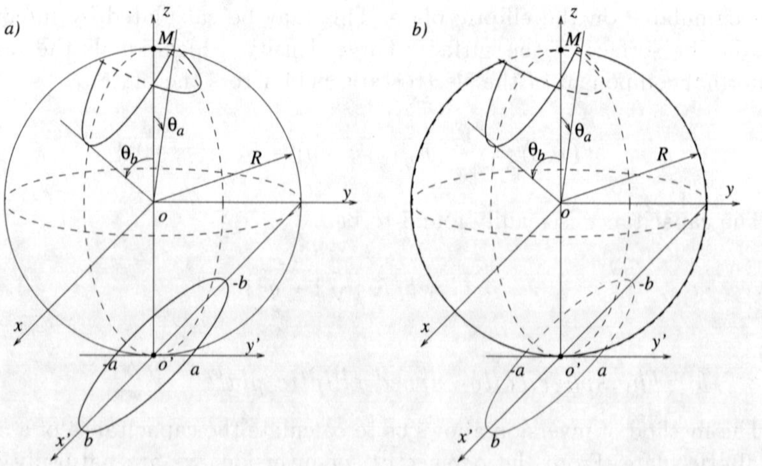

FIGURE 8.3. (a) The spherically conformal elliptic plate; its image under inversion is the elliptic disc. (b) The sphere with elliptic aperture; its image under inversion is the plane with the elliptic disc removed.

The angles

$$\theta_b = 2\arctan\left(\frac{b}{2R}\right), \ \theta_a = 2\arctan\left(\frac{a}{2R}\right) \tag{8.43}$$

corresponding to the image of end-points of the semi-axes of the plane ellipse measure the angular spread of the curved plate.

First consider the conformal plate charged to unit potential. According to Bouwkamp's theorem, the problem to be solved is equivalent to the electrostatic problem for the grounded planar elliptic plate in the presence of a unit negative charge located at M. The free-space potential generated by this charge is

$$\psi^0(x, y, z) = -\left\{x^2 + y^2 + (z - 2R)^2\right\}^{-\frac{1}{2}}. \tag{8.44}$$

Based upon previous results we are led to the following dual integral equations to be solved for the unknown function f,

$$\int_0^\infty d\nu \cos\nu x \int_0^\infty d\mu f(\nu, \mu) \cos\mu x = \left\{x^2 + y^2 + 4R^2\right\}^{-\frac{1}{2}}, \tag{8.45}$$

$$\int_0^\infty d\nu \cos\nu x \int_0^\infty d\mu \sqrt{\nu^2 + \mu^2} f(\nu, \mu) \cos\mu x = 0. \tag{8.46}$$

The first equation holds for points (x, y) lying inside the disc, whilst the second holds for those points outside. Substituting (8.28), we obtain

$$\int_0^\infty d\nu \cos(\nu b\rho\cos\phi) \int_0^\infty d\mu f(\nu,\mu)\cos(\mu a\rho\sin\phi)$$
$$= b^{-1}\left\{\rho^2+\gamma^2-k^2\rho^2\sin^2\phi\right\}^{-\frac{1}{2}}, \qquad 0 \le \rho < 1, \quad (8.47)$$

$$\int_0^\infty d\nu \cos(\nu b\rho\cos\phi) \int_0^\infty d\mu \sqrt{\nu^2+\mu^2} f(\nu,\mu)\cos(\mu a\rho\sin\phi) = 0, \ \rho > 1, \quad (8.48)$$

where $\gamma = 2R/b$ and $k = \sqrt{1-q^2}$; this holds for $0 \le \phi < \frac{1}{2}\pi$.

Expand the right-hand side of (8.47) in a Chebyshev series

$$\left\{\rho^2+\gamma^2-k^2\rho^2\sin^2\phi\right\}^{-\frac{1}{2}} = \sum_{m=0}^{\infty}(2-\delta_{0m})\alpha_{2m}T_{2m}(\cos\phi), \quad (8.49)$$

where

$$\alpha_{2m} = \alpha_{2m}(\rho) = \frac{2}{\pi}\int_0^{\frac{\pi}{2}} \frac{T_{2m}(\cos\phi)d\phi}{\sqrt{\rho^2+\gamma^2-k^2\rho^2\sin^2\phi}}. \quad (8.50)$$

In particular,

$$\alpha_0 = \alpha_0(\rho) = \frac{2}{\pi\sqrt{\rho^2+\gamma^2}} K\left(\frac{k\rho}{\sqrt{\rho^2+\gamma^2}}\right), \quad (8.51)$$

where K denotes the complete elliptic integral of first kind. It is evident that (8.47) and (8.48) imply that

$$\int_0^\infty d\nu \int_0^\infty d\mu f_{2m}(\nu,\mu) J_{2m}(\sqrt{\nu^2+q^2\mu^2}b\rho) = b^{-1}\alpha_{2m}(\rho), \ \rho < 1, \quad (8.52)$$

$$\int_0^\infty d\nu \int_0^\infty d\mu \sqrt{\nu^2+\mu^2} f_{2m}(\nu,\mu) J_{2m}(\sqrt{\nu^2+q^2\mu^2}b\rho) = 0, \ \rho > 1, \quad (8.53)$$

where $m = 0, 1, 2, \ldots$, and

$$f_{2m}(\nu,\mu) = T_{2m}\left(\frac{q\mu}{\sqrt{\nu^2+q^2\mu^2}}\right) f(\nu,\mu). \quad (8.54)$$

To find the solution use the extended form of the representation (8.34)

$$f_{2m}(\nu,\mu) = \left(\nu^2+\mu^2\right)^{-\frac{1}{2}}\left(\nu^2+q^2\mu^2\right)^{-\frac{1}{4}} \sum_{n=0}^{\infty} x_k^m J_{2k+2m+\frac{1}{2}}(\sqrt{\nu^2+q^2\mu^2}b). \quad (8.55)$$

Its substitution in (8.52) and (8.53) produces

$$\varkappa(q)\sum_{k=0}^{\infty} x_k^m \int_0^{\infty} \tau^{-\frac{1}{2}} J_{2k+2m+\frac{1}{2}}(\tau b)\, J_{2m}(\tau b\rho)\, d\tau = b^{-1}\alpha_{2m}, \quad 0 \le \rho < 1, \tag{8.56}$$

$$\sum_{k=0}^{\infty} x_k^m \int_0^{\infty} \tau^{\frac{1}{2}} J_{2k+2m+\frac{1}{2}}(\tau b)\, J_{2m}(\tau b\rho)\, d\tau = 0, \quad \rho > 1. \tag{8.57}$$

We employ the generalisation of the integrals given in (8.23) and (8.24),

$$\int_0^{\infty} \tau^{\frac{1}{2}} J_{2k+2m+\frac{1}{2}}(\tau b)\, J_{2m}(\tau b\rho)\, d\tau = \frac{2^{-2m+\frac{1}{2}}}{b^{\frac{3}{2}}} \frac{\Gamma(k+1)}{\Gamma\left(k+2m+\frac{1}{2}\right)} \frac{P_{2k+2m}^{2m}\left(\sqrt{1-\rho^2}\right)}{\sqrt{1-\rho^2}} H(1-\rho), \tag{8.58}$$

$$\int_0^{\infty} \tau^{-\frac{1}{2}} J_{2k+2m+\frac{1}{2}}(\tau b)\, J_{2m}(\tau b\rho)\, d\tau = \frac{2^{-2m-\frac{1}{2}}}{b^{\frac{1}{2}}} \frac{\Gamma\left(k+\frac{1}{2}\right)}{\Gamma(k+2m+1)} P_{2k+2m}^{2m}\left(\sqrt{1-\rho^2}\right), \quad 0 \le \rho < 1, \tag{8.59}$$

to deduce

$$\sum_{k=0}^{\infty} \frac{\Gamma\left(k+\frac{1}{2}\right)}{\Gamma(k+2m+1)} x_k^m P_{2k+2m}^{2m}\left(\sqrt{1-\rho^2}\right) = \frac{2^{2m+\frac{1}{2}}}{b^{\frac{1}{2}}\varkappa(q)} \alpha_{2m}(\rho), \quad 0 \le \rho < 1, \tag{8.60}$$

where $m = 0, 1, 2, \ldots$. The solution of this equation immediately follows by exploiting the orthogonality property of the associated Legendre functions on $(0,1)$:

$$\int_0^1 \frac{\rho}{\sqrt{1-\rho^2}} P_{2k+2m}^{2m}\left(\sqrt{1-\rho^2}\right) P_{2s+2m}^{2m}\left(\sqrt{1-\rho^2}\right) d\rho = \frac{1}{4k+4m+1} \frac{\Gamma(2k+4m+1)}{\Gamma(2k+1)} \delta_{ks}. \tag{8.61}$$

As a result we obtain

$$x_s^m = \left(\frac{2}{b}\right)^{\frac{1}{2}} 2^{-2m} \frac{(4s+4m+1)\,\Gamma(s+1)}{\Gamma\left(s+2m+\frac{1}{2}\right)\Gamma\left(s+\frac{1}{2}\right)} \frac{\beta_{sm}}{\varkappa(q)}, \tag{8.62}$$

where

$$\beta_{sm} = \int_0^1 \frac{\rho}{\sqrt{1-\rho^2}} \alpha_{2m}(\rho) P_{2s+2m}^{2m}\left(\sqrt{1-\rho^2}\right) d\rho. \tag{8.63}$$

We may now calculate the capacitance C of the spherically conforming elliptic plate. By Bouwkamp's theorem, it is proportional to the value of the induced potential at the centre of inversion M:

$$C = 4R^2\psi(0,0,2R). \tag{8.64}$$

It is readily seen that the calculation of C only requires a knowledge of the function $f_0(\nu,\mu) = f(\nu,\mu)$. Let us now demonstrate the solution of Equations (8.52) and (8.53) (with $m = 0$) by the Abel integral transform method. Based on the results at the beginning of this section (see also (8.55)), let us seek the unknown function f in the form

$$f(\nu,\mu) = \left(\nu^2+\mu^2\right)^{-\frac{1}{2}} F\left(\sqrt{\nu^2+q^2\mu^2}\right). \tag{8.65}$$

After some evident manipulation, we obtain the dual integral equations

$$\int_0^\infty F(\tau)J_0(\tau\rho b)d\tau = \frac{2}{\pi b}\frac{1}{\sqrt{\rho^2+\gamma^2}}\frac{1}{K\left(\sqrt{1-q^2}\right)}K\left(\frac{\sqrt{1-q^2}\rho}{\sqrt{\rho^2+\gamma^2}}\right),$$
$$0 \le \rho < 1, \tag{8.66}$$

$$\int_0^\infty \tau F(\tau)J_0(\tau\rho b)d\tau = 0, \qquad \rho > 1. \tag{8.67}$$

Use the method described in Section 2.7 to transform these dual equations to the Fourier cosine form

$$\int_0^\infty F(\tau)\cos\tau\rho b\, d\tau = \frac{b^{-1}\gamma}{K\left(\sqrt{1-q^2}\right)}\left(\gamma^2+\rho^2\right)^{-\frac{1}{2}}\left(\gamma^2+q^2\rho^2\right)^{-\frac{1}{2}} H(1-\rho), \tag{8.68}$$

and invert this expression to obtain

$$F(\tau) = \frac{2}{\pi}\frac{b^{-1}\gamma}{K\left(\sqrt{1-q^2}\right)}\int_0^1 \frac{\cos\tau\rho b}{\sqrt{\left(\gamma^2+\rho^2\right)\left(\gamma^2+q^2\rho^2\right)}}d\rho. \tag{8.69}$$

According to (8.64) the capacitance of the spherically-conforming elliptic plate is

$$C = 4R^2\int_0^1 \left\{\left(1-t^2\right)\left[1-\left(1-q^2\right)t^2\right]\right\}^{-\frac{1}{2}} \times$$
$$\int_0^\infty F(\tau)e^{-\tau\sqrt{1-(1-q^2)t^2}2R/q}d\tau dt. \tag{8.70}$$

Remarkably, substitution of (8.69) into (8.70) produces the closed form expression

$$C = \frac{2R}{K\left(\sqrt{1-q^2}\right)} \left\{ \frac{\arctan \gamma^{-1} - q \arctan q\gamma^{-1}}{1-q^2} \right\}, \tag{8.71}$$

which may be written in terms of the angles θ_a, θ_b (defined by (8.43)) as

$$C = \frac{R}{K\left(\sqrt{1-q^2}\right)} \left\{ \frac{\theta_b - q\theta_a}{1-q^2} \right\}. \tag{8.72}$$

When the elliptic plate degenerates to a circular disc ($q \to 1$), the conforming plate becomes a spherical cap; since $K(0) = \frac{1}{2}\pi$ and

$$\lim_{q \to 1} \frac{\arctan \gamma^{-1} - q \arctan q\gamma^{-1}}{1-q^2} = \frac{1}{2}\left\{ \arctan \gamma^{-1} + \gamma \left(1+\gamma^2\right)^{-1} \right\}$$
$$= \frac{1}{4}\left\{\theta_b + \sin\theta_b\right\} \tag{8.73}$$

the expression for its capacitance reduces to the well-known value previously calculated for the spherical cap, namely $\pi^{-1}(\theta_b + \sin\theta_b)$.

This completes our discussion of the capacitance of the spherically conforming elliptic plate. The complementary structure – the spherical shell with an elliptic aperture – may be analysed in a similar fashion.

8.3 Polygonal plates

In contrast to the plates with smooth boundaries considered in previous sections, this section examines polygonal plates, particularly regular polygons of N equal sides ($N = 3, 4, \ldots$). As shown in Figure 8.4, the angle subtended by each side at the centre O of the polygon is $2\alpha = 2\pi/N$. If the circle circumscribing the polygon has radius a, the difference in length between an edge AB of the polygon and the circular arc AB of the circumscribing circle is $a\left(2\pi/N - 2\sin\pi/N\right)$; as $n \to \infty$, this difference is $\frac{1}{3}\pi^3/N^3 + O(N^{-5})$, and the circle approximates the polygon in some sense. When the plate is charged, symmetry implies that we may concentrate on the right-angled triangular sector OAC, where the angle $\widehat{OAC} = \alpha$.

The potential on the charged circular plate $S_0 = \left\{(x,y,0) : x^2 + y^2 < a^2\right\}$ is determined by the dual equations of the form (see (8.16))

$$\int_0^\infty d\nu \cos\nu x \int_0^\infty d\mu f(\mu,\nu) \cos\mu y = 1, (x,y) \in S_0, \tag{8.74}$$

$$\int_0^\infty d\nu \cos\nu x \int_0^\infty d\mu \sqrt{\nu^2+\mu^2} f(\mu,\nu) \cos\mu y = 0, (x,y) \notin S_0. \tag{8.75}$$

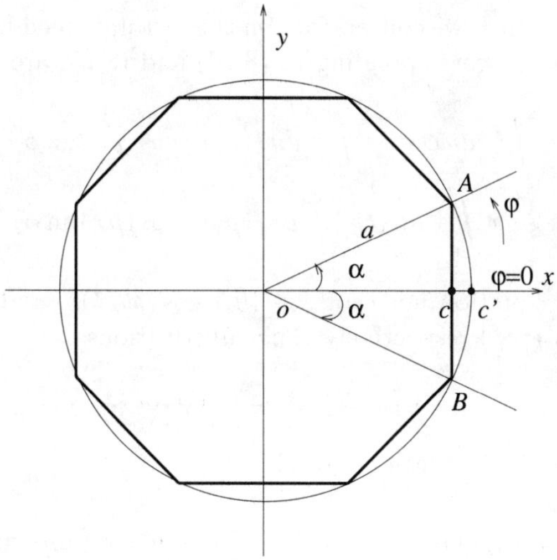

FIGURE 8.4. The polygonal plate and circumscribing circle.

These dual integral equations were solved in Section 8.1. The same equations hold for the polygon charged to unit potential, except the region S_0 is differently defined. It is sufficient to consider the triangular region OAC and the associated unbounded sector defined by angle α.

In the limit when $N \to \infty$, this sector degenerates to a half-line or ray. If we consider the ray $y = 0$, the equations (8.74), (8.75) are

$$\int_0^\infty d\mu \int_0^\infty d\nu f(\mu,\nu)\cos\nu x = 1,\ 0 < x < a, \tag{8.76}$$

$$\int_0^\infty d\mu \int_0^\infty d\nu \sqrt{\nu^2+\mu^2} f(\mu,\nu)\cos\nu x = 0,\ x > a. \tag{8.77}$$

The substitution $\tau = \sqrt{\nu^2+\mu^2}$ leads to the readily solvable equations for the potential distribution on the circular disc,

$$\int_0^\infty \tau f(\tau) J_0(\tau x) d\tau = \frac{2}{\pi},\ 0 < x < a, \tag{8.78}$$

$$\int_0^\infty \tau^2 f(\tau) J_0(\tau x) d\tau = 0,\ x > a. \tag{8.79}$$

We shall solve the potential problem by transforming the dual equations to a form that may be recognised as a perturbation of the equations describing the circular disc.

Setting $y = x \tan \phi$, we concentrate on the sector defined by $\phi \in (0, \pi/N)$. The dual equations corresponding to (8.74) and (8.75) are

$$\int_0^\infty d\nu \cos \nu x \int_0^\infty d\mu f(\mu, \nu) \cos (\mu x \tan \phi) = 1, \quad (8.80)$$

$$\int_0^\infty d\nu \cos \nu x \int_0^\infty d\mu \sqrt{\nu^2 + \mu^2} f(\mu, \nu) \cos (\mu x \tan \phi) = 0, \quad (8.81)$$

where the first equation hods for $x \in (0, a \cos (\pi/N))$, and the second for $x \in (a \cos (\pi/N), \infty)$ respectively. The substitutions

$$\rho = x \sec (\pi/N), \ u = \tan \phi \cot (\pi/N), \quad (8.82)$$

transform these dual equations to

$$\int_0^\infty d\nu \cos \left(\nu \rho \cos \frac{\pi}{N}\right) \int_0^\infty d\mu f(\mu, \nu) \cos \left(\mu \rho u \sin \frac{\pi}{N}\right) = 1, \quad (8.83)$$

$$\int_0^\infty d\nu \cos \left(\nu \rho \cos \frac{\pi}{N}\right) \int_0^\infty d\mu \sqrt{\nu^2 + \mu^2} f(\mu, \nu) \cos \left(\mu \rho u \sin \frac{\pi}{N}\right) = 0, \quad (8.84)$$

where the first equation holds for $\rho \in (0, a), u \in (0, 1)$ and the second for $\rho \in (a, \infty), u \in (0, 1)$. Arguing as in Section 8.1, the dependence upon u in these equations can be eliminated by transformation to the form

$$\int_0^\infty d\nu \cos \left(\nu \rho \cos \frac{\pi}{N}\right) \int_0^\infty d\mu f(\mu, \nu) J_0 \left(\mu \rho \sin \frac{\pi}{N}\right) = 1, \quad (8.85)$$

$$\int_0^\infty d\nu \cos \left(\nu \rho \cos \frac{\pi}{N}\right) \int_0^\infty d\mu \sqrt{\nu^2 + \mu^2} f(\mu, \nu) J_0 \left(\mu \rho \sin \frac{\pi}{N}\right) = 0, \quad (8.86)$$

holding for $\rho \in (0, a)$ and $\rho \in (a, \infty)$, respectively. It should be observed that when $N \to \infty$, Equations (8.85) and (8.86) degenerate to (8.76) and (8.77).

When the plate is a circular or elliptic disc, the dual equations analogous to (8.85) and (8.86) have particularly simple solutions of the form $f(\mu, \nu) = f(\tau)$ (where $\tau = \sqrt{\nu^2 + \mu^2}$ for the circular disc, and $\tau = \sqrt{\nu^2 + q^2 \mu^2}$ for the elliptic disc). It is not obvious *a priori* that the solution $f(\mu, \nu)$ to (8.85) and (8.86) has a solution of a similarly simple form. However, it turns out that the form is exactly the same as that for the circular disc; thus, we shall assume

$$f(\mu, \nu) = f(\tau), \text{ where } \tau = \sqrt{\nu^2 + \mu^2}, \quad (8.87)$$

and justify this assumption retrospectively by showing that the solution so constructed satisfies all equations and associated conditions. With this assumption, the dual equations become

$$\int_0^\infty F(\tau)S_N(\tau\rho)d\tau = 1,\ \rho \in (0,a)\,, \tag{8.88}$$

$$\int_0^\infty \tau F(\tau)S_N(\tau\rho)d\tau = 0,\ \rho \in (a,\infty)\,, \tag{8.89}$$

where $F(\tau) = \tau f(\tau)$ and the kernel S_N is defined by

$$\begin{aligned} S_N(\tau\rho) &= \int_0^\tau \frac{J_0\left(\sqrt{\tau^2-\nu^2}\rho\sin\frac{\pi}{N}\right)}{\sqrt{\tau^2-\nu^2}}\cos\left(\nu\rho\cos\frac{\pi}{N}\right)d\tau \\ &= \int_0^\tau \frac{\cos\left(\sqrt{\tau^2-\nu^2}\rho\cos\frac{\pi}{N}\right)}{\sqrt{\tau^2-\nu^2}}J_0\left(\nu\rho\sin\frac{\pi}{N}\right)d\tau \\ &= \frac{\pi}{2}J_0\left(\tau\rho\cos^2\frac{\pi}{2N}\right)J_0\left(\tau\rho\sin^2\frac{\pi}{2N}\right). \end{aligned} \tag{8.90}$$

When $N \to \infty$, the kernel becomes

$$S_\infty(\tau\rho) = \lim_{N\to\infty} S_N(\tau\rho) = \frac{\pi}{2}J_0(\tau\rho)\,, \tag{8.91}$$

which is identical with that encountered for the circular disc.

This construction justifies our assumption of the form (8.87) for f. We may therefore seek the solution to the dual equations in the form

$$F(\tau) = \int_0^a G(\tau)\cos\left(\tau t\cos\frac{\pi}{N}\right)dt \tag{8.92}$$

where both the function G and its derivative G' are continuous on $(0,a)$. Integrating by parts, F is representable as

$$F(\tau) = \sec\frac{\pi}{N}\left\{G(a)\frac{\sin\left(a\tau\cos\frac{\pi}{N}\right)}{\tau} - \frac{1}{\tau}\int_0^a G'(\tau)\sin\left(\tau t\cos\frac{\pi}{N}\right)dt\right\}. \tag{8.93}$$

Now substitute (8.93) into (8.89) and invert the order of integration. Then when $\rho > a$,

$$\begin{aligned} G(a)\int_0^\infty & S_N(\tau\rho)\sin\left(a\tau\cos\frac{\pi}{N}\right)d\tau - \\ & \int_0^a G'(t)\left\{S_N(\tau\rho)\sin\left(t\tau\cos\frac{\pi}{N}\right)d\tau\right\}dt = 0. \end{aligned} \tag{8.94}$$

However, it is well known (see [19]) that

$$\int_0^\infty J_\nu(ax)J_\nu(bx)\sin xy\,dx = 0, \qquad 0 < y < b-a, \tag{8.95}$$

when $b > a, \operatorname{Re}\nu > -1$, so that the equation (8.94) holds identically.

Following the basic idea of regularisation, we split the kernel S_N as a sum of its limiting value S_∞ and a correction term and analytically invert that part of the equation containing the limiting kernel contribution, corresponding to the circular disc problem. This is most naturally done in the present context by using the result derived from the addition theorem for Bessel functions [14],

$$J_0\left(\tau\rho\cos^2\frac{\pi}{2N}\right) J_0\left(\tau\rho\sin^2\frac{\pi}{2N}\right) = J_0(\tau\rho) - 2\sum_{n=1}^{\infty}(-1)^n J_n\left(\tau\rho\cos^2\frac{\pi}{2N}\right) J_n\left(\tau\rho\sin^2\frac{\pi}{2N}\right). \quad (8.96)$$

We may now construct the representation of the function to be determined. First expand G in a series with Gegenbauer polynomials $C_{2k}^{(\frac{1}{2})} = P_{2k}$,

$$G(t) = \sum_{k=1}^{\infty} b_k C_{2k}^{(\frac{1}{2})}(t/a). \quad (8.97)$$

Substitute this expression in (8.92), invert the order of summation and integration and obtain

$$F(\tau) = \sum_{k=1}^{\infty} b_k \int_0^a \cos\left(\tau t\cos\frac{\pi}{N}\right) C_{2k}^{(\frac{1}{2})}(t/a)dt. \quad (8.98)$$

Using the tabulated integral [14] (Vol. 1)

$$\int_0^a \cos\left(\tau t\cos\frac{\pi}{N}\right) C_{2k}^{(\frac{1}{2})}(t/a)dt = (-1)^k\left(\frac{\pi a}{2}\sec\frac{\pi}{N}\right)^{\frac{1}{2}} \tau^{-\frac{1}{2}} J_{2k+\frac{1}{2}}\left(\tau a\cos\frac{\pi}{N}\right), \quad (8.99)$$

we deduce that

$$F(\tau) = \tau^{-\frac{1}{2}} \sum_{k=1}^{\infty} b_k^* J_{2k+\frac{1}{2}}\left(\tau a\cos\frac{\pi}{N}\right), \quad (8.100)$$

where

$$b_k^* = (-1)^k\left(\frac{\pi a}{2}\sec\frac{\pi}{N}\right)^{\frac{1}{2}} b_k. \quad (8.101)$$

Substitute (8.100) into (8.88) and change the order of integration and summation to obtain

$$\sum_{k=1}^{\infty} b_k^* \int_0^\infty \tau^{-\frac{1}{2}} S_N(\tau\rho) J_{2k+\frac{1}{2}}\left(\tau a\cos\frac{\pi}{N}\right) d\tau = 1, \quad \rho \in (0,a). \quad (8.102)$$

We recall that Equation (8.89) is satisfied automatically with the representation (8.92) or its equivalent form (8.100). After some manipulation, we deduce from (8.100) that

$$\sum_{k=0}^{\infty} b_k^* \frac{\Gamma\left(k+\frac{1}{2}\right)}{\Gamma\left(k+1\right)} P_{2k}\left(\sqrt{1-\rho^2/a^2}\right)$$

$$= \frac{2}{\pi}\left(2a\cos\frac{\pi}{N}\right)^{\frac{1}{2}} + \sum_{k=0}^{\infty} b_k^* \frac{\Gamma\left(k+\frac{1}{2}\right)}{\Gamma\left(k+1\right)} F_k\left(\rho\right), \; \rho \in (0,a), \quad (8.103)$$

where

$$F_k\left(\rho\right) = P_{2k}\left(\sqrt{1-\rho^2/a^2}\right) - \int_0^{\pi} \wp_k\left(\rho, x\right) dx, \quad (8.104)$$

$$\wp_k\left(\rho, x\right) = \begin{cases} \pi^{-1} P_{2k}\left(\sqrt{1-\rho^2/\rho_c^2}\right), & \rho < \rho_c, \\ 2\pi^{-\frac{1}{2}} \arcsin\left(\rho/\rho_c\right) \Gamma\left(k+\frac{1}{2}\right)/\Gamma\left(k+1\right), & \rho > \rho_c, \end{cases} \quad (8.105)$$

and the value of ρ_c is defined by the relation

$$\frac{a}{\rho_c} = \sec\frac{\pi}{N}\left(1 - \sin^2\frac{\pi}{N}\cos^2\frac{\pi}{N}\right)^{\frac{1}{2}}. \quad (8.106)$$

It is evident that as $N \to \infty$, $F_k\left(\rho\right) \to 0$.

Apply the usual principle of orthogonality of Legendre polynomials on the interval $[0,1]$ to obtain the i.s.l.a.e. of the second kind,

$$x_s - \sum_{k=0}^{\infty} \gamma_{ks} x_k = \frac{2}{\pi}\delta_{0s}, \quad (8.107)$$

for $s = 0, 1, 2, \ldots$, where

$$b_k^* = \left(2a\cos\frac{\pi}{N}\right)^{\frac{1}{2}} \frac{\Gamma\left(k+1\right)}{\Gamma\left(k+\frac{1}{2}\right)} \left(4k+1\right)^{\frac{1}{2}} x_k, \quad (8.108)$$

and

$$\gamma_{ks} = \left[\left(4k+1\right)\left(4s+1\right)\right]^{\frac{1}{2}} \int_0^1 \frac{t}{\sqrt{1-t^2}} F_k\left(t\right) P_{2s}\left(\sqrt{1-t^2}\right) dt. \quad (8.109)$$

The solution of this system of equations $\{x_k\}_{k=0}^{\infty}$ is sought in l_2. The computation of the integrals defining the matrix elements is straightforward. Furthermore, as $N \to \infty$, $F_k\left(\rho\right) \to 0$ and estimates of the difference between the potential distribution for a circular disc and a polygonal disc with many vertices ($N \gg 1$) are readily derived from (8.107).

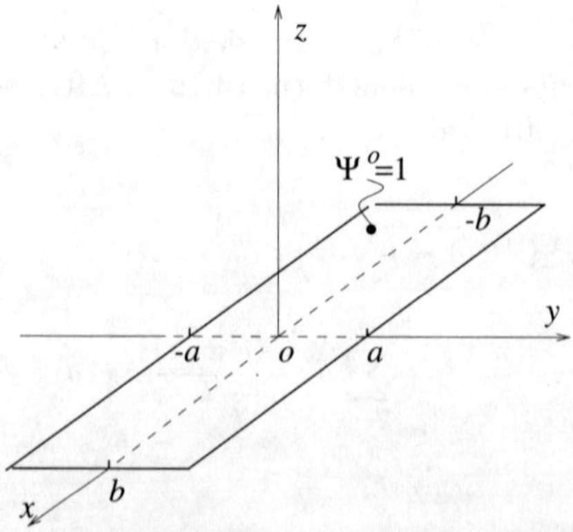

FIGURE 8.5. The finite strip.

8.4 The finite strip

In Section 7.2 we examined the potential associated with charged infinitely long thin strips. Although this two-dimensional problem has its own intrinsic interest, it is worth examining the more physically realistic structure of a finitely long strip. Consider the flat strip of width $2a$ and length $2b > 2a$ lying in the plane $z = 0$ as shown in Figure 8.5. The centre lies at the origin and the edges are aligned with the x and y axes. Suppose the strip is charged to unit potential. The mixed boundary conditions satisfied by the electrostatic potential ψ are

$$\psi(x, y, +0) = \psi(x, y, -0) = 1, \ |x| \le a, |y| \le b, \tag{8.110}$$

and by its normal derivative are

$$\frac{\partial}{\partial z}\psi(x, y, +0) = \frac{\partial}{\partial z}\psi(x, y, +0), \ |x| > a \text{ or } |y| > b. \tag{8.111}$$

The symmetry of the structure leads to the familiar form (8.16) for the solution, and enforcement of the mixed boundary conditions leads to dual integral equations for the unknown function $F = F(\nu, \mu)$,

$$\int_0^\infty d\nu \cos \nu b x' \int_0^\infty d\mu F(\nu, \mu) \cos(\mu a y') = 1, |x'| \le 1, |y'| \le 1, \tag{8.112}$$

$$\int_0^\infty d\nu \cos \nu b x' \int_0^\infty d\mu \sqrt{\nu^2 + \mu^2} F(\nu, \mu) \cos(\mu a y') = 0, |x'| > 1 \text{ or } |y'| > 1 \tag{8.113}$$

where $x' = x/a$, $y' = y/b$.

The distinctive feature of these equations is the apparent lack of coupling between the rescaled variables x' and y'. This dictates a special choice for the form of the solution to be found by the substitution method. In order to satisfy (8.113) automatically, it is sufficient to represent the unknown function F by an expansion in Bessel functions of even order,

$$F(\nu,\mu) = \left(\nu^2+\mu^2\right)^{-\frac{1}{2}} \sum_{n=0}^{\infty}\sum_{m=0}^{\infty} x_{nm} J_{2n}(\nu b) J_{2m}(\mu a), \tag{8.114}$$

where the coefficients x_{nm} are to be determined. Substitution of this form in (8.113) leads to

$$\sum_{n=0}^{\infty}\sum_{m=0}^{\infty} x_{nm} \int_0^{\infty} d\nu \cos \nu b x' J_{2n}(\nu b) \int_0^{\infty} d\mu \cos\left(\mu a y'\right) J_{2m}(\mu a) = 0, \tag{8.115}$$

when $|x'| > 1$ or $|y'| > 1$. The product of integrals occurring in (8.115) vanish because [19]

$$\int_0^{\infty} J_{2n}(\alpha x) \cos xy dx = (-1)^n \left(\alpha^2 - y^2\right)^{-\frac{1}{2}} T_{2n}(y/\alpha) H\left(\alpha^2 - y^2\right). \tag{8.116}$$

Moreover, it is apparent from (8.116) that the behaviour of the surface charge density $\sigma(x', y')$ near the edges will be in accord with physical expectation, namely

$$\sigma(x', y') \backsim \sigma_0 \left(b^2 - x^2\right)^{-\frac{1}{2}} \left(a^2 - y^2\right)^{-\frac{1}{2}} \qquad (\sigma_0 \text{ constant}).$$

Now substitute (8.114) into (8.112). Using the expansions

$$\begin{aligned}
\cos\left(\nu b x'\right) &= \sum_{s=0}^{\infty} (-1)^s \left(2-\delta_{0s}\right) J_{2s}(\nu b) T_{2s}(x'), \\
\cos\left(\mu a y'\right) &= \sum_{p=0}^{\infty} (-1)^p \left(2-\delta_{0p}\right) J_{2p}(\mu a) T_{2p}(y'),
\end{aligned}$$

and the orthogonality of the Chebyshev polynomials on $[0,1]$, we obtain the i.s.l.a.e. for the unknowns $x_{nm} (n, m = 0, 1, 2, \ldots)$,

$$\sum_{n=0}^{\infty}\sum_{m=0}^{\infty} x_{nm} R_{nmsp} = \delta_{0s}\delta_{0p}, \tag{8.117}$$

where $s, p = 0, 1, 2, \dots$, and matrix elements R_{nmsp} are given by

$$(-1)^{s+p} \int_0^\infty \int_0^\infty d\nu d\mu \left(\nu^2+\mu^2\right)^{-\frac{1}{2}} J_{2n}(\nu b) J_{2m}(\mu a) J_{2s}(\nu b) J_{2p}(\mu a)$$
$$= \frac{(-1)^{s+p}}{b} \int_0^\infty du J_{2n}(u) J_{2s}(u) \int_0^\infty dv \left(u^2+v^2\right)^{-\frac{1}{2}} J_{2m}(qv) J_{2p}(qv), \tag{8.118}$$

with $q = a/b$. This reduction to the i.s.l.a.e. (8.117) is a very formal procedure. The representation (8.118) of the matrix elements R_{nmsp} in terms of slowly convergent iterated integrals makes numerical procedures problematic.

Let us transform (8.118), where for convenience we will set $b = 1$. Making use of the representation for the product of Bessel functions [14]

$$J_{2n}(u) J_{2s}(u) = \frac{2}{\pi} \int_0^{\frac{\pi}{2}} J_{2n+2s}(2u\cos\theta) \cos\left[(2s-2n)\,\theta\right] d\theta, \tag{8.119}$$

valid when $\mathrm{Re}\,(\nu+\mu) > -1$, and the tabulated integral [14]

$$\int_0^\infty \frac{J_\nu(cx)dx}{\sqrt{x^2+z^2}} = I_{\frac{1}{2}\nu}\left(\frac{cz}{2}\right) K_{\frac{1}{2}\nu}\left(\frac{cz}{2}\right), \tag{8.120}$$

valid when $c > 0, \mathrm{Re}\, z > 0, \mathrm{Re}\,\nu > -1$, the expression for the matrix element R_{nmsp} becomes

$$R_{nmsp} = \frac{2}{\pi}(-1)^{s+p} \int_0^{\frac{\pi}{2}} \cos\left(2n-2s\right)\phi \times$$
$$\left\{\int_0^\infty J_{2m}(qv) J_{2p}(qv) I_{n+s}\left(v\cos\phi\right) K_{n+s}(v\cos\phi) dv\right\} d\phi. \tag{8.121}$$

Using the Mellin transform one may represent the product of modified Bessel functions occurring in (8.121) in the form (see [61])

$$I_{n+s}\left(v\cos\phi\right) K_{n+s}(v\cos\phi) =$$
$$= \frac{1}{8\pi^{\frac{3}{2}} i} \int_{c-i\infty}^{c+i\infty} \frac{\Gamma\left(n+s+\frac{t}{2}\right)\Gamma\left(\frac{t}{2}\right)\Gamma\left(\frac{1}{2}-\frac{t}{2}\right)}{\Gamma\left(n+s+1-\frac{t}{2}\right)} \cos^{-t}\phi v^{-t} dt, \tag{8.122}$$

where $0 < c < 1$. After substitution of (8.122) into (8.121) and some obvious rearrangement, the expression for the matrix element takes the form

$$R_{nmsp} = \frac{(-1)^{s+p}}{4\pi^{\frac{5}{2}} i} \int_{c-i\infty}^{c+i\infty} \frac{\Gamma\left(n+s+\frac{t}{2}\right)\Gamma\left(\frac{t}{2}\right)\Gamma\left(\frac{1}{2}-\frac{t}{2}\right)}{\Gamma\left(n+s+1-\frac{t}{2}\right)} A_{ns}\left(t\right) B_{mp}\left(t\right) dt \tag{8.123}$$

where

$$A_{ns}(t) = \int_0^{\frac{\pi}{2}} \cos^{-t}\phi \cos(2n-2s)\phi d\phi, \tag{8.124}$$

$$B_{mp}(t) = \int_0^{\infty} v^{-t} J_{2m}(qv) J_{2p}(qv) dv. \tag{8.125}$$

Both integrals occurring in (8.124) and (8.125) are tabulated in [19], and so

$$A_{ns}(t) = \frac{\sqrt{\pi}}{2} \frac{\Gamma\left(\frac{1}{2}-\frac{t}{2}\right)\Gamma\left(1-\frac{t}{2}\right)}{\Gamma\left(s-n+1-\frac{t}{2}\right)\Gamma\left(n-s+1-\frac{t}{2}\right)}, \tag{8.126}$$

$$B_{mp}(t) = \frac{q^{t-1}}{2\sqrt{\pi}} \frac{\Gamma\left(\frac{t}{2}\right)\Gamma\left(\frac{1}{2}+\frac{t}{2}\right)\Gamma\left(p+m+\frac{1}{2}-\frac{t}{2}\right)}{\Gamma\left(m-p+\frac{1}{2}+\frac{t}{2}\right)\Gamma\left(p+m+\frac{1}{2}+\frac{t}{2}\right)\Gamma\left(p-m+\frac{1}{2}+\frac{t}{2}\right)}. \tag{8.127}$$

Insert (8.126) and (8.127) into (8.123), make the substitution $t = 2r + 1$, and replace r by t, to obtain

$$\begin{aligned} R_{nmsp} = \frac{(-1)^{s+p}}{4\pi^{\frac{3}{2}}} \frac{1}{2\pi i} \int_L & \frac{\Gamma^2\left(\frac{1}{2}+t\right)\Gamma^2(-t)\Gamma\left(\frac{1}{2}-t\right)\Gamma(1+t)}{\Gamma\left(n+s+\frac{1}{2}-t\right)\Gamma\left(s-n+\frac{1}{2}-t\right)} \times \\ & \frac{\Gamma\left(n+s+\frac{1}{2}+t\right)\Gamma(p+m-t)}{\Gamma\left(n-s+\frac{1}{2}-t\right)\Gamma(m-p+1+t)} \times \\ & \frac{1}{\Gamma(p+m+1+t)\Gamma(p-m+1+t)} q^{2t} dt \end{aligned} \tag{8.128}$$

where the contour L runs from $-i\infty$ to $+i\infty$, intersecting the real axis at a point t_0 satisfying the inequality $-\frac{1}{2} < t_0 < 0$. It is evident that all the poles of $\Gamma(-t)$ and $\Gamma(\lambda - t)$ lie to the right of L, whereas all the poles of $\Gamma(1+t), \Gamma\left(\frac{1}{2}+t\right)$ and $\Gamma(\mu+t)$ lie to the left of L. We may express the contour integral in terms of *Meijer's G-function*, as defined in [14], via

$$R_{nmsp} = \frac{(-1)^{s+p}}{4\pi^{\frac{3}{2}}} G^{4,4}_{7,7}\left(q^2 \left| \begin{matrix} \frac{1}{2}, \frac{1}{2}, 0, -n-s+\frac{1}{2}, n+s+\frac{1}{2}, s-n+\frac{1}{2}, n-s+\frac{1}{2} \\ 0, 0, \frac{1}{2}, \quad p+m, \quad p-m, \quad -p-m, \quad m-p \end{matrix} \right.\right). \tag{8.129}$$

When $s = n$ and $p = m$, simple identities satisfied by Meijer's G-function show that the "diagonal" matrix elements R_{nmnm} are given by

$$R_{nmnm} = \frac{(-1)^{n+m}}{4\pi^{\frac{3}{2}}} G^{3,3}_{5,5}\left(q^2 \left| \begin{matrix} \frac{1}{2}, & \frac{1}{2}, & -2n+\frac{1}{2}, & 2n+\frac{1}{2}, & \frac{1}{2} \\ 0, & 0, & 2m, & 0, & -2m \end{matrix} \right.\right). \tag{8.130}$$

Using the well-known relations for the Gamma function (see Appendix, (B.3))

$$\begin{aligned}\Gamma(-t)\Gamma(1+t) &= -\frac{\pi}{\sin(\pi t)},\\ \Gamma(\lambda-t)\Gamma(-\lambda+1+t) &= -(-1)^{\lambda}\frac{\pi}{\sin(\pi t)},\\ \Gamma\left(\mu+\frac{1}{2}-t\right)\Gamma\left(-\mu+\frac{1}{2}+t\right) &= (-1)^{\mu}\frac{\pi}{\cos(\pi t)},\end{aligned}$$

we may derive the expression

$$R_{nmsp} = -\frac{(-1)^{p+m}}{4\sqrt{\pi}}\frac{1}{2\pi i}\int_L \frac{\cos^2(\pi t)}{\sin^3(\pi t)}\frac{\Gamma\left(\frac{1}{2}+t\right)}{\Gamma(1+t)}\frac{\Gamma\left(n+s+\frac{1}{2}+t\right)}{\Gamma(-p+m+1+t)}\times$$
$$\frac{\Gamma\left(-n-s+\frac{1}{2}+t\right)\Gamma\left(-s+n+\frac{1}{2}+t\right)\Gamma\left(s-n+\frac{1}{2}+t\right)}{\Gamma(m-p+1+t)\Gamma(p+m+1+t)\Gamma(p-m+1+t)}q^{2t}\,dt. \quad (8.131)$$

Evaluation of the contour is thus reduced to the evaluation of residues at the poles $t = 0, 1, 2, \ldots$. After some manipulation, this yields

$$R_{nmsp} = \frac{(-1)^{\mu}}{8\pi}\sum_{k=0}^{\infty}\frac{\left(\frac{1}{2}\right)_k\left(\kappa+\frac{1}{2}\right)_k\left(-\kappa+\frac{1}{2}\right)_k\left(\lambda+\frac{1}{2}\right)_k\left(-\lambda+\frac{1}{2}\right)_k}{k!\,(\mu+k)!\,(\nu+k)!}\times$$
$$q^{2k}N^k_{\kappa\lambda\mu\nu} \quad (8.132)$$

where $\kappa = n+s, \lambda = s-n, \mu = p+m, \nu = p-m$, and the coefficients $N^k_{\kappa\lambda\mu\nu}$ are defined as follows. When $\mu \le k$ and $\nu \le k$,

$$\begin{aligned}N^k_{\kappa\lambda\mu\nu} = \frac{(-1)^k}{(-\mu+k)!\,(-\nu+k)!}\{\pi^2 &+ \psi'(1+k) +\\ &\psi'(\mu+1+k)+\psi'(-\mu+1+k)+\\ &\psi'(\nu+1+k)+\psi'(-\nu+1+k)-\\ \psi'\left(\frac{1}{2}+k\right)&-\psi'\left(\kappa+\frac{1}{2}+k\right)-\psi'\left(-\kappa+\frac{1}{2}+k\right)-\\ &\psi'\left(\lambda+\frac{1}{2}+k\right)-\psi'\left(-\lambda+\frac{1}{2}+k\right)-\\ (\psi\left(\frac{1}{2}+k\right)&+\psi\left(\kappa+\frac{1}{2}+k\right)+\psi\left(-\kappa+\frac{1}{2}+k\right)+\\ \psi\left(\lambda+\frac{1}{2}+k\right)&+\psi\left(-\lambda+\frac{1}{2}+k\right)+2\log q-\psi(1+k)-\\ \psi(\mu+1+k)-\psi(-\mu+1+k)&-\psi(\nu+1+k)-\psi(-\nu+1+k))^2\}.\end{aligned} \quad (8.133)$$

When $\mu > k$ and $\nu \leq k$, or $\mu \leq k$ and $\nu > k$, it is necessary to remove the indeterminacy which appears in this formula arising from the product of zero and infinite terms by use of the formulae

$$\left.\frac{\psi(x)}{\Gamma(x)}\right|_{x=-j} = -\frac{d}{dx}\left[\frac{1}{\Gamma(x)}\right]\Big|_{x=-j} = (-1)^{j-1}\Gamma(j+1), \tag{8.134}$$

$$\left.\frac{\psi'(x)-\psi^2(x)}{\Gamma(x)}\right|_{x=-j} = 2(-1)^j\Gamma(j+1)\psi(j+1) \tag{8.135}$$

where $j = 0, 1, 2, \ldots$. Thus when $\mu > k$ and $\nu \leq k$, the expression becomes

$$N^k_{\kappa\lambda\mu\nu} = 2(-1)^\mu \frac{\Gamma(\mu-k)}{\Gamma(-\nu+1+k)}\left\{F^k_{\kappa\lambda\mu\nu} - \psi(\mu-k) - \psi(-\nu+1+k)\right\} \tag{8.136}$$

where

$$\begin{aligned} F^k_{\kappa\lambda\mu\nu} = {} & \psi\left(\frac{1}{2}+k\right) + \psi\left(\kappa+\frac{1}{2}+k\right) + \psi\left(-\kappa+\frac{1}{2}+k\right) + \\ & \psi\left(\lambda+\frac{1}{2}+k\right) + \psi\left(-\lambda+\frac{1}{2}+k\right) + 2\log q - \\ & \psi(1+k) - \psi(\mu+1+k) - \psi(\nu+1+k); \end{aligned} \tag{8.137}$$

when $\mu \leq k$ and $\nu > k$, the expression becomes

$$N^k_{\kappa\lambda\mu\nu} = 2(-1)^\nu \frac{\Gamma(\nu-k)}{\Gamma(-\mu+1+k)}\left\{F^k_{\kappa\lambda\mu\nu} - \psi(\nu-k) - \psi(-\mu+1+k)\right\}.$$

Finally when $\mu > k$ and $\nu > k$, Formulae (8.135) may be used to show that

$$N^k_{\kappa\lambda\mu\nu} = 2(-1)^{k+1}\Gamma(\nu-k)\Gamma(\mu-k).$$

From this final form it may be shown that the diagonal terms of the matrix elements R_{nmsp} dominate so that the system (8.117) is satisfactory for computation. It may be verified that for narrow strips at least ($q \ll 1$), this i.s.l.a.e. is non-singular, and an analytic solution can be developed. In the general case ($0 < q < 1$) numerical techniques may be employed. This completes our regularisation of the dual integral equations associated with the finite strip.

8.5 Coupled charged conductors: the spherical cap and circular disc

In Section 2.9, we briefly described techniques for calculating the potential distribution surrounding coupled charged conductors with components that

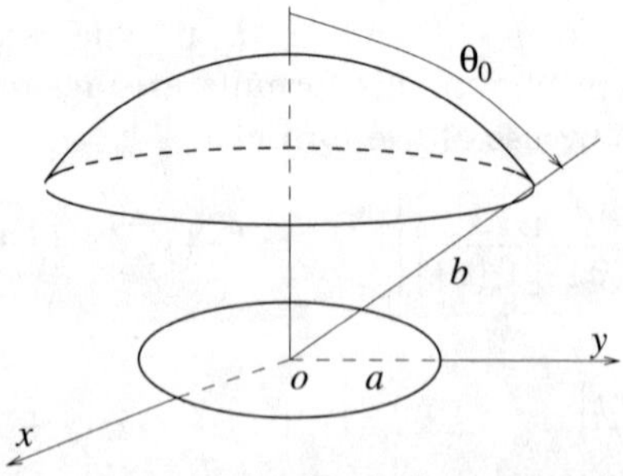

FIGURE 8.6. The coupled disc and spherical cap.

are parts of coordinate surfaces belonging to *different* coordinate systems. One of the simplest examples is the combination of a spherical cap and circular disc.

Suppose the circular disc of radius a is located in the plane $z = 0$ with centre at the origin O; the spherical cap also has its centre at O, subtends an angle $\theta_0 \leq \frac{1}{2}\pi$ at O, and has radius $b > a$; let $q = a/b$. (See Figure 8.6.) Both disc and cap are charged to unit potential.

Following the usual principle of superposition, the total potential U may be expressed as the sum of two contributions

$$U = U^c + U^d, \tag{8.138}$$

where the cap contribution may be represented in the form

$$U^c = \sum_{n=0}^{\infty} x_n P_n(\cos\theta) \begin{cases} (r/b)^n, & r < b, \\ (r/b)^{-n-1}, & r > b, \end{cases} \tag{8.139}$$

whilst the disc contribution may be represented as

$$U^d = \int_0^\infty G(\nu) J_0(\nu\rho) e^{-\nu|z|} d\nu. \tag{8.140}$$

The unknown coefficients $\{x_n\}_{n=0}^{\infty}$ and function G are to be found.

The obvious boundary conditions to be enforced are

$$U(b,\theta) = 1, \quad \theta \in (0,\theta_0), \tag{8.141}$$

$$\left[\frac{\partial}{\partial r} U(r,\theta)\right]_{r=b-0}^{r=b+0} = 0, \quad \theta \in (\theta_0,\pi), \tag{8.142}$$

$$U(\rho,0) = 1, \quad \rho \in (0,a), \tag{8.143}$$

$$\left[\frac{\partial}{\partial z} U(\rho,z)\right]_{r=b-0}^{r=b+0} = 0, \quad \rho \in (a,\infty), \tag{8.144}$$

where (8.141) and (8.142) have been expressed in terms of the standard spherical coordinate system (r,θ,ϕ) centred at O, whereas (8.143) and

(8.144) have been expressed in terms of the standard cylindrical coordinate system (ρ, ϕ, z) centred at O (so that $\rho = r\sin\theta, z = r\cos\theta$).

Enforcement of the boundary conditions (8.141)–(8.144) leads to the integro-series equations for the unknowns

$$\sum_{n=0}^{\infty} x_n P_n(\cos\theta) = 1 - \int_0^{\infty} G(\nu) J_0(\nu b \sin\theta) e^{-\nu b|\cos\theta|} d\nu, \quad \theta \in (0, \theta_0), \tag{8.145}$$

$$\sum_{n=0}^{\infty} (2n+1)\, x_n P_n(\cos\theta) = 0, \qquad \theta \in (\theta_0, \pi), \tag{8.146}$$

$$\int_0^{\infty} G(\nu) J_0(\nu\rho) d\nu = 1 - \sum_{n=0}^{\infty} x_{2n} P_{2n}(0) (\rho/b)^{2n}, \quad \rho \in (0, a), \tag{8.147}$$

$$\int_0^{\infty} \nu G(\nu) J_0(\nu\rho) d\nu = 0, \qquad \rho \in (a, \infty). \tag{8.148}$$

In deriving (8.147), the property $P_{2n+1}(0) = 0$ was used. Using the method described in Section 2.6, and noting the value

$$P_{2n}(0) = (-1)^n \frac{(2n-1)!!}{(2n)!!},$$

we may transform (8.147) and (8.148) to the form

$$\int_0^{\infty} G(\nu) \cos\nu\rho \, d\nu = \left\{ 1 - \sum_{n=0}^{\infty} (-1)^n x_{2n} (\rho/b)^{2n} \right\} H(a-\rho). \tag{8.149}$$

The application of an inverse cosine Fourier transform to (8.149) produces our first integro-series equation in algebraic form,

$$G(\nu) + \sum_{n=0}^{\infty} x_{2n} R_n(\nu) = \frac{2}{\pi} \frac{\sin\nu a}{\nu}, \tag{8.150}$$

where

$$R_n(\nu) = \frac{a}{\pi} (-1)^n \frac{q^{2n}}{2n+1} \, {}_1F_1(2n+1; 2n+2; i\nu a) + \frac{a}{\pi} (-1)^n \frac{q^{2n}}{2n+1} \, {}_1F_1(2n+1; 2n+2; -i\nu a). \tag{8.151}$$

The sum of the Kummer functions may be simplified to

$$
\begin{aligned}
&{}_1F_1(2n+1;2n+2;i\nu a)+{}_1F_1(2n+1;2n+2;-i\nu a) \\
&= -i\,(-1)^n\,(2n+1)!\sum_{k=0}^{2n}\frac{i^k\,(\nu a)^{k-2n-1}}{k!}\left\{(-1)^k\,e^{i\nu a}-e^{-i\nu a}\right\}. \qquad (8.152)
\end{aligned}
$$

Before turning to the analysis of (8.145), we expand that part of the integrand appearing in (8.145) in a series of Legendre polynomials (see [14])

$$
J_0(\nu b\sin\theta)e^{\pm\nu b\cos\theta}=\sum_{n=0}^{\infty}\frac{(\nu b)^n}{n!}(-1)^n\,P_n(\cos\theta). \qquad (8.153)
$$

By the methods developed in Section 2.1, Equations (8.145) and (8.146) may be transformed to

$$
\begin{aligned}
&\sum_{n=0}^{\infty}x_n\cos\left(n+\frac{1}{2}\right)\theta \\
&=\begin{cases}\cos\frac{1}{2}\theta-\int_0^{\infty}G(\nu)e^{-\nu b\cos\theta}\sin\left(\frac{1}{2}\theta-\nu b\sin\theta\right)d\nu, & \theta<\theta_0\\ 0, & \theta>\theta_0\end{cases} \qquad (8.154)
\end{aligned}
$$

where we have used the series (derived the generating function, see Appendix, (B.138))

$$
\sum_{n=0}^{\infty}\frac{(-1)^n}{n!}(\nu b)^n\cos\left(n+\frac{1}{2}\right)\theta=e^{-\nu b\cos\theta}\sin\left(\frac{1}{2}\theta-\nu b\sin\theta\right). \qquad (8.155)
$$

From (8.154) we may derive the companion integro-series equation in algebraic form

$$
x_m+\int_0^{\infty}G(\nu)S_m(\nu)d\nu=Q_{0m}(\theta_0), \qquad (8.156)
$$

where $m=0,1,2\ldots$, and

$$
S_m(\nu)=\frac{2}{\pi}\int_0^{\theta_0}e^{-\nu b\cos\theta}\sin\left(\frac{1}{2}\theta-\nu b\sin\theta\right)\cos\left(m+\frac{1}{2}\right)\theta\,d\theta. \qquad (8.157)
$$

The structure of Equations (8.150) and (8.156) is interesting. If the contribution from the functions R_m and S_m are neglected, then the closed form solutions are precisely those previously obtained for the isolated disc and isolated spherical cap, respectively. The contribution from the functions R_m and S_m may be regarded as perturbation terms (though, as we shall see, not necessarily small in magnitude).

The simultaneous solution of Equations (8.150) and (8.156) provides the potential of the coupled two-component structure. It is clear that a second-kind Fredholm equation for G may be obtained by elimination of the terms involving x_n; equally, a second-kind i.s.l.a.e. for the sequence $\{x_n\}_{n=0}^{\infty}$ may be obtained by elimination of the function G. Using (8.150) to eliminate G, this i.s.l.a.e. is

$$x_m - \sum_{n=0}^{\infty} x_{2n}\alpha_{nm}(q,\theta_0) = Q_{0m}(\theta_0) - \beta_m(q,\theta_0), \tag{8.158}$$

where $m = 0, 1, 2, \ldots$, and

$$\alpha_{nm}(q,\theta_0) = \frac{4(-1)^n}{\pi^2}\int_0^{\theta_0} \cos\left(m+\frac{1}{2}\right)\theta \left\{\int_0^q t^{2n}\frac{t^2\sin\frac{3}{2}\theta - \sin\frac{1}{2}\theta}{t^4+2t^2\cos 2\theta+1}dt\right\}d\theta, \tag{8.159}$$

$$\beta_m(q,\theta_0) = \frac{1}{\pi^2}\int_0^{\theta_0} \cos\left(m+\frac{1}{2}\right)\theta \times \left\{2\arctan\left[\frac{2q\cos\theta}{1-q^2}\right]\sin\frac{1}{2}\theta - \cos\frac{1}{2}\theta\ln\left[\frac{1+2q\cos\theta+q^2}{1-2q\cos\theta+q^2}\right]\right\}d\theta. \tag{8.160}$$

In a similar way, we may deduce that G satisfies the second-kind integral equation

$$G(\mu) - \int_0^{\infty} G(\nu)H(\nu,\mu;q,\theta_0) = \frac{2}{\pi}\frac{\sin\mu a}{\mu} - L(\mu;q,\theta_0), \tag{8.161}$$

where

$$H(\nu,\mu;q,\theta_0) = \frac{4}{\pi^2}b\int_0^q dt\cos\mu bt \times \int_0^{\theta_0} d\theta e^{-\nu b\cos\theta}\sin\left(\frac{1}{2}\theta - \nu b\sin\theta\right)\frac{t^2\cos\frac{3}{2}\theta+\cos\frac{1}{2}\theta}{t^4+2t^2\cos 2\theta+1}, \tag{8.162}$$

and

$$L(\mu;q,\theta_0) = -\frac{b}{\pi^2}\int_0^q dt\cos\mu bt \times \left\{\theta_0 + \arctan\left[\frac{1-t^2}{1+t^2}\tan\theta_0\right] + \frac{1}{2t}\ln\left[\frac{1+2t\sin\theta_0+t^2}{1-2t\sin\theta_0+t^2}\right]\right\}. \tag{8.163}$$

When the disc is much smaller than the radius of curvature of the cap ($q \ll 1$), it is possible to obtain an approximate analytical solution. In this limiting case the matrix elements can be factored as

$$\alpha_{nm} = \xi_n\beta_m^*(q,\theta_0)\left(1+O(q^3)\right), \tag{8.164}$$

where

$$\xi_n = (-1)^n \frac{q^{2n}}{2n+1}, \; \beta_m^*(q,\theta_0) = \frac{2}{\pi^2} q \left[\frac{\cos(m+1)\theta_0 - 1}{m+1} - \frac{\cos m\theta_0 - 1}{m} \right]. \tag{8.165}$$

It should be noted that $\beta_m = \beta_m^*(q,\theta_0) + O(q^3)$. For the given approximation ($q \ll 1, \theta_0$ arbitrary), the solution of the i.s.l.a.e. (8.158) is

$$x_m = (C-1)\beta_m^*(q,\theta_0) + Q_{0m}(\theta_0) + O(q^3), \tag{8.166}$$

where $C = \sum_{n=0}^{\infty} x_{2n}\xi_n$. The value of C is readily computed from (8.166); the final solution is

$$x_m = Q_{0m}(\theta_0) - \frac{2}{\pi^2} q \frac{1 - Q_{00}(\theta_0)}{1 + \frac{4}{\pi^2} q \sin^2 \frac{1}{2}\theta_0} \left[\frac{\cos(m+1)\theta_0 - 1}{m+1} - \frac{\cos m\theta_0 - 1}{m} \right]. \tag{8.167}$$

The total charge Q accumulated on both components is the sum of that accumulated on the disc (Q^d) and of that accumulated on the cap (Q^c); these are simply

$$Q^c = b x_0, \; Q^d = G(0). \tag{8.168}$$

Using (8.167) and (8.150) at $\nu = 0$, we deduce

$$Q^c = b \left\{ Q_{00}(\theta_0) + \frac{2}{\pi^2} q \frac{1 - \cos\theta_0}{1 + \frac{2}{\pi^2} q (1 - \cos\theta_0)} + O(q^3) \right\} \tag{8.169}$$

and

$$\begin{aligned} Q^d = \frac{2}{\pi} a \left\{ 1 - Q_{00}(\theta_0) - \frac{2}{\pi^2} q (1 - Q_{00}(\theta_0)) (1 - \cos\theta_0) \right\} \\ + \frac{2}{\pi} a q^2 \left\{ \frac{4}{\pi^2} (1 - \cos\theta_0)^2 (1 - Q_{00}(\theta_0)) + \frac{1}{3} Q_{02}(\theta_0) \right\} \\ + O(q^3). \end{aligned} \tag{8.170}$$

Some results of calculation based on these approximate formulae for $q = 0.1$ are shown in Table 8.1. The results for an isolated spherical cap are shown for comparison in brackets in the first column of the table.

The presence of the charged disc has a discernible effect on the spherical cap even when it is small, increasing the charge on the cap. Rather more noticeable is the decrease in charge on the disc as the cap size increases; as the angle θ_0 increases, the disc is increasingly shielded by the larger charged conductor, and its surface charge distribution is correspondingly modified.

θ_0 (deg.)	Q^c/b	Q^d/a	Q/b
0°	0	0.6366	0.0636
10°	0.1111(0.1108)	0.5661	0.1677
20°	0.2209(0.2200)	0.4963	0.2705
30°	0.3276(0.3258)	0.4285	0.3722
40°	0.4294(0.4268)	0.3636	0.4683
50°	0.5248(0.5216)	0.3028	0.5582
60°	0.6125(0.6090)	0.2467	0.6407
70°	0.6916(0.6880)	0.1961	0.7147
80°	0.7613(0.7579)	0.1514	0.7797
90°	0.8213(0.8183)	0.1131	0.8354

TABLE 8.1. Normalised value of total charge $Q/b = (Q^c + Q^d)/b$. The parameter $q = a/b = 0.1$.

More generally, whatever the values of the parameters q and θ_0, the regularised second-kind Equations (8.158) and (8.161) are readily solved by standard numerical methods, and the behaviour of the coupled disc-cap structure can be determined as a function of the parameters. If recursion formulae for the coefficients α_{nm} and β_m are exploited, a highly efficient computational algorithm can be obtained for computation.

Appendix A

Notation

The *Kronecker symbol* is defined by

$$\delta_{nm} = \begin{cases} 1, n = m \\ 0, n \neq m \end{cases}.$$

The *order notation* $f(x) = O(g(x))$ as $x \to a$, means that $|f(x)/g(x)|$ remains bounded as $x \to a$. (This includes the possibilities $a = \pm\infty$.) Similarly, the notation $a_n = O(b_n)$ as $n \to \infty$ means $|a_n/b_n|$ remains bounded as $n \to \infty$.

The *Heaviside function* is defined by

$$H(x) = \begin{cases} 1, x > 0 \\ 0, x < 0 \end{cases}.$$

Appendix B
Special Functions

Only the most important relations for the special functions employed in this book are included in this section. For more detailed information, the reader is referred to standard works on the special functions including, for example, [59, 1, 57, 58], and a summary treatment in [27].

B.1 The Gamma function

The Gamma function Γ defined by

$$\Gamma(z) = \int_0^\infty t^{z-1} e^{-t} dt, \qquad \mathrm{Re}(z) > 0 \tag{B.1}$$

is a generalization of the factorial: when n is a nonnegative integer

$$\Gamma(n+1) = n!$$

The *recurrence formula* for the factorial is

$$\Gamma(z+1) = z\Gamma(z), \tag{B.2}$$

and the *reflection formula* is

$$\Gamma(z)\Gamma(1-z) = \frac{\pi}{\sin(\pi z)}, \tag{B.3}$$

from which it follows that $\Gamma(\frac{1}{2}) = \sqrt{\pi}$; the *duplication formula* is

$$\Gamma(2z) = (2\pi)^{-\frac{1}{2}} 2^{2z-\frac{1}{2}} \Gamma(z)\Gamma(z+\frac{1}{2}). \tag{B.4}$$

Two *asymptotic formulae* are widely used. Stirling's formula states

$$\Gamma(z) \backsim e^{-z} z^{z-\frac{1}{2}} (2\pi)^{\frac{1}{2}} \left[1 + \frac{1}{12z} + \frac{1}{288z^2} - \frac{139}{51840z^3} - \frac{571}{2488320z^4} + \ldots\right], \tag{B.5}$$

when $z \to \infty$ in $|\arg z| < \pi$; Field's formula states that the ratio of Gamma functions has an asymptotic expansion of the form for suitable c_n,

$$\frac{\Gamma(z+a)}{\Gamma(z+b)} \backsim z^{a-b} \sum_{n=0}^{\infty} c_n \frac{\Gamma(b-a+n)}{\Gamma(b-a)} \frac{1}{z^n}, \tag{B.6}$$

when $z \to \infty$ and $z \neq -a, -a-1, \ldots; z \neq -b, -b-1, \ldots$. The first few terms in the expansion are

$$\frac{\Gamma(z+a)}{\Gamma(z+b)} = z^{a-b}(1 + \frac{(a-b)(a+b-1)}{2z} + \frac{1}{12}\binom{a-b}{2}(3(a+b-1)^2 - a + b - 1)\frac{1}{z^2} + \ldots). \tag{B.7}$$

Closely connected with the Gamma function is the Beta function defined for $\mathrm{Re}(p) > 0, \mathrm{Re}(q) > 0$; it equals

$$B(p,q) = \int_0^1 t^{p-1}(1-t)^{q-1} dt = \frac{\Gamma(p)\Gamma(q)}{\Gamma(p+q)}. \tag{B.8}$$

B.2 Hypergeometric functions

The generalised hypergeometric function is defined by

$${}_pF_q(a_1, \ldots, a_p; b_1, \ldots, b_q; z) \equiv \sum_{k=0}^{\infty} \frac{(a_1)_k (a_2)_k \ldots (a_p)_k}{(b_1)_k (b_2)_k \ldots (b_q)_k} \cdot \frac{z^k}{k!} \tag{B.9}$$

where the notation for the *Pochhammer symbol*

$$(a)_k \stackrel{def}{=} a(a+1)\ldots(a+k-1); (a)_0 \stackrel{def}{=} 1 \tag{B.10}$$

has been used; the *upper parameters* $\overrightarrow{a} = (a_1, \ldots, a_p)$ are unrestricted, whereas the *lower parameters* $\overrightarrow{b} = (b_1, \ldots, b_q)$ are restricted so that $b_j \neq 0, -1, -2, \ldots$. Note that when $a \neq 0, -1, -2, \ldots$,

$$(a)_k = \frac{\Gamma(a+k)}{\Gamma(a)}. \tag{B.11}$$

When $p \leq q$, the series converges for all complex z. When $p = q + 1$, the series has radius of convergence 1, converging inside the unit disc $|z| < 1$; it converges on the unit disc $|z| = 1$ provided

$$\operatorname{Re}\left(\sum_{k=1}^{q} b_k - \sum_{j=1}^{q+1} a_j\right) > 0, \tag{B.12}$$

or alternatively, it converges everywhere on the unit disc, except at the point $z = 1$, provided

$$-1 < \operatorname{Re}\left(\sum_{k=1}^{q} b_k - \sum_{j=1}^{q+1} a_j\right) \leq 0. \tag{B.13}$$

If the one of upper parameters is equal to zero or a negative integer, then the series terminates and is a *hypergeometric* polynomial.

The function ${}_1F_1(a; b; z) \equiv M(a, b, z)$ is known as Kummer's function; many special functions are expressible as Kummer's function with particular parameters [1, 59].

The Gaussian hypergeometric series is a special case of the hypergeometric function with $p = 2, q = 1$,

$${}_2F_1(a,b; c; z) = \sum_{k=0}^{\infty} \frac{(a)_k (b)_k}{(c)_k} \cdot \frac{z^k}{k!}. \tag{B.14}$$

It satisfies *the differential equation*

$$z(1-z)\frac{d^2U}{dz^2} + [c - (a+b+1)z]\frac{dU}{dz} - abU = 0. \tag{B.15}$$

When a or b is equal to a negative integer, then the series (B.14) terminates and is a *hypergeometric polynomial;* if $a = -m$ (m a positive integer),

$$F(-m, b; c; z) = \sum_{n=0}^{m} \frac{(-m)_n (b)_n}{(c)_n} \frac{z^n}{n!}. \tag{B.16}$$

This formula is also well defined when $c = -m - l$, $l = 0, 1, 2, \ldots$

$$F(-m, b; -m-l; z) = \sum_{n=0}^{m} \frac{(-m)_n (b)_n}{(-m-l)_n} \frac{z^n}{n!}. \tag{B.17}$$

Many special functions are particular examples of the Gaussian hypergeometric series (B.14) with appropriate arguments, including the Jacobi polynomials discussed in the next section. Hypergeometric functions satisfy a great number of transformation rules (see [1]) that provide many interesting and useful connections between the various special functions.

B.3 Orthogonal polynomials: Jacobi polynomials, Legendre polynomials

Jacobi polynomials and Legendre polynomials are two families of *classical orthogonal polynomials* whose properties are extensively described in [58].

For each fixed (α, β) with $\alpha > -1, \beta > -1$, the *Jacobi polynomials* $P_n^{(\alpha,\beta)}$ are polynomials of degree n $(= 0, 1, 2, \ldots)$, and are orthogonal with respect to the *weighted scalar product* on $[-1, 1]$ employing the weight function $w_{\alpha,\beta}(x) = (1-x)^{\alpha}(1+x)^{\beta}$:

$$(P_n^{(\alpha,\beta)}, P_m^{(\alpha,\beta)}) = \int_{-1}^{1} (1-x)^{\alpha}(1+x)^{\beta} P_n^{(\alpha,\beta)}(x) P_m^{(\alpha,\beta)}(x)dx = h_n^{(\alpha,\beta)}\delta_{nm}. \tag{B.18}$$

The polynomials are *normalised* by their value at $x = 1$,

$$P_n^{(\alpha,\beta)}(1) = \binom{n+\alpha}{n} = \frac{\Gamma(n+\alpha+1)}{\Gamma(n+1)\Gamma(\alpha+1)}, \tag{B.19}$$

so that their squared norm is

$$h_n^{(\alpha,\beta)} = \left\| P_n^{(\alpha,\beta)} \right\|^2 = \frac{2^{\alpha+\beta+1}}{2n+\alpha+\beta+1} \frac{\Gamma(n+\alpha+1)\Gamma(n+\beta+1)}{n!\Gamma(n+\alpha+\beta+1)}. \tag{B.20}$$

Jacobi polynomials may also be normalised by the requirement that the weighted scalar product be equal to unity when $n = m$; the members of this orthonormal family are denoted $\hat{P}_n^{(\alpha,\beta)} = \left\{h_n^{(\alpha,\beta)}\right\}^{-\frac{1}{2}} P_n^{(\alpha,\beta)}$.

In common with all the families of classical orthogonal polynomials, the Jacobi polynomials satisfy a *recurrence relation* of form

$$p_{n+1} - (a_n x + b_n)p_n + c_n p_{n-1} = 0, \quad n = 1, 2, \ldots \tag{B.21}$$

For the Jacobi polynomials $p_n = P_n^{(\alpha,\beta)}$, the coefficients a_n, b_n, c_n and the two lowest degree polynomials are

$$\begin{aligned} a_n &= \frac{(2n+\alpha+\beta+1)(2n+\alpha+\beta+2)}{(2n+2)(n+\alpha+\beta+1)}, \\ b_n &= \frac{(2n+\alpha+\beta+1)(\alpha^2-\beta^2)}{(2n+2)(n+\alpha+\beta+1)(2n+\alpha+\beta)}, \\ c_n &= \frac{2(n+\alpha)(n+\beta)(2n+\alpha+\beta+2)}{(2n+2)(n+\alpha+\beta+1)(2n+\alpha+\beta)}, \end{aligned} \tag{B.22}$$

and

$$P_0^{(\alpha,\beta)}(x) = 1, P_1^{(\alpha,\beta)}(x) = \frac{1}{2}(\alpha-\beta) + \left[1 + \frac{1}{2}(\alpha+\beta)\right] x. \tag{B.23}$$

They satisfy the *differential equation*

$$(1-x^2)\frac{d^2y}{dx^2}+[\beta-\alpha-(\alpha+\beta+2)x]\frac{dy}{dx}+n(n+\alpha+\beta+1)y=0. \tag{B.24}$$

Comparing this with the differential Equation (B.15) for the Gaussian hypergeometric series and making the transformation $z=\frac{1}{2}(1-x)$ leads us to make the identification

$$a=-n,\ b=n+\alpha+\beta+1,\ c=\alpha+1$$

and to recognise that $P_n^{(\alpha,\beta)}(x)$ is the *hypergeometric polynomial*

$$P_n^{(\alpha,\beta)}(x)=\binom{n+\alpha}{n}\,{}_2F_1\left(-n,n+\alpha+\beta+1;\alpha+1;\frac{1-x}{2}\right). \tag{B.25}$$

Thus, an explicit form for $P_n^{(\alpha,\beta)}(x)$ is

$$\frac{\Gamma(n+\alpha+1)}{n!\Gamma(\alpha+1)}\sum_{m=0}^{n}\frac{(-n)_m(n+\alpha+\beta+1)_m}{m!(\alpha+1)_m}\left(\frac{1-x}{2}\right)^m$$

$$=\frac{\Gamma(n+\alpha+1)}{n!\Gamma(n+\alpha+\beta+1)}\sum_{k=0}^{n}\frac{\Gamma(n+1)\Gamma(n+k+\alpha+\beta+1)}{\Gamma(k+1)\Gamma(n-k+1)\Gamma(k+\alpha+1)}\left(\frac{x-1}{2}\right)^k.$$

From *the symmetry property*

$$P_n^{(\alpha,\beta)}(-x)=(-1)^nP_n^{(\beta,\alpha)}(x), \tag{B.26}$$

one obtains the alternative representation

$$P_n^{(\alpha,\beta)}(x)=(-1)^n\binom{n+\beta}{n}\,{}_2F_1\left(-n,n+\alpha+\beta+1;\beta+1;\frac{1+x}{2}\right). \tag{B.27}$$

Many other representations are possible because of the great number of transformation relations that the hypergeometric function satisfies.

The Jacobi polynomials satisfy *Rodrigues' formula*

$$P_n^{(\alpha,\beta)}(x)=\frac{(-1)^n}{2^nn!}\frac{1}{(1-x)^\alpha(1+x)^\beta}\left(\frac{d}{dx}\right)^n\left[(1-x)^{\alpha+n}(1+x)^{\beta+n}\right], \tag{B.28}$$

from which follows the useful relation

$$-2n(1-x)^\alpha(1+x)^\beta P_n^{(\alpha,\beta)}(x)=\frac{d}{dx}\left[(1-x)^{\alpha+1}(1+x)^{\beta+1}P_{n-1}^{(\alpha+1,\beta+1)}(x)\right]. \tag{B.29}$$

The *differential relation* expresses derivatives in terms of polynomials of the same parameters (α, β) :

$$(2n+\alpha+\beta)(1-x^2)\frac{d}{dx}P_n^{(\alpha,\beta)}(x)$$
$$= n\left[\alpha-\beta-(2n+\alpha+\beta)x\right]P_n^{(\alpha,\beta)}(x)+2(n+\alpha)(n+\beta)P_{n-1}^{(\alpha,\beta)}(x). \tag{B.30}$$

Other *recurrence relations* connect polynomials with indices (α, β) to those with indices $(\alpha+1, \beta)$ and $(\alpha, \beta+1)$,

$$\left(n+\frac{\alpha}{2}+\frac{\beta}{2}+1\right)(1-x)P_n^{(\alpha+1,\beta)}(x)$$
$$= (n+\alpha+1)P_n^{(\alpha,\beta)}(x)-(n+1)P_{n+1}^{(\alpha,\beta)}(x), \tag{B.31}$$

$$\left(n+\frac{\alpha}{2}+\frac{\beta}{2}+1\right)(1+x)P_n^{(\alpha,\beta+1)}(x)$$
$$= (n+\beta+1)P_n^{(\alpha,\beta)}(x)+(n+1)P_{n+1}^{(\alpha,\beta)}(x), \tag{B.32}$$

$$2P_n^{(\alpha,\beta)}(x)=(1-x)P_n^{(\alpha+1,\beta)}(x)+(1+x)P_n^{(\alpha,\beta+1)}(x); \tag{B.33}$$

also *recurrence relations* between polynomials with indices (α, β) and those with indices $(\alpha-1, \beta)$ and $(\alpha, \beta-1)$

$$(2n+\alpha+\beta)\,P_n^{(\alpha-1,\beta)}(x)=(n+\alpha+\beta)P_n^{(\alpha,\beta)}(x)-(n+\beta)P_{n-1}^{(\alpha,\beta)}(x), \tag{B.34}$$

$$(2n+\alpha+\beta)\,P_n^{(\alpha,\beta-1)}(x)=(n+\alpha+\beta)P_n^{(\alpha,\beta)}(x)+(n+\alpha)P_{n-1}^{(\alpha,\beta)}(x), \tag{B.35}$$

$$P_n^{(\alpha,\beta-1)}(x)-P_n^{(\alpha-1,\beta)}(x)=P_{n-1}^{(\alpha,\beta)}(x). \tag{B.36}$$

These relations may be used to extend the definition of Jacobi polynomials for parameters (α, β) where $\alpha \leq -1$ or $\beta \leq 1$; in the text, the most commonly encountered examples are

$$P_n^{(-1,0)}(x) = \frac{1}{2}\left(P_n(x)-P_{n-1}(x)\right), \tag{B.37}$$

$$P_n^{(0,-1)} = \frac{1}{2}\left(P_n(x)+P_{n-1}(x)\right). \tag{B.38}$$

The *generating function* is

$$F(z,x) = \sum_{n=0}^{\infty} P_n^{(\alpha,\beta)}(x) z^n = 2^{\alpha+\beta} R^{-1}(1-z+R)^{-\alpha}(1+z+R)^{-\beta}, \tag{B.39}$$

where $R = \sqrt{1-2xz+z^2}$, the branch being fixed by specifying $R = 1$ when $z = 0$; the power series is convergent when $|z| < 1$. For particular values of α, β there are other generating functions.

An asymptotic formula with α, β, x fixed and $n \to \infty$ is

$$P_n^{(\alpha,\beta)}(\cos\theta) = \frac{\cos\left(\left[n+\frac{1}{2}(\alpha+\beta+1)\right]\theta - \frac{\pi}{4}(2\alpha+1)\right)}{\sqrt{\pi n}\left(\sin\frac{1}{2}\theta\right)^{\alpha+\frac{1}{2}}\left(\cos\frac{1}{2}\theta\right)^{\beta+\frac{1}{2}}} + O\left(n^{-\frac{3}{2}}\right) \tag{B.40}$$

where $0 < \theta < \pi$.

Many of the classical orthogonal polynomials are particular examples of Jacobi polynomials, including the Legendre polynomials $P_n = P_n^{(0,0)}$, the Chebyshev polynomials of first kind

$$T_n = \frac{\Gamma\left(\frac{1}{2}\right)\Gamma(n+1)}{\Gamma\left(n+\frac{1}{2}\right)} P_n^{\left(-\frac{1}{2},-\frac{1}{2}\right)}, \tag{B.41}$$

the Chebyshev polynomials of second kind

$$U_n = \frac{\Gamma\left(\frac{3}{2}\right)\Gamma(n+1)}{\Gamma\left(n+\frac{3}{2}\right)} P_n^{\left(\frac{1}{2},\frac{1}{2}\right)}, \tag{B.42}$$

and the Gegenbauer polynomials

$$C_n^{\gamma} = \frac{(2\gamma)_n}{\left(\gamma+\frac{1}{2}\right)_n} P_n^{\left(\gamma-\frac{1}{2},\gamma-\frac{1}{2}\right)}. \tag{B.43}$$

Thus if n is a nonnegative integer,

$$\cos n\theta = \frac{\Gamma\left(\frac{1}{2}\right)\Gamma(n+1)}{\Gamma\left(n+\frac{1}{2}\right)} P_n^{\left(-\frac{1}{2},-\frac{1}{2}\right)}(\cos\theta), \tag{B.44}$$

$$\sin n\theta = \frac{\Gamma\left(\frac{3}{2}\right)\Gamma(n+1)}{\Gamma\left(n+\frac{1}{2}\right)} \sin\theta\, P_{n-1}^{\left(\frac{1}{2},\frac{1}{2}\right)}(\cos\theta). \tag{B.45}$$

Explicit forms for other trigonometric functions are

$$\cos(n+\frac{1}{2})\theta = \frac{\Gamma\left(\frac{1}{2}\right)\Gamma(n+1)}{\Gamma\left(n+\frac{1}{2}\right)} \cos\frac{1}{2}\theta\, P_n^{\left(-\frac{1}{2},\frac{1}{2}\right)}(\cos\theta), \tag{B.46}$$

$$\sin(n+\frac{1}{2})\theta = \frac{\Gamma\left(\frac{1}{2}\right)\Gamma(n+1)}{\Gamma\left(n+\frac{1}{2}\right)} \sin\frac{1}{2}\theta\, P_n^{\left(\frac{1}{2},-\frac{1}{2}\right)}(\cos\theta). \tag{B.47}$$

B.3.1 The associated Legendre polynomials.

When $n \geqslant m$, the relationship between the associated Legendre functions P_n^m and the Jacobi polynomials $P_{n-m}^{(m,m)}$ is

$$P_n^m(\cos\theta) = 2^{-m}\sin^m\theta\frac{\Gamma(n+m+1)}{\Gamma(n+1)}P_{n-m}^{(m,m)}(\cos\theta) \tag{B.48}$$

and the connection with Legendre polynomials is

$$P_n^m(x) = \left(1-x^2\right)^{\frac{m}{2}}\frac{d^m}{dx^m}P_n(x). \tag{B.49}$$

Another orthonormal family of Jacobi polynomials ($n \geq k$, k fixed), considered in Chapter 2 has the form

$$\hat{P}_{n-k}^{(k-\frac{1}{2},k+\frac{1}{2})}(\cos\theta) = \frac{(-1)^k}{\sqrt{\pi}}\left\{\frac{(n-k)!}{(n+k)!}\right\}^{\frac{1}{2}}\left(\frac{1}{\sin\theta}\frac{d}{d\theta}\right)^k\left[\frac{\cos\left(n+\frac{1}{2}\right)\theta}{\cos\frac{1}{2}\theta}\right]. \tag{B.50}$$

B.3.2 The Legendre polynomials.

The Legendre polynomials $P_n(x)$ form a subclass of the associated Legendre functions $P_\nu^m(z)$ (where $m = 0$, $\nu = n = 0, 1, 2, \ldots$ and $z = x$ is real, $-1 \leq x \leq 1$) that are considered in the next subsection and so all properties of these functions are valid for the Legendre polynomials. In the context of classical orthogonal polynomials, the Legendre polynomials are the Jacobi polynomials with $\alpha = \beta = 0$. Thus, they are orthogonal with respect to the constant (unit) function, are normalised by the condition $P_n(1) = 1$, and have square norm where $h_n = h_n^{(0,0)} = \|P_n\|^2 = 2\,(2n+1)^{-1}$. They satisfy the *recurrence relation*

$$(n+1)P_{n+1}(x) - (2n+1)xP_n(x) + nP_{n-1}(x) = 0, \quad n = 1, 2, \ldots \tag{B.51}$$

where $P_0(x) = 1$, $P_1(x) = x$. Thus $P_2(x) = \frac{3}{2}x^2 - \frac{1}{2}$. They satisfy the *differential equation*

$$(1-x^2)\frac{d^2y}{dx^2} - 2x\frac{dy}{dx} + n(n+1)y = 0, \tag{B.52}$$

and have the *hypergeometric polynomial* representation

$$P_n(x) = {}_2F_1\left(-n, n+1; 1; \frac{1-x}{2}\right). \tag{B.53}$$

The *Rodrigues' formula* is simply

$$P_n(x) = \frac{1}{2^n n!}\left(\frac{d}{dx}\right)^n\left[\left(x^2-1\right)^n\right]. \tag{B.54}$$

Useful *differential and integration relations* are

$$\begin{aligned}(1-x^2)\frac{d}{dx}P_n(x) &= n\left[P_{n-1}(x)-xP_n(x)\right] \\ &= (n+1)\left[xP_n(x)-P_{n+1}(x)\right] \qquad (B.55)\\ nP_n(x) &= x\frac{d}{dx}P_n(x)-\frac{d}{dx}P_{n-1}(x) \qquad (B.56)\\ (n+1)P_n(x) &= \frac{d}{dx}P_{n+1}(x)-x\frac{d}{dx}P_n(x) \qquad (B.57)\\ (2n+1)\int P_n(x)dx &= P_{n+1}(x)-P_{n-1}(x). \qquad (B.58)\end{aligned}$$

Two *generating functions* are

$$\sum_{n=0}^{\infty} P_n(x)z^n = (1-2xz+z^2)^{-1}, \quad -1<x<1, |z|<1, \tag{B.59}$$

$$\sum_{n=0}^{\infty} \frac{1}{n!}P_n(\cos\theta)z^n = e^{z\cos\theta}J_0(z\sin\theta) \tag{B.60}$$

The asymptotic formula for the Legendre polynomials when $n\to\infty$ is

$$P_n(\cos\theta) = \frac{\Gamma(n+1)}{\Gamma(n+\frac{3}{2})}(\frac{1}{2}\pi\sin\theta)^{-\frac{1}{2}}\cos\left[\left(n+\frac{1}{2}\right)\theta-\frac{\pi}{4}\right]+O\left(n^{-1}\right) \tag{B.61}$$

where $x=\cos\theta$ is fixed and $0<\theta<\pi$.

B.4 Associated Legendre functions

Associated Legendre functions of degree ν and order μ are solutions of complex argument z of the differential equation

$$(1-z^2)\frac{d^2y}{dz^2}-2z\frac{dy}{dz}+\left[\nu(\nu+1)-\frac{\mu^2}{1-z^2}\right]y=0. \tag{B.62}$$

The constants ν and μ are in general arbitrary complex numbers. The singularities of the differential equation are located at $z=\pm1,\infty$ and are regular. We shall consider first the ordinary Legendre functions of degree ν corresponding to the choice $\mu=0$, and subsequently consider the associated Legendre functions of nonzero order μ, restricting it to be integral.

B.4.1 Ordinary Legendre functions

When $\mu = 0$, the differential equation becomes

$$(1-z^2)\frac{d^2y}{dz^2} - 2z\frac{dy}{dz} + \nu(\nu+1)y = 0. \tag{B.63}$$

A pair of linearly independent solutions is the first-kind and second-kind Legendre functions denoted $P_\nu(z)$ and $Q_\nu(z)$; they are entire functions of z in the plane cut along $(-\infty, 1]$. The first-kind function is defined by

$$P_\nu(z) = {}_2F_1\left(-\nu, \nu+1; 1; \frac{1-z}{2}\right), \quad |\arg(z+1)| < \pi. \tag{B.64}$$

It possesses the symmetry property $P_{-\nu-1} = P_\nu$. An alternative representation for P_ν that is useful for large z is

$$P_\nu(z) = \frac{(2z)^{-\nu-1}\,\Gamma(-\frac{1}{2}-\nu)}{\sqrt{\pi}\Gamma(-\nu)}\,{}_2F_1\left(\frac{\nu}{2}+1, \frac{\nu+1}{2}; \nu+\frac{3}{2}; \frac{1}{z^2}\right) + \frac{(2z)^{\nu}\,\Gamma(\nu+\frac{1}{2})}{\Gamma(\nu+1)}\,{}_2F_1\left(\frac{1-\nu}{2}, -\frac{\nu}{2}; \frac{1}{2}-\nu; \frac{1}{z^2}\right), \tag{B.65}$$

valid when $|z| > 1$, $|\arg z| < \pi$, $\nu \neq \pm\frac{1}{2}, \pm\frac{3}{2}, \ldots$. Another useful representation is

$$P_\nu(z) = \frac{\Gamma\left(\frac{\nu}{2}+\frac{1}{2}\right)}{\sqrt{\pi}\Gamma\left(\frac{\nu}{2}+1\right)}\cos\frac{\nu\pi}{2}\,{}_2F_1\left(\frac{1+\nu}{2}, -\frac{\nu}{2}; \frac{1}{2}-\nu; z^2\right) + \frac{2\Gamma\left(\frac{\nu}{2}+1\right)}{\sqrt{\pi}\Gamma\left(\frac{\nu}{2}+\frac{1}{2}\right)}\sin\frac{\nu\pi}{2} z\,{}_2F_1\left(\frac{1-\nu}{2}, \frac{\nu}{2}+1; \frac{3}{2}; z^2\right), \tag{B.66}$$

valid when $|z| < 1$, and ν is arbitrary.

The second-kind Legendre function is defined by

$$Q_\nu(z) = \frac{\sqrt{\pi}\Gamma(\nu+1)}{\Gamma\left(\nu+\frac{3}{2}\right)(2z)^{\nu+1}}\,{}_2F_1\left(\frac{\nu}{2}+1, \frac{\nu}{2}+\frac{1}{2}; \nu+\frac{3}{2}; z^{-2}\right), \tag{B.67}$$

where $\nu \neq -1, -2, \ldots$; it possesses an analytic continuation in the entire complex plane, excluding the points $z = \pm 1$, with a branch cut along $(-\infty, 1]$. Another useful expansion is

$$Q_\nu(z) = e^{\mp i\nu\pi/2}\frac{\sqrt{\pi}\Gamma\left(\frac{\nu}{2}+1\right)}{\Gamma\left(\frac{\nu}{2}+\frac{1}{2}\right)} z\,{}_2F_1\left(\frac{1-\nu}{2}, 1+\frac{\nu}{2}; \frac{3}{2}; z^2\right) \mp e^{\mp i\nu\pi/2} i \frac{\sqrt{\pi}\Gamma\left(\frac{\nu}{2}+\frac{1}{2}\right)}{2\Gamma\left(\frac{\nu}{2}+1\right)}\,{}_2F_1\left(\frac{1+\nu}{2}, -\frac{\nu}{2}; \frac{1}{2}; z^2\right), \tag{B.68}$$

valid when $|z| < 1, \nu \neq -1, -2, ...$, the upper sign being taken when $\operatorname{Im} z > 0$, and the lower sign when $\operatorname{Im} z < 0$.

The *Wronskian* is

$$W\left\{P_\nu(z), Q_\nu(z)\right\} = P_\nu'(z)Q_\nu(z) - P_\nu(z)Q_\nu'(z) = \left(1 - z^2\right)^{-1}. \tag{B.69}$$

The following formulae are particularly useful for estimation of the asymptotically small parameters encountered in Chapters 3 and 4.

$$Q_\nu(\cosh\alpha) = \frac{\sqrt{\pi}\Gamma(\nu+1)}{\Gamma(\nu+\frac{3}{2})} e^{-(\nu+1)\alpha} F\left(\nu+1, \frac{1}{2}; \nu+\frac{3}{2}; e^{-2\alpha}\right) \tag{B.70}$$

$$\begin{aligned} P_\nu(\cosh\alpha) = {} & \frac{\Gamma(\nu+1)}{\sqrt{\pi}\Gamma(\nu+\frac{3}{2})} \tan(\nu\pi)\, e^{-(\nu+1)\alpha} F\left(\nu+1, \frac{1}{2}; \nu+\frac{3}{2}; e^{-2\alpha}\right) \\ & + \frac{\Gamma(\nu+\frac{1}{2})}{\sqrt{\pi}\Gamma(\nu+1)} e^{\nu\alpha} F\left(-\nu, \frac{1}{2}; \frac{1}{2}-\nu; e^{-2\alpha}\right), \end{aligned} \tag{B.71}$$

where $\nu \neq \pm\frac{1}{2}, \pm\frac{3}{2}, \ldots$.

Asymptotic expansions valid when $|\nu| \to \infty, |\arg\nu| \leq \frac{\pi}{2} - \delta$, and α is fixed $(0 < \alpha < \infty)$ are

$$P_\nu(\cosh\alpha) = \frac{e^{(\nu+\frac{1}{2})\alpha}}{\sqrt{2\nu\pi\sinh\alpha}} \left[1 + O\left(|\nu|^{-1}\right)\right] \tag{B.72}$$

$$Q_\nu(\cosh\alpha) = \frac{\sqrt{\pi}}{\sqrt{2\nu\sinh\alpha}} e^{-(\nu+\frac{1}{2})\alpha} \left[1 + O\left(|\nu|^{-1}\right)\right]; \tag{B.73}$$

when ν is real and $\nu \to \infty$, and θ is fixed in the interval $\delta \leq \theta \leq \pi - \delta$ (for some $\delta > 0$),

$$P_\nu(\cos\theta) = \sqrt{\frac{2}{\nu\pi\sin\theta}} \sin\left[(\nu+\frac{1}{2})\theta + \frac{1}{4}\pi\right] \left[1 + O\left(|\nu|^{-1}\right)\right], \tag{B.74}$$

$$Q_\nu(\cos\theta) = \sqrt{\frac{2}{\nu\pi\sin\theta}} \cos\left[(\nu+\frac{1}{2})\theta + \frac{1}{4}\pi\right] \left[1 + O\left(|\nu|^{-1}\right)\right]. \tag{B.75}$$

Explicit expressions are

$$P_0(z) = 1, \quad Q_0(z) = \frac{1}{2}\ln\left(\frac{z+1}{z-1}\right), \tag{B.76}$$

$$P_1(z) = z, \quad Q_1(z) = \frac{z}{2}\ln\left(\frac{z+1}{z-1}\right) - 1; \tag{B.77}$$

these are valid when z takes real values $x \in (-1,1)$.

$P_{\pm\frac{1}{2}}$ and $Q_{\pm\frac{1}{2}}$ are closely related to *complete elliptic integrals* of the first kind

$$K(k) = \int_0^{\frac{\pi}{2}} \frac{d\theta}{\sqrt{1-k^2\sin^2\theta}} \tag{B.78}$$

and of the second kind

$$E(k) = \int_0^{\frac{\pi}{2}} \sqrt{1-k^2\sin^2\theta}\, d\theta, \tag{B.79}$$

the properties of which are discussed in [59, 1]; in particular [14]

$$P_{-\frac{1}{2}}(z) = \frac{2}{\pi}\sqrt{\frac{2}{z+1}} K\left(\sqrt{\frac{z-1}{z+1}}\right), \tag{B.80}$$

$$Q_{-\frac{1}{2}}(z) = \sqrt{\frac{2}{z+1}} K\left(\sqrt{\frac{2}{z+1}}\right), \tag{B.81}$$

$$P_{\frac{1}{2}}(z) = \frac{2}{\pi}\left(z+\sqrt{z^2-1}\right)^{\frac{1}{2}} E\left(\sqrt{\frac{2(z^2-1)^{1/2}}{z+(z^2-1)^{1/2}}}\right), \tag{B.82}$$

$$Q_{\frac{1}{2}}(z) = z\sqrt{\frac{2}{z+1}} K\left(\sqrt{\frac{2}{z+1}}\right) - \sqrt{2(z+1)} E\left(\sqrt{\frac{2}{z+1}}\right). \tag{B.83}$$

When $z = x$ is real and $-1 < x < 1$, these become

$$P_{-\frac{1}{2}}(x) = \frac{2}{\pi} K\left(\sqrt{\frac{1-x}{2}}\right), \tag{B.84}$$

$$Q_{-\frac{1}{2}}(x) = K\left(\sqrt{\frac{1+x}{2}}\right), \tag{B.85}$$

$$P_{\frac{1}{2}}(x) = \frac{2}{\pi}\left[2E\left(\sqrt{\frac{1-x}{2}}\right) - K\left(\sqrt{\frac{1-x}{2}}\right)\right], \tag{B.86}$$

$$Q_{\frac{1}{2}}(x) = K\left(\sqrt{\frac{1+x}{2}}\right) - 2E\left(\sqrt{\frac{1+x}{2}}\right). \tag{B.87}$$

When $z = \cosh\alpha$ is real and exceeds 1, these become

$$\begin{aligned}
P_{-\frac{1}{2}}(\cosh\alpha) &= \left(\frac{\pi}{2}\cosh\frac{\alpha}{2}\right)^{-1} K\left(\tanh\frac{\alpha}{2}\right), && \text{(B.88)}\\
Q_{-\frac{1}{2}}(\cosh\alpha) &= 2e^{-\alpha/2} K\left(e^{-\alpha}\right), && \text{(B.89)}\\
P_{\frac{1}{2}}(\cosh\alpha) &= \frac{2}{\pi} e^{\alpha/2} E\left(\sqrt{1-e^{-2\alpha}}\right), && \text{(B.90)}\\
Q_{\frac{1}{2}}(\cosh\alpha) &= \left(2\cosh\frac{\alpha}{2} - \operatorname{sech}\frac{\alpha}{2}\right) K\left(\operatorname{sech}\frac{\alpha}{2}\right) \\
&\quad -2\cosh\frac{\alpha}{2} E\left(\operatorname{sech}\frac{\alpha}{2}\right), && \text{(B.91)}
\end{aligned}$$

Another useful result is

$$Q_{-\frac{1}{2}}(\cosh 2\sigma) = \operatorname{sech}\sigma K(\operatorname{sech}\sigma). \tag{B.92}$$

Integral representations valid for any complex ν and $\operatorname{Re}\cosh\alpha > 0$ are

$$P_\nu(\cosh\alpha) = \int_0^\alpha \frac{\cosh\left(\nu+\frac{1}{2}\right)\theta}{\sqrt{2\cosh\alpha - 2\cosh\theta}} d\theta, \tag{B.93}$$

and the Mehler-Dirichlet formula ([55])

$$P_\nu(\cos\beta) = \frac{2}{\pi}\int_0^\beta \frac{\cos\left(\nu+\frac{1}{2}\right)\theta}{\sqrt{2\cos\theta - 2\cos\beta}} d\theta. \tag{B.94}$$

When $\alpha > 0$, and $-1 < \operatorname{Re}\nu < 1$,

$$P_\nu(\cosh\alpha) = \frac{2}{\pi}\cot\left(\nu+\frac{1}{2}\right)\pi \int_\alpha^\infty \frac{\sinh\left(\nu+\frac{1}{2}\right)\theta}{\sqrt{2\cosh\theta - 2\cosh\alpha}} d\theta. \tag{B.95}$$

Also, when $\operatorname{Re}\nu > -1$,

$$Q_\nu(\cosh\alpha) = \int_\alpha^\infty \frac{e^{-\left(\nu+\frac{1}{2}\right)\theta}}{\sqrt{2\cosh\theta - 2\cosh\alpha}} d\theta. \tag{B.96}$$

A definite integral that frequently occurs is

$$\int_{-1}^{z_0} Q_{-\frac{1}{2}}(z)\, P_m(z)\, dz = \\ \frac{1-z_0^2}{\left(m+\frac{1}{2}\right)^2}\left\{P_m(z_0)\, Q'_{-\frac{1}{2}}(z_0) - P'_m(z_0)\, Q_{-\frac{1}{2}}(z_0)\right\}. \tag{B.97}$$

(It is evaluated using integration by parts and the defining differential equations for these functions.)

B.4.2 Conical functions

The Legendre functions $P_{-\frac{1}{2}+i\tau}$ and $Q_{-\frac{1}{2}+i\tau}$ with real τ occur in boundary value problems in conical geometry. The function $P_{-\frac{1}{2}+i\tau}(\cos\phi) = P_{-\frac{1}{2}-+i\tau}(\cos\phi)$ is real for real ϕ, as may be seen from its hypergeometric representation derived from (B.64),

$$P_{-\frac{1}{2}+i\tau}(\cos\phi) = {}_2F_1\left(\frac{1}{2}+i\tau, \frac{1}{2}-i\tau; 1; \sin^2\frac{1}{2}\phi\right) \tag{B.98}$$

Although $P_{-\frac{1}{2}+i\tau}$ and $Q_{-\frac{1}{2}+i\tau}$ are linearly independent solutions of the differential equation, the functions $P_{-\frac{1}{2}+i\tau}(x)$ and $P_{-\frac{1}{2}+i\tau}(-x)$ are also linearly independent. The Wronskians are

$$\begin{aligned} &W\left(P_{-\frac{1}{2}+i\tau}(x), P_{-\frac{1}{2}+i\tau}(-x)\right) \\ &\quad = P_{-\frac{1}{2}+i\tau}(x)P'_{-\frac{1}{2}+i\tau}(-x) - P_{-\frac{1}{2}+i\tau}(-x)P'_{-\frac{1}{2}+i\tau}(x) \\ &\quad = \frac{2}{\pi}\cosh(\pi\tau)\, W\left(P_{-\frac{1}{2}+i\tau}(x), Q_{-\frac{1}{2}+i\tau}(x)\right) \\ &\quad = \frac{2}{\pi}\cosh(\pi\tau)\left(1-x^2\right)^{-1}. \end{aligned} \tag{B.99}$$

$P_{-\frac{1}{2}+i\tau}$ has the integral representation

$$\begin{aligned} P_{-\frac{1}{2}+i\tau}(\cosh x) &= \frac{\sqrt{2}}{\pi}\int_0^x \frac{\cos\tau t\, dt}{\sqrt{\cosh x - \cosh t}}, \\ &= \frac{\sqrt{2}}{\pi}\coth(\pi\tau)\int_x^\infty \frac{\sin\tau t\, dt}{\sqrt{\cosh t - \cosh x}}. \end{aligned} \tag{B.100}$$

When as $\tau \to \infty$,

$$P_{-\frac{1}{2}+i\tau}(\cos\theta) \backsim \frac{e^{\tau\theta}}{\sqrt{2\pi\tau\sin\theta}}, \tag{B.101}$$

uniformly in the sector $\delta \le \theta \le \pi - \delta$.

B.4.3 Associated Legendre functions of integer order

The conventional choice for a pair of linearly independent solutions to the differential Equation (B.62) employs the first-kind and second-kind associated Legendre functions denoted $P_\nu^\mu(z)$ and $Q_\nu^\mu(z)$ and defined by

$$P_\nu^\mu(z) = \frac{1}{\Gamma(1-\mu)}\left(\frac{z+1}{z-1}\right)^{\frac{1}{2}\mu} {}_2F_1\left(-\nu, \nu+1; 1-\mu; \frac{1-z}{2}\right), \tag{B.102}$$

and

$$Q_\nu^\mu(z) = \frac{\sqrt{\pi}e^{\mu\pi i}\Gamma(\nu+\mu+1)\left(z^2-1\right)^{\frac{1}{2}\mu}}{2^{\nu+1}z^{\nu+\mu+1}\Gamma\left(\nu+\frac{3}{2}\right)} \times$$
$${}_2F_1\left(\frac{1}{2}\nu+\frac{1}{2}\mu+1, \frac{1}{2}\nu+\frac{1}{2}\mu+\frac{1}{2}; \nu+\frac{3}{2}; z^{-2}\right). \quad \text{(B.103)}$$

This is valid for the complex plane with a branch cut along $(-\infty, 1]$. When μ is a positive integer, the Gamma function factor creates some difficulty; in this case the definitions of the associated Legendre functions of degree $m\,(=1,2,\ldots)$ are taken to be

$$P_\nu^m(z) = \left(z^2-1\right)^{\frac{1}{2}m}\frac{d^m}{dz^m}P_\nu(z) \quad \text{(B.104)}$$

and

$$Q_\nu^m(z) = \left(z^2-1\right)^{\frac{1}{2}m}\frac{d^m}{dz^m}Q_\nu(z). \quad \text{(B.105)}$$

When $z = x \in (-1,1)$ is real, it is convenient to modify these definitions in the fashion described in [27]. P_ν^m, Q_ν^m are generalisations of the Legendre polynomials P_n, Q_n, reducing to them when $m=0$ and $\nu = n = 0,1,2,\ldots$.

$P_\nu^m(z)$ is an entire function of ν, while $Q_\nu^m(z)$ is a meromorphic function of ν with poles at the points $\nu = -1,-2,\ldots$. They have *the hypergeometric function* representations

$$P_\nu^m(z) = \frac{\Gamma(\nu+m+1)}{2^m\Gamma(m+1)\Gamma(\nu-m+1)}(z^2-1)^{\frac{1}{2}m} \times$$
$${}_2F_1\left(m-\nu, \nu+1+m; m+1; \frac{1-z}{2}\right), \quad \text{(B.106)}$$

valid when $|z-1|<2$, $|\arg(z-1)|<\pi$, and ν is arbitrary, and

$$Q_\nu^m(z) = \frac{(-1)^m\sqrt{\pi}\Gamma(\nu+m+1)(z^2-1)^{\frac{1}{2}m}}{2^{\nu+1}z^{\nu+m+1}\Gamma(\nu+\frac{3}{2})} \times$$
$${}_2F_1\left(\frac{\nu+m}{2}+1, \frac{\nu+m+1}{2}; \nu+\frac{3}{2}; \frac{1}{z^2}\right), \quad \text{(B.107)}$$

valid when $|z| >$; $|\arg(z\pm 1)|<\pi$, and $\nu \neq -1,-2,\ldots$. When $x \in (-1,1)$ is real,

$$P_\nu^m(x) = \frac{(-1)^m\Gamma(\nu+m+1)}{2^m\Gamma(m+1)\Gamma(\nu-m+1)}(1-x^2)^{\frac{1}{2}m} \times$$
$${}_2F_1\left(m-\nu, \nu+m+1; m+1; \frac{1-x}{2}\right). \quad \text{(B.108)}$$

Both functions $P_\nu^\mu(z)$ and $Q_\nu^\mu(z)$ satisfy the same *recurrence relations:*

$$P_\nu^{\mu+1}(z) = (z^2-1)^{-\frac{1}{2}}\left[(\nu-\mu)zP_\nu^\mu(z) - (\nu+\mu)P_{\nu-1}^\mu(z)\right], \tag{B.109}$$

$$\begin{aligned}(\nu-\mu+1)P_{\nu+1}^\mu(z) &= (2\nu+1)zP_\nu^\mu(z) - (\nu+\mu)P_{\nu-1}^\mu(z), && \text{(B.110)}\\ (z^2-1)\frac{dP_\nu^\mu(z)}{dz} &= \nu zP_\nu^\mu(z) - (\nu+\mu)P_{\nu-1}^\mu(z). && \text{(B.111)}\end{aligned}$$

Transformation formulae relate negative and positive indices:

$$\begin{aligned}P_{-\nu-1}^\mu(z) &= P_\nu^\mu(z),\\ P_{-\nu-1}^\mu(x) &= P_\nu^\mu(x), \quad -1<x<1; && \text{(B.112)}\end{aligned}$$

$$\begin{aligned}P_\nu^{-m}(z) &= \frac{\Gamma(\nu-m+1)}{\Gamma(\nu+m+1)}P_\nu^m(z),\\ P_\nu^{-m}(x) &= (-1)^m\frac{\Gamma(\nu-m+1)}{\Gamma(\nu+m+1)}P_\nu^m(x), \quad -1<x<1; && \text{(B.113)}\end{aligned}$$

$$Q_{-\nu-1}^\mu(z) = \frac{1}{\sin\pi(\nu-\mu)}\left[-\pi e^{\mu\pi i}\cos\nu\pi P_\nu^\mu(z) + \sin\pi(\nu+\mu)Q_\nu^\mu(z)\right]; \tag{B.114}$$

$$\begin{aligned}Q_\nu^{-\mu}(z) &= e^{-2\mu\pi i}\frac{\Gamma(\nu-\mu+1)}{\Gamma(\nu+\mu+1)}Q_\nu^\mu(z), && \text{(B.115)}\\ Q_\nu^{-m}(x) &= (-1)^m\frac{\Gamma(\nu-m+1)}{\Gamma(\nu+m+1)}Q_\nu^m(x). && \text{(B.116)}\end{aligned}$$

The Formulae (B.114)–(B.116) require that $-1<x<1$ and $\nu \neq m-1, m-2, \ldots$. Finally we note that when $m>n$,

$$P_n^m(z) = P_n^m(x) = 0. \tag{B.117}$$

Also

$$P_n^m(-x) = (-1)^{m+n}P_n^m(x), \quad -1<x<1. \tag{B.118}$$

The *Wronskian* is

$$W\{P_\nu^m(z), Q_\nu^m(z)\} = \frac{(-1)^m}{(1-z^2)}\frac{\Gamma(\nu+m+1)}{\Gamma(\nu-m+1)}, \tag{B.119}$$

$$W\{P_\nu^m(x), Q_\nu^m(x)\} = \frac{1}{(1-x^2)}\frac{\Gamma(\nu+m+1)}{\Gamma(\nu-m+1)}, \quad -1<x<1. \tag{B.120}$$

Some explicit expressions are

$$P_0^{-1}(z) = P_{-1}^{-1}(z) = \sqrt{\frac{z-1}{z+1}}, \tag{B.121}$$

$$P_1^1(x) = -\sqrt{1-x^2}, \; P_2^1(x) = -3x\sqrt{1-x^2}. \tag{B.122}$$

For fixed $z \notin (-\infty, -1) \cup (1, \infty)$ and fixed μ, as $\mathrm{Re}(\nu) \to \infty$

$$\begin{aligned} P_\nu^\mu(z) = & \frac{1}{\sqrt{2\pi}(z^2-1)^{1/4}} \frac{\Gamma(\nu+\mu+1)}{\Gamma(\nu+\frac{3}{2})} \left[z+\sqrt{z^2-1}\right]^{\nu+\frac{1}{2}} \times \\ & {}_2F_1\left(\frac{1}{2}+\mu, \frac{1}{2}-\mu; \frac{3}{2}+\nu; \frac{z+\sqrt{z^2-1}}{2\sqrt{z^2-1}}\right) + \\ & \frac{1}{\sqrt{2\pi}(z^2-1)^{1/4}} \frac{\Gamma(\nu+\mu+1)}{\Gamma(\nu+\frac{3}{2})} ie^{-i\mu\pi} \left[z-\sqrt{z^2-1}\right]^{\nu+\frac{1}{2}} \times \\ & {}_2F_1\left(\frac{1}{2}+\mu, \frac{1}{2}-\mu; \frac{3}{2}+\nu; \frac{-z+\sqrt{z^2-1}}{2\sqrt{z^2-1}}\right), \end{aligned} \tag{B.123}$$

and for fixed $z \notin (-\infty, -1)$ and fixed μ, as $\mathrm{Re}(\nu) \to \infty$

$$\begin{aligned} Q_\nu^\mu(z) = & e^{i\mu\pi}\sqrt{\frac{\pi}{2}} \frac{1}{(z^2-1)^{1/4}} \frac{\Gamma(\nu+\mu+1)}{\Gamma(\nu+\frac{3}{2})} \left[z-\sqrt{z^2-1}\right]^{\nu+\frac{1}{2}} \times \\ & {}_2F_1\left(\frac{1}{2}+\mu, \frac{1}{2}-\mu; \frac{3}{2}+\nu; \frac{-z+\sqrt{z^2-1}}{2\sqrt{z^2-1}}\right). \end{aligned} \tag{B.124}$$

B.5 Bessel functions

The commonly employed solutions of Bessel's differential equation

$$z^2\frac{d^2w}{dz^2} + z\frac{dw}{dz} + (z^2-\nu^2)w = 0 \tag{B.125}$$

are the Bessel functions of the first kind $J_\nu(z)$, of the second kind $Y_\nu(z)$ (also called the Neumann function), and of the third kind $H_\nu^{(1)}(z)$, $H_\nu^{(2)}(z)$ (also called the Hankel functions of the first and second kind, respectively), defined below; ν, z are in general complex. The classic treatise is Watson [73]. Each is a regular (holomorphic) function of z in the entire z - plane cut along the negative real axis; for fixed z ($\neq 0$) each is an entire function of ν. When ν is integral, $J_\nu(z)$ has no branch point and is an entire function of z.

The series representation for J_ν is

$$J_\nu(z) = \frac{1}{\Gamma(\nu+1)}\left(\frac{z}{2}\right)^\nu {}_0F_1\left(\nu+1; -\frac{z^2}{4}\right) = \sum_{k=0}^{\infty} \frac{(-1)^k (z/2)^{2k+\nu}}{k!\Gamma(k+\nu+1)}. \tag{B.126}$$

When $\nu = -n$ is a negative integer, and $J_{-n}(z) = (-1)^n J_n(z)$, for all z. The Neumann function is defined by

$$Y_\nu(z) = \frac{1}{\sin(\nu\pi)}\left[J_\nu(z)\cos(\nu\pi) - J_{-\nu}(z)\right] \tag{B.127}$$

where the right-hand side of this equation is replaced by its limiting value if ν is an integer or zero. When $\nu = n$ is a nonnegative integer,

$$\begin{aligned} Y_n(z) = \frac{2}{\pi} J_n(z) \ln\left(\frac{z}{2}\right) - \frac{1}{\pi}\left(\frac{z}{2}\right)^{-n} \sum_{k=0}^{n-1} \frac{(n-k-1)!}{k!}\left(\frac{z}{2}\right)^{2k} - \\ \frac{1}{\pi}\left(\frac{z}{2}\right)^n \sum_{k=0}^{\infty} \left[\psi(k+1) + \psi(n+k+1)\right] \frac{(-1)^k}{k!(n+k)!}\left(\frac{z}{2}\right)^{2k} \end{aligned} \tag{B.128}$$

where $\psi(k) = -\gamma + \sum_{n=0}^{\infty} \left(1/(n+1) - 1/(k+n)\right)$; also $Y_{-n}(z) = (-1)^n Y_n(z)$. The Hankel functions are defined to be

$$H_\nu^{(1)}(z) = J_\nu(z) + iY_\nu(z), \; H_\nu^{(2)}(z) = J_\nu(z) - iY_\nu(z). \tag{B.129}$$

The set $\{J_\nu, Y_v\}$ is a linearly independent pair of solutions of Bessel's differential equation. The same is true of the pair $\left\{H_\nu^{(1)}, H_\nu^{(2)}\right\}$. The *Wronskians* are

$$W\{J_\nu(z), Y_\nu(z)\} = J_\nu'(z)Y_\nu(z) - J_\nu(z)Y_\nu'(z) = \frac{2}{\pi z} \tag{B.130}$$

and

$$W\left\{H_\nu^{(1)}(z), H_\nu^{(2)}(z)\right\} = H_\nu^{(1)\prime}(z)H_\nu^{(2)}(z) - H_\nu^{(1)}(z)H_\nu^{(2)\prime}(z) = -\frac{4i}{\pi z}. \tag{B.131}$$

The functions $J_\nu, Y_\nu, H_\nu^{(1)}, H_\nu^{(2)}$ all satisfy the same *recurrence relations*

$$zF_{\nu-1}(z) + zF_{\nu+1}(z) = 2\nu F_\nu(z) \tag{B.132}$$

$$2\frac{d}{dz}F_\nu(z) = F_{\nu-1}(z) - F_{\nu+1}(z) \tag{B.133}$$

$$z\frac{d}{dz}F_\nu(z) = \pm\nu F_\nu(z) \mp zF_{\nu\pm1}(z) \tag{B.134}$$

$$\frac{d}{dz}\left[z^{\pm\nu}F_\nu(z)\right] = \pm z^{\pm\nu}F_{\nu\mp1}(z) \tag{B.135}$$

and the *differentiation formulae*

$$\left(\frac{1}{z}\frac{d}{dz}\right)^m \left[z^{\pm\nu} F_\nu(z)\right] = (\pm 1)^m z^{\pm\nu - m} F_{\nu \mp m}(z) \tag{B.136}$$

$$\frac{d^m}{dz^m} F_\nu(z) = \frac{1}{2^m} \sum_{k=0}^{m} (-1)^k \binom{m}{k} F_{\nu-m+2k}(z). \tag{B.137}$$

In particular, $J_0' = -J_1$, $Y_0' = -Y_1$ and $H_0^{(i)\prime}(z) = -H_1^{(i)}(z), (i = 1, 2)$.
The *generating function* is

$$\exp\left[\left(t - t^{-1}\right)\frac{z}{2}\right] = \sum_{n=-\infty}^{\infty} t^n J_n(z) \tag{B.138}$$

from which is derived

$$\cos(z\sin\theta) = J_0(z) + 2\sum_{k=1}^{\infty} J_{2k}(z)\cos(2k\theta) \tag{B.139}$$

$$\sin(z\sin\theta) = 2\sum_{k=0}^{\infty} J_{2k+1}(z)\sin\{(2k+1)\theta\} \tag{B.140}$$

$$\cos(z\cos\theta) = J_0(z) + 2\sum_{k=1}^{\infty} (-1)^k J_{2k}(z)\cos(2k\theta) \tag{B.141}$$

$$\sin(z\cos\theta) = 2\sum_{k=0}^{\infty} (-1)^k J_{2k+1}(z)\cos(2k+1)\theta \tag{B.142}$$

Asymptotics. When $|z| \to 0$ with ν fixed, the power series expansions (B.126)–(B.128) serve as asymptotic relations,

$$J_\nu(z) \sim \left(\frac{z}{2}\right)^\nu \frac{1}{\Gamma(\nu+1)}, \nu \neq -1, -2, \ldots \tag{B.143}$$

and when $\mathrm{Re}(\nu) > 0$,

$$Y_\nu(z) \sim -iH_\nu^{(1)}(z) \sim iH_\nu^{(2)}(z) \sim -\frac{1}{\pi}\Gamma(\nu)\left(\frac{z}{2}\right)^{-\nu}. \tag{B.144}$$

When z is fixed and $\nu \to \infty$,

$$J_\nu(z) \sim \frac{1}{\sqrt{2\pi\nu}}\left(\frac{ez}{2\nu}\right)^\nu, Y_\nu(z) \sim -\sqrt{\frac{2}{\pi\nu}}\left(\frac{ez}{2\nu}\right)^{-\nu}. \tag{B.145}$$

When ν is fixed and $|z| \to \infty$,

$$J_\nu(z) = \sqrt{\frac{2}{\pi z}}\left\{\cos\left(z - \frac{1}{2}\nu\pi - \frac{1}{4}\pi\right) + O\left(|z|^{-1}\right)\right\}, |\arg z| < \pi \tag{B.146}$$

$$Y_\nu(z) = \sqrt{\frac{2}{\pi z}}\left\{\sin\left(z - \frac{1}{2}\nu\pi - \frac{1}{4}\pi\right) + O\left(|z|^{-1}\right)\right\}, |\arg z| < \pi \tag{B.147}$$

$$H_\nu^{(1)}(z) \sim \sqrt{\frac{2}{\pi z}}\exp\left[i\left(z - \frac{1}{2}\nu\pi - \frac{1}{4}\pi\right)\right], -\pi < \arg z < 2\pi. \tag{B.148}$$

B.5.1 Spherical Bessel functions

The spherical Bessel functions $j_n, y_n, h_n^{(1,2)}$ are defined for integral n to be

$$\begin{aligned} j_n(z) &= \sqrt{\frac{\pi}{2z}}J_{n+1/2}(z), \\ y_n(z) &= \sqrt{\frac{\pi}{2z}}Y_{n+1/2}(z), \\ h_n^{(1,2)}(z) &= \sqrt{\frac{\pi}{2z}}H_{n+1/2}^{(1,2)}(z), \end{aligned} \tag{B.149}$$

and can be expressed in terms of elementary functions as

$$j_n(z) = (-z)^n\left(\frac{1}{z}\frac{d}{dz}\right)^n\left(\frac{\sin z}{z}\right), \tag{B.150}$$

$$y_n(z) = -(-z)^n\left(\frac{1}{z}\frac{d}{dz}\right)^n\left(\frac{\cos z}{z}\right). \tag{B.151}$$

B.5.2 Modified Bessel functions

Bessel functions with argument $\pm iz$ are known as modified Bessel functions and are solutions of the *differential equation*

$$z^2\frac{d^2w}{dz^2} + z\frac{dw}{dz} - (z^2 + \nu^2)w = 0. \tag{B.152}$$

The first-kind and second-kind modified Bessel functions are defined by

$$I_\nu(z) = e^{-\frac{1}{2}\nu\pi i}J_\nu\left(ze^{\frac{1}{2}\pi i}\right), \quad -\pi < \arg z \le \frac{1}{2}\pi, \tag{B.153}$$

$$I_\nu(z) = e^{\frac{3}{2}\nu\pi i}J_\nu\left(ze^{-\frac{3}{2}\pi i}\right), \quad \frac{1}{2}\pi < \arg z \le \pi, \tag{B.154}$$

and

$$K_\nu(z) = \frac{1}{2}\pi i e^{\frac{1}{2}\nu\pi i} H_\nu^{(1)}\left(ze^{\frac{1}{2}\pi i}\right), \quad -\pi < \arg z \le \frac{1}{2}\pi, \tag{B.155}$$

$$K_\nu(z) = -\frac{1}{2}\pi i e^{-\frac{1}{2}\nu\pi i} H_\nu^{(2)}\left(ze^{-\frac{1}{2}\pi i}\right), \quad \frac{1}{2}\pi < \arg z \le \pi. \tag{B.156}$$

Each is a regular function of z throughout the z-plane cut along the negative real axis, and for fixed z ($\neq 0$) each is an entire function of ν; when ν is integral, $I_\nu(z)$ is an entire function of z. They constitute a linearly independent pair of solutions to the differential equation with *Wronskian*

$$W\{I_\nu(z), K_\nu(z)\} = -\frac{1}{z}. \tag{B.157}$$

Also

$$K_\nu(z) = \frac{\pi}{2\sin(\nu\pi)}\left[I_{-\nu}(z) - I_\nu(z)\right] \tag{B.158}$$

where the right of this equation is replaced by its limiting value if ν is an integer or zero. The series expansions are

$$I_\nu(z) = \sum_{k=0}^{\infty} \frac{(z/2)^{2k+\nu}}{k!\Gamma(k+\nu+1)}, \tag{B.159}$$

and

$$K_n(z) = (-1)^{n+1} I_n(z) \ln\left(\frac{z}{2}\right) + \frac{1}{2}\left(\frac{z}{2}\right)^{-n} \sum_{k=0}^{n-1} \frac{(n-k-1)!}{k!}\left(\frac{z}{2}\right)^{2k} +$$

$$(-1)^n \frac{1}{2}\left(\frac{z}{2}\right)^n \sum_{k=0}^{\infty} \frac{[\psi(k+1) + \psi(n+k+1)]}{k!(n+k)!}\left(\frac{z}{2}\right)^{2k}, \tag{B.160}$$

where $\psi(k)$ was defined above. Also

$$I_{-n}(z) = I_n(z), K_{-\nu}(z) = K_\nu(z). \tag{B.161}$$

Recurrence relations satisfied by modified Bessel functions include

$$2\nu I_\nu(z) = zI_{\nu-1}(z) - zI_{\nu+1}(z), \tag{B.162}$$

$$2\nu K_\nu(z) = -zK_{\nu-1}(z) + zK_{\nu+1}(z). \tag{B.163}$$

Asymptotics. When ν is fixed and $z \to \infty$,

$$I_\nu(z) \sim \frac{1}{\sqrt{2\pi z}} e^z \sum_{n=0}^{\infty} (-1)^n (2z)^{-n} \frac{\Gamma(\frac{1}{2}+\nu+n)}{n!\Gamma(\frac{1}{2}+\nu-n)}, |\arg z| < \frac{\pi}{2}, \tag{B.164}$$

and

$$K_\nu(z) \sim \sqrt{\frac{\pi}{2z}} e^{-z} \sum_{n=0}^{\infty} (2z)^{-n} \frac{\Gamma(\frac{1}{2}+\nu+n)}{n!\Gamma(\frac{1}{2}+\nu-n)}, |\arg z| < \frac{3\pi}{2}. \tag{B.165}$$

B.6 The incomplete scalar product

The incomplete scalar product for the family of Jacobi polynomials is defined by

$$Q_{sn}^{(\alpha,\beta)}(t) = \int_t^1 (1-x)^\alpha (1+x)^\beta \, P_s^{(\alpha,\beta)}(x) P_n^{(\alpha,\beta)}(x) dx, \tag{B.166}$$

whilst its normalised counterpart is defined by

$$\hat{Q}_{sn}^{(\alpha,\beta)}(t) = \int_t^1 (1-x)^\alpha (1+x)^\beta \, \hat{P}_s^{(\alpha,\beta)}(x) \hat{P}_n^{(\alpha,\beta)}(x) dx. \tag{B.167}$$

Elementary properties of the normalised incomplete scalar product valid for all $s, n = 0, 1, \ldots$ are

$$\hat{Q}_{sn}^{(\alpha,\beta)}(1) = 0, \tag{B.168}$$

an index symmetry

$$\hat{Q}_{sn}^{(\alpha,\beta)}(t) = \hat{Q}_{ns}^{(\alpha,\beta)}(t), \tag{B.169}$$

and

$$\hat{Q}_{sn}^{(\alpha,\beta)}(-t) = \delta_{sn} - (-1)^{s-n} \, \hat{Q}_{sn}^{(\beta,\alpha)}(t). \tag{B.170}$$

Two other relationships frequently used are

$$\begin{aligned} \hat{Q}_{sn}^{(\alpha,\beta)}(t) = {} & \frac{(1-t)^{\alpha+1}(1+t)^\beta}{[(s+\alpha+1)(s+\beta)]^{\frac{1}{2}}} \hat{P}_s^{(\alpha+1,\beta-1)}(t) \hat{P}_n^{(\alpha,\beta)}(t) \\ & + \left[\frac{(n+\alpha+1)(n+\beta)}{(s+\alpha+1)(s+\beta)}\right]^{\frac{1}{2}} \hat{Q}_{sn}^{(\alpha+1,\beta-1)}(t), \end{aligned} \tag{B.171}$$

valid when $\alpha > -1, \beta > 0$, and

$$\begin{aligned} \hat{Q}_{sn}^{(\alpha,\beta)}(t) = {} & \frac{-(1-t)^{\alpha}(1+t)^{\beta+1}}{[(s+\alpha)(s+\beta+1)]^{\frac{1}{2}}} \hat{P}_s^{(\alpha-1,\beta+1)}(t) \hat{P}_n^{(\alpha,\beta)}(t) \\ & + \left[\frac{(n+\alpha)(n+\beta+1)}{(s+\alpha)(s+\beta+1)}\right]^{\frac{1}{2}} \hat{Q}_{sn}^{(\alpha-1,\beta+1)}(t), \end{aligned} \tag{B.172}$$

valid when $\alpha > 0, \beta > -1$. Formulae (B.171) and (B.172) are deduced from the relationships (1.173) and (1.174). Finally, the property

$$\sum_{l=0}^{\infty} \hat{Q}_{sl}^{(\alpha,\beta)}(t) \, \hat{Q}_{ln}^{(\alpha,\beta)}(t) = \hat{Q}_{sn}^{(\alpha,\beta)}(t) \tag{B.173}$$

allows us to interpret the matrix operator $K(t)$ with elements $\hat{Q}_{sn}^{(\alpha,\beta)}(t)$ as a projection operator on l_2.

Employ the following differentiation formulae, which follow from the index recurrence relations and the differential recurrence relations

$$-\frac{d}{dz}\left[(1-z)^{\alpha+1}\hat{P}_n^{(\alpha+1,\beta-1)}(z)\right] \\ = \sqrt{(n+\alpha+1)(n+\beta)}(1-z)^{\alpha}\hat{P}_n^{(\alpha,\beta)}(z), \quad \text{(B.174)}$$

$$\frac{d}{dz}\left[(1+z)^{\beta+1}\hat{P}_n^{(\alpha-1,\beta+1)}(z)\right] \\ = \sqrt{(n+\beta+1)(n+\alpha)}(1+z)^{\beta}\hat{P}_n^{(\alpha,\beta)}(z), \quad \text{(B.175)}$$

and apply integration by parts to obtain two equivalent expressions for the incomplete scalar product, valid when $s \neq l$,

$$\hat{Q}_{sl}^{(\alpha,\beta)}(z_0) = \frac{(1-z_0)^{\alpha+1}(1+z_0)^{\beta}}{[(s+\alpha+1)(s+\beta)-(l+\alpha+1)(l+\beta)]} \times \\ \left\{\sqrt{(s+\alpha+1)(s+\beta)}\hat{P}_s^{(\alpha+1,\beta-1)}(z_0)\hat{P}_l^{(\alpha,\beta)}(z_0) - \right. \\ \left. \sqrt{(l+\alpha+1)(l+\beta)}\hat{P}_s^{(\alpha,\beta)}(z_0)\hat{P}_l^{(\alpha+1,\beta-1)}(z_0)\right\} \quad \text{(B.176)}$$

and

$$\hat{Q}_{sl}^{(\alpha,\beta)}(z_0) = \frac{-(1-z_0)^{\alpha}(1+z_0)^{\beta+1}}{[(s+\beta+1)(s+\alpha)-(l+\beta+1)(l+\alpha)]} \times \\ \left\{\sqrt{(s+\beta+1)(s+\alpha)}\hat{P}_s^{(\alpha-1,\beta+1)}(z_0)\hat{P}_l^{(\alpha,\beta)}(z_0) - \right. \\ \left. \sqrt{(l+\beta+1)(l+\alpha)}\hat{P}_s^{(\alpha,\beta)}(z_0)\hat{P}_l^{(\alpha-1,\beta+1)}(z_0)\right\}. \quad \text{(B.177)}$$

Thus, when $s \neq l$, the incomplete scalar products $\hat{Q}_{sl}^{(\alpha,\beta)}(z_0)$ may be calculated in terms of the normalized Jacobi polynomials $\hat{P}_n^{(\alpha,\beta)}$. These polynomials are efficiently evaluated by a normalised form of the recurrence relation (B.21) on the polynomial order:

$$\hat{P}_{n+1}^{(\alpha,\beta)}(x) = (\hat{b}_n + x\hat{a}_n)\hat{P}_n^{(\alpha,\beta)}(x) - \hat{c}_n\hat{P}_{n-1}^{(\alpha,\beta)}(x) \quad \text{(B.178)}$$

with initialisation

$$\hat{P}_0^{(\alpha,\beta)}(x) = \left\{h_0^{(\alpha,\beta)}\right\}^{-\frac{1}{2}},$$

$$\hat{P}_1^{(\alpha,\beta)}(x) = \frac{1}{2}\left\{h_1^{(\alpha,\beta)}\right\}^{-\frac{1}{2}}[\alpha-\beta+x(\alpha+\beta+2)].$$

The recurrence coefficients are defined by

$$\hat{a}_n = a_n \left(h_n^{(\alpha,\beta)} / h_{n+1}^{(\alpha,\beta)} \right)^{\frac{1}{2}}, \tag{B.179}$$

$$\hat{b}_n = b_n \left(h_n^{(\alpha,\beta)} / h_{n+1}^{(\alpha,\beta)} \right)^{\frac{1}{2}}, \tag{B.180}$$

$$\hat{c}_n = c_n \left(h_{n-1}^{(\alpha,\beta)} / h_{n+1}^{(\alpha,\beta)} \right)^{\frac{1}{2}}. \tag{B.181}$$

The ratio of norm values in (B.179), (B.180), and (B.181) are simple rational expressions in n, α, and β; also

$$\hat{b}_n = \frac{(\alpha^2 - \beta^2)\hat{a}_n}{(2n+\alpha+\beta)(2n+\alpha+\beta+2)}. \tag{B.182}$$

When $s = l$, the following recurrence relation for the incomplete scalar product may be employed. Consider (B.178) with $n = s$ and $n = l$:

$$\hat{P}_{s+1}^{(\alpha,\beta)}(x) = (\hat{b}_s + x\hat{a}_s)\hat{P}_s^{(\alpha,\beta)}(x) - \hat{c}_s \hat{P}_{s-1}^{(\alpha,\beta)}(x), \tag{B.183}$$

$$\hat{P}_{l+1}^{(\alpha,\beta)}(x) = (\hat{b}_l + x\hat{a}_l)\hat{P}_l^{(\alpha,\beta)}(x) - \hat{c}_l \hat{P}_{l-1}^{(\alpha,\beta)}(x). \tag{B.184}$$

Multiply (B.183) by $\hat{a}_l \hat{P}_l^{(\alpha,\beta)}(x)$, (B.184) by $\hat{a}_s \hat{P}_s^{(\alpha,\beta)}(x)$ and subtract to eliminate the term containing x. Now multiply by the factor $(1-x)^\alpha(1+x)^\beta$ and integrate over $(z_0, 1)$ to deduce the following recurrence relation:

$$\begin{aligned} &\hat{a}_l \hat{Q}_{s+1,l}^{(\alpha,\beta)}(z_0) - \hat{a}_s \hat{Q}_{l+1,s}^{(\alpha,\beta)}(z_0) \\ &\quad = (\hat{a}_l \hat{b}_s - \hat{a}_s \hat{b}_l)\hat{Q}_{sl}^{(\alpha,\beta)}(z_0) - \hat{c}_s \hat{a}_l \hat{Q}_{s-1,l}^{(\alpha,\beta)}(z_0) + \hat{c}_l \hat{a}_s \hat{Q}_{l-1,s}^{(\alpha,\beta)}(z_0). \end{aligned} \tag{B.185}$$

Setting $s = l+1$ in (B.185) produces a recurrence formula involving $\hat{Q}_{ll}^{(\alpha,\beta)}$, and three other incomplete scalar products of form $\hat{Q}_{nm}^{(\alpha,\beta)}$ with $n \neq m$.

$$\begin{aligned} \hat{Q}_{l+1,l+1}^{(\alpha,\beta)}(z_0) = {} & \frac{\hat{a}_l}{\hat{a}_{l+1}} \hat{Q}_{l+2,l}^{(\alpha,\beta)}(z_0) + \left(\hat{b}_l - \hat{b}_{l+1}\frac{\hat{a}_l}{\hat{a}_{l+1}} \right) \hat{Q}_{l+1,l}^{(\alpha,\beta)}(z_0) \\ & + \hat{c}_{l+1}\frac{\hat{a}_l}{\hat{a}_{l+1}} \hat{Q}_{ll}^{(\alpha,\beta)}(z_0) - \hat{c}_l \hat{Q}_{l-1,l+1}^{(\alpha,\beta)}(z_0). \end{aligned} \tag{B.186}$$

It may be initialised by the value

$$\hat{Q}_{00}^{(\alpha,\beta)}(z_0) = \left\{ h_0^{(\alpha,\beta)} \right\}^{-1} \int_{z_0}^{1} (1-x)^\alpha (1+x)^\beta \, dx. \tag{B.187}$$

Special cases commonly encountered are

$$\hat{Q}_{nm}^{(-\frac{1}{2},-\frac{1}{2})}(\cos\theta_0) = \frac{1}{\pi}\left[\frac{\sin(n-m)\,\theta_0}{n-m} + \frac{\sin(n+m)\,\theta_0}{n+m}\right], \tag{B.188}$$

$$\hat{Q}_{nm}^{(\frac{1}{2},\frac{1}{2})}(\cos\theta_0) = \frac{1}{\pi}\left[\frac{\sin(n-m)\,\theta_0}{n-m} - \frac{\sin(n+m)\,\theta_0}{n+m}\right], \tag{B.189}$$

$$\hat{Q}_{nm}^{(-\frac{1}{2},\frac{1}{2})}(\cos\theta_0) = \frac{1}{\pi}\left[\frac{\sin(n-m)\,\theta_0}{n-m} + \frac{\sin(n+m+1)\,\theta_0}{n+m+1}\right], \tag{B.190}$$

$$\hat{Q}_{n-1,m-1}^{(\frac{1}{2},-\frac{1}{2})}(\cos\theta_0) = \frac{1}{\pi}\left[\frac{\sin(n-m)\,\theta_0}{n-m} - \frac{\sin(n+m+1)\,\theta_0}{n+m+1}\right]. \tag{B.191}$$

These are valid when $n \neq m$; when $n = m$, the term

$$\frac{\sin(n-m)\,\theta_0}{n-m}$$

occurring in (B.188)–(B.191) is replaced by θ_0.

Appendix C

Elements of Functional Analysis

C.1 Hilbert spaces

In this section we collect some concepts from functional analysis. There are many standard introductory texts on this material, including [34, 33, 78, 10].

A Hilbert space is a vector space H over a field of either real or complex scalars, endowed with an inner product. The inner product is a bilinear map that associates to each pair of elements f, g in H a complex number denoted (f, g) with the following properties: (1) $(\alpha_1 f_1 + \alpha_2 f_2, g) = \alpha_1 (f_1, g) + \alpha_2 (f_2, g)$ for all $f_1, f_2, g \in H$, and scalars α_1, α_2; (2) $(f, g) = \overline{(g, f)}$ for all $f, g \in H$, where the bar denotes complex conjugate; and (3) $(f, f) \geq 0$ and $(f, f) = 0 \Leftrightarrow f = 0$. We normally deal with real Hilbert spaces with a real inner product. The third property allows us to define the norm of an element $f \in H$ to be $||f|| = (f, f)^{\frac{1}{2}}$. It satisfies the properties (1) $||f|| \geq 0$ and $||f|| = 0 \Leftrightarrow f = 0$; (2) $||\alpha f|| = |\alpha|\, ||f||$ for all scalars α; and (3) $||f + g|| = ||f|| + ||g||$ for all $f, g \in H$. Moreover, the *Cauchy-Schwarz inequality* $|(f, g)| \leq ||f||\, ||g||$ holds. The Hilbert space H is *complete* with respect to this norm, i.e., every sequence $\{f_n\}_{n=1}^{\infty}$ in H that is Cauchy (so that $||f_n - f_m|| \to 0$ as $n, m \to \infty$) is also convergent to an element f of H ($||f_n - f|| \to 0$ as $n \to \infty$).

A basis for H is a set of elements $\{e_1, e_2, \ldots\}$ of H such that every element f of H is a unique linear combination of the basis elements: there exist scalars $\alpha_1, \alpha_2, \ldots$ such that

$$f = \sum_n \alpha_n e_n. \tag{C.1}$$

If the basis can be ordered as a countably infinite sequence $\{e_n\}_{n=1}^{\infty}$ H is called separable, and the sum (C.1) is interpreted to mean that

$$\left\| f - \sum_{n=1}^{N} \alpha_n e_n \right\| \to 0 \text{ as } N \to \infty. \tag{C.2}$$

(If the basis is not countable, then only countably many scalars in the sum (C.1) may be nonzero and the sum is interpreted in the sense of (C.2) for the nonzero scalar elements sequentially ordered.) The basis is orthogonal if $(f_n, f_m) = h_n \delta_{nm}$, where $h_n = \|f_n\|^2$ is necessarily positive. If $h_n = 1$ for all n, the basis is orthonormal; this may always be arranged by replacing each basis element f_n by $f_n / \|f_n\|$.

Examples of Hilbert spaces.

1. Let l_2 denote the space of (real or complex) sequences $\{a_n\}_{n=1}^{\infty}$ such that $\sum_{n=1}^{\infty} |a_n|^2$ converges. It is a Hilbert space with the inner product of sequences $a = \{a_n\}_{n=1}^{\infty}$ and $b = \{b_n\}_{n=1}^{\infty}$ defined to be

$$(a, b) = \sum_{n=1}^{\infty} a_n \overline{b_n}. \tag{C.3}$$

An orthonormal basis is the set of sequences $S = \{e_n, n = 1, 2, \ldots\}$ where $e_n = \{\delta_{nm}\}_{m=1}^{\infty}$.

2. Let $w = \{w_n\}_{n=1}^{\infty}$ be a positive real sequence, and define $l_2(w)$ to be space of (real or complex) sequences $\{a_n\}_{n=1}^{\infty}$ such that $\sum_{n=1}^{\infty} w_n |a_n|^2$ converges. It is a Hilbert space with the inner product of sequences $a = \{a_n\}_{n=1}^{\infty}$ and $b = \{b_n\}_{n=1}^{\infty}$ defined to be

$$(a, b) = \sum_{n=1}^{\infty} w_n a_n \overline{b_n}. \tag{C.4}$$

The set S defined above is an orthogonal basis, and is orthonormal only if $w_n = 1$ for all n. A particular example of interest is the choice $w_n = n^{\mu}$ where μ is a fixed real number; this space is denoted $l_2(\mu)$.

3. Let $L_2(a, b)$ denote the set of (real or complex) valued functions f defined on the interval (a, b) such that $\int_a^b |f|^2$ converges. It is a separable Hilbert space with the inner product of functions f, g defined to be

$$(f, g) = \int_a^b f \overline{g}. \tag{C.5}$$

The Lebesgue integral is used for this purpose with the understanding that two functions f, g are regarded as equal if they differ only on a set of Lebesgue measure zero (f, g are said to be *equal almost everywhere*); this allows us to assert that the only function of norm zero is the function that is zero almost everywhere.

4. Let w be a real valued positive function defined on (a, b). Let $L_{2,w}(a, b)$ denote the set of (real or complex) valued functions f defined on (a, b) such that $\int_a^b w|f|^2$ converges. It is a separable Hilbert space with the inner product of functions f, g defined to be

$$(f, g) = \int_a^b w f \overline{g}, \tag{C.6}$$

with derived norm

$$||f|| = \left(\int_a^b w|f|^2 \right)^{\frac{1}{2}}. \tag{C.7}$$

If α and β are real numbers exceeding -1, and w is defined by $w(x) = (1-x)^{\alpha}(1+x)^{\beta}$, then the Jacobi polynomials $\left\{ P_n^{(\alpha,\beta)} \right\}_{n=1}^{\infty}$ form an orthogonal basis for $L_{2,w}(-1, 1)$, and the normalised Jacobi polynomials $\left\{ \hat{P}_n^{(\alpha,\beta)} \right\}_{n=1}^{\infty}$ form an orthonormal basis. The cosine functions $\{\cos n\theta\}_{n=1}^{\infty}$ and the complex exponential functions $\{e^{in\theta}\}_{n=1}^{\infty}$ form orthogonal bases for $L_2(0, \pi)$ and $L_2(0, 2\pi)$, respectively.

C.2 Operators

A linear operator T on H is a function $T : H \to H$ that is linear: $T(\alpha_1 f_1 + \alpha_2 f_2) = \alpha_1 T(f_1) + \alpha_2 T(f_2)$ for all $f_1, f_2, g \in H$, and scalars α_1, α_2. T is bounded if there exists a positive constant M such that $||T(f)|| \leq M||f||$ for all $f \in H$; the norm of the operator is then defined to be

$$||T|| = \sup_{f \neq 0} \frac{||T(f)||}{||f||} = \sup_{||f||=1} ||T(f)||. \tag{C.8}$$

The null space $N(T)$ of T is the set $\{f \in H : T(f) = 0\}$; the range of T is the image $T(H)$ of H under the action of T.

An example is the integral operator K formed from a real or complex valued kernel function k of two variables defined on $(a, b) \times (a, b)$ via

$$K(f)(x) = \int_a^b k(x, t) f(t) dt \tag{C.9}$$

for each function $f \in L_2(a,b)$; the condition

$$\int_a^b \int_a^b |k(x,t)|^2 \, dx dt < \infty \qquad \text{(C.10)}$$

ensures that K is a bounded linear operator on $L_2(a,b)$ with norm $\|K\|$ not exceeding $\left(\int_a^b \int_a^b |k(x,t)|^2 \, dx dt \right)^{\frac{1}{2}}$. A discrete analogue is the operator K with associated matrix $(k_{nm})_{n,m=1}^{\infty}$ defined via

$$(Ka)_n = \sum_{m=1}^{\infty} k_{nm} a_m, \qquad (m = 1, 2, \ldots), \qquad \text{(C.11)}$$

for each sequence $\{a_n\}_{n=1}^{\infty}$ in l_2; the condition

$$\sum_{m=1}^{\infty} \sum_{n=1}^{\infty} |k_{nm}|^2 < \infty \qquad \text{(C.12)}$$

ensures that K is a bounded linear operator on l_2 with norm $\|K\|$ not exceeding $\left(\sum_{m=1}^{\infty} \sum_{n=1}^{\infty} |k_{nm}|^2 \right)^{\frac{1}{2}}$.

Of particular importance in numerical methods are projection operators P that may be characterised by the requirement that

$$P^2 = P.$$

In practice, such an operator is often associated with a finite dimensional space and is used to convert operator equations of the form $Kf = g$ to systems of finitely many linear equations; the relation between the (computed) solution to the finite system and the original (infinite dimensional) system is important in determining the success of numerical solution methods (see below).

The adjoint K^* of a linear operator K on H is uniquely defined by the requirement that

$$(K^* f, g) = (f, Kg) \qquad \text{(C.13)}$$

for all $f, g \in H$. The adjoint of the integral operator defined in (C.9) is an integral operator of the same form with kernel h defined by

$$h(x,t) = \overline{k(t,x)}. \qquad \text{(C.14)}$$

The adjoint of the matrix operator defined in (C.11) is a matrix operator of the same form with matrix h defined by

$$h_{nm} = \overline{k}_{mn}, \qquad \text{(C.15)}$$

for all $n, m = 1, 2, \ldots$

The operator K on H is compact (also called completely continuous) if for every bounded sequence $\{f_n\}_{n=1}^{\infty}$ in H, the image sequence $\{K(f_n)\}_{n=1}^{\infty}$ has a convergent subsequence (in H). Bounded finite rank operators (those with finite dimensional range) are necessarily compact. The integral operator and matrix operator defined by (C.9) and (C.11) are compact. By contrast, the identity operator I is never compact in infinite dimensional spaces. If $\{e_n\}_{n=1}^{\infty}$ is a basis for H, and $\{\lambda_n\}_{n=1}^{\infty}$ is a sequence of scalars, the diagonal operator defined by

$$K(e_n) = \lambda_n e_n \tag{C.16}$$

for all n is compact if and only if $\lambda_n \to 0$ as $n \to \infty$.

Properties of compact operators are discussed in [34, 33]. In particular, the set of eigenvalues of a compact operator K (those values of λ for which the equation $(\lambda I - K)x = 0$ has nontrivial solutions x) is countable (perhaps finite or even empty); 0 is the only possible point of accumulation of this set. The Abel integral operator A defined on $L_2(0,1)$ by

$$A(f)(x) = \int_0^x \frac{f(t)\,dt}{\sqrt{x^2 - t^2}}, \quad x \in (0,1) \tag{C.17}$$

has norm $\|A\| = \frac{\pi}{2}$ and is not compact; for, as observed in [4], the functions $f_\alpha(t) = t^\alpha$ (with $\alpha \geq 0$), are eigenfunctions of A satisfying $Af_\alpha = \lambda_\alpha f_\alpha$, where the eigenvalues λ_α vary continuously between 0 and $\frac{\pi}{2}$ as α ranges from 0 to ∞, so that A cannot be compact.

The dimension of each eigenspace of K is finite; for each $\lambda \neq 0$, there is a unique smallest integer r so that the null spaces satisfy

$$N((\lambda I - K)^r) = N\left((\lambda I - K)^{r+1}\right) = N\left((\lambda I - K)^{r+1}\right) = \ldots \tag{C.18}$$

and the range spaces satisfy

$$(\lambda I - K)^r H = (\lambda I - K)^{r+1} H = (\lambda I - K)^{r+1} H = \ldots. \tag{C.19}$$

The space H has the orthogonal decomposition

$$H = N((\lambda I - K)^r) \oplus (\lambda I - K)^r H \tag{C.20}$$

(every element of H is a unique sum of two orthogonal elements lying in $N((\lambda I - K)^r)$ and $(\lambda I - K)^r H$).

C.3 The Fredholm alternative and regularisation

The following result, known as the *Fredholm alternative*, is very important in establishing the solubility of second-kind equations of the form

$(\lambda I - K)\,x = y$, where λ is a scalar and K is a compact operator on a Hilbert space H ($\lambda^{-1}K$ is a compact perturbation of the identity operator I). We consider the four equations

$$(\lambda I - K)\,x = y \tag{C.21}$$
$$(\lambda I - K)\,x = 0 \tag{C.22}$$
$$(\lambda I - K^*)\,u = v \tag{C.23}$$
$$(\lambda I - K^*)\,u = 0 \tag{C.24}$$

where y and v are given elements of H.

Theorem 12 *(The Fredholm alternative.) The Equation (C.21) has a solution $x \in H$ if and only if $(y, u) = 0$ for all solutions u of the homogeneous Equation (C.24). Thus if the zero solution $u = 0$ is the only solution of (C.24), then for every y, the Equation (C.21) is solvable, i.e., the range of $\lambda I - K$ is H; the solution x depends continuously on y. Likewise, Equation (C.23) has a solution $u \in H$ if and only if $(x, v) = 0$ for all solutions x of the homogeneous Equation (C.22). Equations (C.22) and (C.24) have the same number of linearly independent solutions.*

These and allied properties of second-kind equations permit the construction of relatively simple numerical methods that are stable and well-conditioned and for which error analyses are possible. Atkinson's book [4] is a comprehensive survey of methods particularly appropriate to integral equations, paying attention to error estimates. In a similar way, Kantorovich [30] discusses error estimates for second-kind matrix systems that are solved by the truncation method; Kress [33] also discusses such estimates in the context of projection methods.

By contrast, first-kind equations, such as

$$Kx = y \tag{C.25}$$

where K is a compact operator (for example the matrix operator defined by (C.9) or the integral operator given by (C.11)), are generally unstable, and simple numerical methods are ill-conditioned and yield poor results. It is necessary to employ some method of regularising the equation. One such method is Tikhonov regularisation, that consists of replacing (C.25) by

$$\left(\varepsilon^2 I + K^* K\right) x = K^* y. \tag{C.26}$$

For small ε, solutions to (C.26) approximately equal those of (C.25) (and are identical when $\varepsilon = 0$), but the precise selection of ε is rather problem dependent and requires some care in achieving acceptably accurate numerical solutions [22].

Many problems of diffraction theory and potential theory give rise to systems of matrix equations or integral equations of the form

$$Ax = y, \tag{C.27}$$

which are singular in the sense that they are not of the second kind involving a compact operator. From a theoretical point of view it can be difficult to establish whether such equations have solutions, even though there may be good physical reasons to expect the existence of a solution. Moreover, the continuous dependence of the solution x on y is not obvious, though clearly necessary for any physically plausible model of potential or diffraction. From a computational point of view, the equation is likely to be unstable, i.e., small perturbations to y result in large (and physically implausible) changes in the computed solution x. It is not difficult to see how this effect arises for the first-kind Equation (C.25) when the compact operator K is given by (C.16).

It is therefore desirable, wherever possible, to convert the singular Equation (C.27) to one of second kind with a compact operator for which the Fredholm alternative holds so that the associated benefits described above are obtained. This process is known as (analytical) regularisation. It may be described formally as follows. The bounded linear operator R is called a (left) regulariser of A if

$$RA = I - K$$

where K is a compact operator on H. Some general properties of regularisers are described in [33]. Application of the regulariser R to (C.27) produces an equation of the desired format:

$$(I - K)\, x = Ry.$$

In general, the construction of R may be difficult, if not impossible. However, the dual series equations arising from the potential problems and diffraction problems considered in this book and its companion volume can indeed be regularised; the regularisation process is explicitly described in Section 2.1, although the regulariser appears only implicitly in the analytical treatment of the dual series equations. The regularised equations enjoy all the advantages of second-kind equations for which the Fredholm alternative holds, including precise estimates of the error or difference of any solution computed to a truncated system, from the true solution (as a function of truncation number N_{tr}). The error decays to zero as $N_{tr} \to \infty$ (and in practice quite rapidly beyond a certain cutoff point, usually related to the electrical size of the body in diffraction problems).

The same remarks apply to triple series equations, as well as to the dual and triple integral equations arising from the mixed boundary value problems associated with Laplace's equation, the Helmholtz equation, and Maxwell's equations for the various canonical structures described in these volumes.

Appendix D

Transforms and Integration of Series

D.1 Fourier and Hankel transforms

The Fourier transform of the function f defined on $(-\infty, \infty)$ is

$$F(y) = \int_{-\infty}^{\infty} f(x) e^{-2\pi i x y} dx, \tag{D.1}$$

and its inverse is given by

$$f(x) = \int_{-\infty}^{\infty} F(y) e^{2\pi i x y} dy. \tag{D.2}$$

Precise conditions on the validity of the inversion formula is given in [9]; a particular useful class for which it holds is $L^p(-\infty, \infty)$ with $1 \le p \le 2$.

The Hankel transform of the function f defined on $(0, \infty)$ is

$$F(y) = \int_0^{\infty} J_\nu(xy)\, f(x)\, (xy)^{\frac{1}{2}}\, dx, \tag{D.3}$$

and its inverse is given by

$$f(x) = \int_0^{\infty} J_\nu(xy)\, F(y)\, (xy)^{\frac{1}{2}}\, dy. \tag{D.4}$$

The inversion formula is valid for parameter $\nu \geqslant -\frac{1}{2}$ when f is integrable on $(0, \infty)$ and of bounded variation near the point x, and is continuous at x; if f has a jump discontinuity at x, the left-hand side of (D.4) is replaced by $\frac{1}{2}(f(x+0) + f(x-0))$ (see [61]).

D.2 Integration of series

In this section we present some results on the validity of term-by-term integration of series.

Theorem 13 *Let* $\{f_n\}_{n=1}^{\infty}$ *be a sequence in* $L_2(a,b)$, *converging to* f *in the* L_2 *norm, i.e.,*

$$\|f - f_n\| = \left(\int_a^b |f - f_n|^2\right)^{\frac{1}{2}} \to 0, \quad \text{as } n \to \infty.$$

Let g *be a function in* $L_2(a,b)$ *and define*

$$h(x) = \int_a^x fg, \qquad h_n(x) = \int_a^x f_n g.$$

Then h_n *converges uniformly to* h *on* $[a,b]$.

Proof. *Fix* $x \in [a,b]$; *from the Cauchy-Schwarz inequality,*

$$\left(\int_a^x |f - f_n|\,|g|\right)^2 \leq \int_a^x |f - f_n|^2 \int_a^x |g|^2.$$

Let $A = 1 + \int_a^b |g|^2$. *Then, given* $\varepsilon > 0$, *there exists* N *such that when* $n > N$,

$$\int_a^b |f - f_n|^2 < \varepsilon^2/A, \text{ so that } \int_a^x |f - f_n|\,|g| < \varepsilon.$$

Thus, h_n *converges uniformly to* h *on* $[a,b]$.

Corollary 14 *Let* $\sum_{n=1}^{\infty} f_n$ *be a series with* $f_n \in L_2(a,b)$ *and converging to* f *in the* L_2 *norm, i.e.,*

$$\left\| f - \sum_{r=1}^{n} f_r \right\| = \left(\int_a^b \left| f - \sum_{r=1}^{n} f_r \right|^2\right)^{\frac{1}{2}} \to 0, \quad \text{as } n \to \infty.$$

Then the series

$$\sum_{n=1}^{\infty} \int_a^x f_n g$$

is uniformly convergent to $\int_a^x fg$ *on* $[a,b]$.

■

In particular, the Fourier series of any function in $L_2(a,b)$ can be integrated term-by-term over the interval $[a,x]$.

The series $\sum_{n=1}^{\infty} a_n$ of real terms is *Abel-summable* if

$$\lim_{r\to 1-0} \sum_{n=1}^{\infty} a_n r^n$$

exists. The series $\sum_{n=1}^{\infty} f_n$ of real valued functions on $[a,b]$ is *uniformly Abel-summable* on $[a,b]$ to the function f, if for all $\varepsilon > 0$, there is some $\delta > 0$ such that for all $x \in [a,b]$,

$$\left| \sum_{n=1}^{\infty} f_n(x) r^n - f(x) \right| < \varepsilon \text{ for } 1-\delta < r < 1.$$

For each fixed r with $0 < r < 1$, the power series $\sum_{n=1}^{\infty} f_n(x) r^n$ is uniformly convergent on $[a,b]$ to its sum, and may be integrated term by term. It immediately follows that term-by-term integration of a *uniformly Abel-summable series* is justified.

References

[1] Abramowitz, M. and Stegun, I.A., *Handbook of Mathematical Functions*, Dover, (1965).

[2] Akhiezer, N.I. and Glazman, I.M., *Theory of Linear Operators in Hilbert Space*, Vols. 1 & 2, Pitman Publishing Co. (1981).

[3] Ashour, A.A., "Note on the relations between series equations and integral equations," *Proc. Camb. Phil. Soc.*, **61**(3), 695–696 (1965).

[4] Atkinson, K.E., *The Numerical Solution of Integral Equations of the Second Kind,* C.U.P. (1997).

[5] Boersma, J. and Danicki, E., "On the solution of an integral equation arising in potential problems for circular and elliptic discs," *SIAM J. Appl. Math.*, **53**(4), 931–941 (1993).

[6] Bowman, J.J., Senior, T.B.A., and Uslenghi, P.L.E., *Electromagnetic and Acoustic Scattering by Simple Shapes*, Hemisphere Publishing Corp., revised printing (1987).

[7] Bouwkamp, C.J., "A simple method of calculating electrostatic capacity," *Physika*, **XXIV** (6), 538–542 (1958).

[8] Busbridge, I.W., "Dual integral equations," *Proc. London Math. Soc.*, **44**, 115 (1938).

[9] Champeney, D.C., *A Handbook of Fourier Theorems*, C.U.P. (1987).

[10] Churchill, R.V., *Fourier Series and Boundary Value Problems*, McGraw-Hill (1948).

[11] Collins, W.D., "On the solution of some axisymmetric boundary value problems by means of integral equations. I. Some electrostatic and hydrodynamic problems for a spherical cap," *Quart. J. Mech. Appl. Math.,* **12**(2), 232–241 (1959).

[12] Collins, W.D., "On some dual series equations and their application to electrostatic problems for spheroidal caps," *Proc. Camb. Phil. Soc.*, **57**(2), 367–384 (1963).

[13] Courant, D. and Hilbert, R., *Methods of Mathematical Physics*, Vols. 1 & 2, Wiley, Reprint (1989).

[14] Erdelyi, A., Magnus, W., Oberhittinger, F., and Tricomi, F.G., *Higher Transcendental Functions,* Vols. 1–3. *Bateman Manuscript Project,* McGraw-Hill (1953).

[15] Erdelyi, A., Magnus, W., Oberhittinger, F., and Tricomi, F.G., *Tables of Integral Transforms,* Vols. 1 & 2. *Bateman Manuscript Project,* McGraw-Hill (1954).

[16] Flammer, C., *Spheroidal Wave Functions*, Stanford University Press (1957).

[17] Folland, G.B., *Introduction to Partial Differential Equations,* Princeton University Press (1976).

[18] Gakhov, F.D., *Boundary Value Problems,* Pergamom Press (1966).

[19] Gradshteyn, I.S. and Rhyzik, I.M., *Tables of Integrals, Series and Products*, 5th edition (edited by A. Jeffrey), Academic Press (1994).

[20] Grant, I.S. and Philips, W.R., *Electromagnetism*, 2nd Edition, Wiley (1990).

[21] Gordon, A.N., "Dual integral equations," *J. London Math. Soc.*, **29,** 360–363 (1954).

[22] Hansen, P.C., "Analysis of discrete ill-posed problems by means of the L-curve," *SIAM Review*, **34**(4), 561–580 (1992).

[23] Hobson, E.W., *Spherical and Spheroidal Harmonics*, C.U.P. (1931).

[24] Hochstadt, H., *Integral Equations*, Wiley (1973).

[25] Homentcovchi, D., "On the electrostatic potential problem for a spherical cap," *ZAMM*, **60**(1), 636-637 (1980).

[26] Iossel' Yu.Ya., Kochanov E.S., and Strunskii M.G., *Raschet electricheskoi emkosti (Calculation of electric capacity)*, Energoizdat, Moskva-Leningrad (1981) (in Russian).

[27] Jones, D.S., *The Theory of Electromagnetism*, Pergamom Press (1964).

[28] Jones, D.S., *The Theory of Distributions*, O.U.P. (1988).

[29] Kapitsa, P.L., Fock, V.A. and Weinstein, L.A., "Static boundary problems for a hollow cylinder of finite length," *Sov. Phys. Techn. Physics*, **29** (10), 1177 (1959) (in Russian).

[30] Kantorovich, L.V. and Akilov, G.P., *Functional analysis in normed spaces*, Pergamom Press (1974).

[31] Kanwal, R.P., *Linear Integral Equations: Theory and Technique*, Academic Press (1971).

[32] Kellogg, O.D., *Foundations of Potential Theory*, Dover (1954).

[33] Kress, R., *Linear Integral Equations*, Springer-Verlag (1995).

[34] Kreyzig, E., *Introduction to Functional Analysis*, Wiley (1978).

[35] Lamé, G., *Journ. de l'Ecole Polyt.*, **14** , 191 (1834).

[36] Lebedev, N.N., *Special Functions and their Applications*, Prentice-Hall (1965).

[37] Lowndes, J.S., "Some dual and triple series equations," *Proc. Edin. Math. Soc.*, **18** (4), 273–280 (1969).

[38] Litvinenko, L.N. and Salnikova, L.P., *Numerical investigation of electrostatic fields for periodic structures*, Naukova Dumka, Kiev (1986) (in Russian).

[39] Martin, P.A. "The spherical-cap crack revisited," *Int. J. Solids & Structures*, to appear (2001).

[40] Meixner, J. and Schafke, F.W., *Mathieusche Funktionen und Sphäroidfunktionen*, Berlin (1954).

[41] Miles, J.W., "Potential and Rayleigh scattering theory for a spherical cap," *Quart. Appl. Math.*, **29** (1), 109–123 (1971).

[42] Minkov, I.M., "Electrostatic field due to the cut spherical condensor," *Sov. Phys. Techn. Physics*, **32** (12), 1409-1412 (1962) (in Russian).

[43] Monna, A.F., *Dirichlet's Principle: A Mathematical Comedy of Errors and its Influence on the Development of Analysis*, Oosthoek, Scheltema & Holkema (1975).

[44] Morse, P.M. and Feschbach, H., *Methods of Mathematical Physics*, Vols. 1 & 2, McGraw-Hill (1953).

[45] Muskhelishvili, N.I., *Singular Integral Equations*, Noordhoff, Croningen (1953).

[46] Noble, B., "The solution of Bessel function dual integral equations by a multiplying factor method," *Proc. Camb. Phil. Soc.*, **59**(2), 351–362 (1963).

[47] Noble, B., "Some dual series equations involving Jacobi polynomials," *Proc. Camb. Phil. Soc.*, **59**(2), 363–371 (1963).

[48] Parihar, K.S., "Some trigonometric series equations and their applications," *Proc. Roy. Soc. Edin.* A, **69**(3), 255–265 (1971).

[49] Pinkus, A. and Zafrany, S., *Fourier Series and Integral Transforms*, C.U.P. (1997).

[50] Porter, D. and Stirling, D.S.G., *Integral Equations*, C.U.P. (1990).

[51] P. Pyati Vital, "Capacitance of a spindle," *Proc. IEEE*, **61**(4), 505–506 (1973).

[52] Shestopalov, V.P., Tuchkin, Yu.A., Poedinchuk, A.E., and Sirenko, Yu.K., *Novie metodi resheniya pryamih i obratnih zadach teorii difrakcii (New methods of solving direct and inverse problems of diffraction theory)*, Osnova, Kharkiv (1997) (in Russian).

[53] Shestopalov, V.P., *The Riemann-Hilbert Problem in Diffraction Theory and Electromagnetic Wave Propagation*, Kharkov University Press, Kharkov (1971) (in Russian).

[54] Smythe, W.R. *Static and Dynamic Electricity*, 3rd edition, revised printing, Hemisphere Publishing Corp. (1989).

[55] Sneddon, I.N., *Mixed Boundary Value Problems in Potential Theory*, North-Holland Publishing Company (1966).

[56] Sneddon, I.N., *The Use of Integral Transforms*, McGraw-Hill (1972).

[57] Sneddon, I.N., *Special Functions of Mathematical Physics and Chemistry,* 3rd edition, Longman (1980).

[58] Szegö, G., *Orthogonal polynomials*, American Mathematical Society Colloquium Publication, **23** (1939).

[59] Temme, N.M., *Special Functions: an Introduction to the Classical Functions of Mathematical Physics*, Wiley (1996).

[60] Tikhonov, A.N. and Samarskii, A.A., *Equations of Mathematical Physics*, Pergamom Press (1963).

[61] Titchmarsh, E.C., *Introduction to the Theory of Fourier Integrals*, 2nd edition, Clarendon Press, Oxford (1948).

[62] Tranter, C.J., "Dual trigonometric series," *Proc. Glasgow Math. Assoc.,* **4**, 49–57 (1960).

[63] Tricomi, F.G., *Integral Equations*, Dover (1985).

[64] Tuchkin, Yu.A. and Shestopalov, V.P., "One model class of boundary problems in electrodynamics," *Differential Equations*, **18** (4), 663–673 (1982) (in Russian).

[65] Tuchkin, Yu.A., Wave scattering by open cylindrical screens of arbitrary profile with Dirichlet boundary conditions, *Soviet Physics Doklady,* (English trans.) **30**(12), 1027–1029 (1985).

[66] Van Bladel, J., *Electromagnetic Fields,* Hemisphere Publishing Corp. (1985).

[67] Vinogradov, S.S., Tuchkin, Yu.A., and Shestopalov, V.P., "An effective solution of paired summation equations with kernels in the form of associated Legendre functions," *Sov. Phys. Doklady*, **23** (9), 650–651 (1978).

[68] Vinogradov, S.S., Tuchkin, Yu.A., and Shestopalov, V.P., "Summator equations with kernels in the form of Jacobi polynomials," *Sov. Phys. Doklady*, **25** (7), 531–532 (1980).

[69] Vinogradov, S.S., Tuchkin, Yu.A., and Shestopalov, V.P., "On the theory of scattering of waves by nonclosed screens of spherical shape," *Sov. Phys. Doklady*, **26** (3), 314–316 (1981).

[70] Vinogradov, S.S., "To solving electrostatic problems for unclosed spherical conductors. Part 2: Non-symmetrical conductors," *Sov. Phys. Techn. Physics (English Trans.)*, **55** (11), 314–316 (1981).

[71] Vinogradov, S.S., "On the method of solving diffraction problems for a thin disc," *Doklady Akademii Nauk Ukrainskii SSR*, Ser. **A,** N6, 37–40 (1983) (in Russian).

[72] Vinogradov, S.S., "Method of the Abel Integral Transform in the Problems of Potential Theory and Diffraction," Dissertation for Degree of Doctor of Science in the Physical and Mathematical Sciences, Kharkhov State University Press (1988).

[73] Watson, G.N., *The Theory of Bessel Functions*, C.U.P. (1944).

[74] Weatherburn, C.E., *Differential Geometry of Three Dimensions*, Vols. 1 & 2, C.U.P. (1939).

[75] Whittaker, E.T. and Watson, G.N., *Modern Analysis*, 4th edition, C.U.P. (1940).

[76] Williams, W.E., "The solution of dual series and dual integral equations," *Proc. Glasgow Math. Assoc.*, **6**(3), 123–129 (1964).

[77] Williams, W.E. , "Note on the reduction of dual and triple series equations to dual and triple integral equations," *Proc. Camb. Phil. Soc.*, **59**(4), 731–734 (1965).

[78] Young, N., *Introduction to Hilbert Space*, C.U.P. (1988).

[79] Zygmund, A., *Trigonometric Series*, Vols. 1 & 2, C.U.P. (1958).

Index

Index